Biological Woman—
The Convenient Myth

Edited by Ruth Hubbard,
Mary Sue Henifin and Barbara Fried

Biological Woman— The Convenient Myth

A Collection of Feminist Essays
and a Comprehensive Bibliography

Copyright © 1982

Schenkman Books, Inc.
Main Street
Rochester, Vermont 05767

*Co/Res
HQ
1206
.B44
1982*

Library of Congress Cataloging in Publication Data

Main entry under title:

Biological woman—the convenient myth.

Bibliography: p.
1. Feminism—Addresses, essays, lectures. 2. Sex discrimination
against women—Addresses, essays, lectures. 3. Sex role—Addresses,
essays, lectures. 4. Women—Health and hygiene—Addresses, essays,
lectures. 5. Women—Health and hygiene—Bibliography. I. Hubbard,
Ruth, 1924-   II. Henifin, Mary Sue, 1953-   III. Fried,
Barbara, 1951-
HQ1206.B44 1982   305.4'2   82-10781
ISBN 0-87073-702-3
ISBN 0-87073-703-1 (pbk.)

Printed in the United States of America.

To the many women, past and present, who have constricted their aspirations to fit within what they were told were the limitations of their biology.

The Brain—is wider than the Sky—
For—put them side by side—
The one the other will contain
With ease—and You—beside—

The Brain is deeper than the sea—
For—hold them—Blue to Blue—
The one the other will absorb—
As Sponges—Buckets—do—

The Brain is just the weight of God—
For—Heft them—Pound for Pound—
And they will differ—if they do—
As Syllable from Sound—

—Emily Dickinson, c. 1862

Contents

List of Figures

Preface and Acknowledgments

This book originated from our participation—one as teacher, two as students— in a seminar on Biology and Women's Issues held at Radcliffe College in the fall of 1975. An important and initially surprising finding was the lack of adequate information on women's biology. Apparently, the wrong questions had been asked, inappropriate methods were used, and the answers were value-laden.

Recognizing the ramifications of this, we came to understand that women's biology not only is not destiny, but is often not even biology. Our biology was not created by God the Father, but by his human sons. As a result, it contains a number of convenient myths that bolster sexist social practices. We decided to communicate some of our discoveries in a first collection of essays, some written by seminar participants and others by colleagues subsequently contacted, which we called *Women Look at Biology Looking at Women: A Collection of Feminist Critiques.* Since its publication in 1979 much new work has been done, and specifically in areas that speak to the diversity among women. We have therefore decided to assemble this new collection which retains six articles from the previous book and to which six new articles have been added. In addition, the bibliography that concluded the earlier book has been greatly expanded.

The inspiration for the first book came from all the seminar participants. Susan Leigh Star and Vicki Druss were part of the initial editorial group, but other demands necessitated their moving too far away to sustain the daily working relationship that was required. The new book has been compiled and edited by the same three editors, with the initial participation of Joan Cindy Amatniek.

When we first decided to publish a collection on "Biology and Women's Issues," we selected as its title a modification of the first line of Emily Dickinson's poem, "The Brain is Wider than the Sky...," because it conveys our feelings about our human capacity to think, to dream, to describe the world. Although it was later impressed on us that such a title was too vague for our purposes, we want to share the poem with our readers. Therefore, we begin the book with it as an epigraph and thank the Harvard University Press for permission to reprint it.

We also wish to thank both the Committee on Instruction of Currier House at Radcliffe College and the Harvard Committee on General Education for sponsoring the seminar, and the staff of Currier House for their hospitality and support. We are grateful for financial assistance from the office of Dr. Matina Horner, President of Radcliffe College; the Radcliffe Union of Students; Radcliffe Education for Action; and the Milton Fund of Harvard University. Much help and advice has come from our publishers. Many friends and colleagues have helped each and all of us; although they remain unnamed, they are not forgotten.

Cambridge, Massachusetts
September, 1978
October, 1981

Figure I. Untitled, 1981.
(Photo: Catherine Allport)

Introduction

". . . what is a woman? I assure you, I do not know. I do not believe that you know. I do not believe that anyone can know until she has expressed herself in all the arts and professions open to human skill."
—Virginia Woolf, Professions for Women

I.

Can you find what is wrong with this description?

> The chief distinction in the intellectual powers of the two sexes is shown by man's attaining to a higher eminence, in whatever he takes up, than can woman—whether requiring deep thought, reason, or imagination, or merely the use of the senses and hands. If two lists were made of the most eminent men and women in poetry, painting, sculpture, music (inclusive both of composition and performance), history, science, and philosophy, with half-a-dozen names under each subject, the two lists would not bear comparison. We may also infer . . . that if men are capable of a decided preeminence over women in many subjects, the average of mental power in man must be above that of women. . . .[1]

In modern literary jargon, the problem might be described as Catch-22. Among feminists, it is termed "androcentric (i.e., ·male-centered) bias." In a court of law, one might call it "double jeopardy," since women are tried twice for the offense of being born female: at birth when assigned to a life of mental ineptitude, and at death when the lives thus lived are judged inept. In biological circles, however, this "problem" is sometimes attributed to sexual selection, a concept formulated by Charles Darwin that has become a cornerstone of modern evolutionary theory.

While many people have long acknowledged an unavoidable subjectivity in perceptions of reality, most continue to attribute objectivity to scientists and science. But science is the result of a process in which nature is filtered through a coarse-meshed sieve: only items that scientists consider worthy of notice are retained. Since scientists are a rather small group of people—mostly economically and socially privileged, university-educated Caucasians, and predominantly male—there is every reason to assume that, like other human productions, science reflects the outlook and interests of its producers.

Scientists do not ask all possible questions that are amenable to their methodology: only those arousing their curiosity and interest, or the curiosity and interest of supporting organizations. They do not accept all possible answers: only those congruent with the implicit assumptions that form the basis of their understanding of the world (an understanding shared with most of their "educated" contemporaries). Furthermore, the very methodology of science limits its applicability to repeatable and measureable phenomena. This discounts vast areas of human experience, indeed most facets of our relationships with ourselves, fellow humans, other living beings, and the inanimate world. When scientific knowledge is held superior to other ways of knowing, it serves to devalue or invalidate much of people's daily experience.

Recent critiques have exposed ideological biases in psychology, anthropology, and other social sciences, documenting the ways in which these disciplines often serve the narrow interests of disciples and social peer groups.[2] However, the myth of scientific objectivity has minimized similar criticism of the natural sciences. The time has come to evaluate the "interesting" questions in biology, and to ask why androcentric scientists find them interesting. For instance, among billions of animal species, why have certain ones been studied repeatedly and in great detail, while others have been ignored? Until very recently, greater attention has been focused on the social structure among Savannah baboons than on chimpanzees or gibbons. Could this be because it has been easy to stereotype baboon social behavior as hierarchical, with relatively rigid sex roles? Could it be because chimpanzees have very fluid relationships with one another that are difficult to stereotype by sex except for the fact that females nourish the unweaned young?

Is it an accident that among billions of insect species, those whose social behavior easily conforms to rigid roles are the ones that have caught the imaginations of naturalists from the nineteenth century onward? The "scientific" language of the last century is still in use—ant and bee societies still contain slaves and queens, as well as workers and soldiers. Yet we hear almost nothing about the behavior of insects whose social arrangements do not lend themselves to analogies reinforcing human social arrangements that many people think of as "natural."

Turning to studies of our own species, is it an accident that scientists have been primarily interested in exploring contraceptive techniques that tamper with the *female* reproductive system, following the curious logic that because "fertility in women depends upon so many finely balanced factors . . . it should be easy to interfere with the process at many different stages. . . ?"[3] Would it not be more sensible to conclude that it is more difficult and riskier to tamper with a woman's reproductive system than a man's *because* the woman's system is made up of "so many finely balanced factors?"

These examples suggest a few of the ways in which the objects selected for scientific study and the manner of study are used to reinforce the interests or preconceptions of the studiers. There is clearly not enough time or money for scientists to ask all possible questions. So one asks only those that promise to lead somewhere. The question is *who gets to decide where?*

This book is our beginning to an answer. Each essay stands alone. Read together, one detects certain recurring themes. The first of them concerns how we use language, or more accurately, how language uses us.

For a long time, biologists, anthropologists, psychologists, and physicians have allocated a great deal of attention to the question, *why are males masculine and females feminine?* They have searched busily for *the* answer: just the right ratio of nature to nurture required to produce our present schizophrenic state of affairs. Nature, it appears, works in ways mysterious and widely varying, depending upon which man of science we accept. Darwin thought that nature enabled us to distinguish boys from girls by endowing the latter with inferior traits at birth—a difference established through sexual selection. John Money contends that nature controls behavior through our hormones and the differing spatial orientation of our genitals (also a favorite with psychologist Erik Erikson). Others have located nature's imprint in a differential development of the left and right sides of our brains. And the sociobiologists locate it in our genes, which supposedly endow women with an updated version of the "mothering instinct"—the evolutionary adaptation that equips women to guard parental investment in our offspring.

While proclaiming erudite and sometimes contradictory answers, they have drawn attention away from the absurdity of the original question:

Why are males masculine and females feminine?

Why? Because that is what we call them. If we began to call "prototypic" females Amazons, before long we would have by definition a race of Amazons. A useful tool, if one gets to pick the names. Having designated the disease of "femininity," we can easily prove its existence by innumerable symptoms of "feminine" behavior which females display to be worthy of their given name. When men display similar behavior, the bimodal model is preserved by saying that they, too, can occasionally be "feminine" (persistence, however, indicates "abnormality").

The fact is that people vary a great deal: they come in a wide range of sizes, shapes, tempers, and talents. We can find examples to support almost any bimodal system we choose to construct: fat people are good humored, thin people, nervous; short people are aggressive, tall people, happy-go-lucky; poor people are shiftless, rich people, circumspect; blacks (jews) are warm, whites (gentiles) reserved. As Humpty Dumpty says, the important thing is who has power to choose the names.[4]

The limits of our language present the limits of "reality" as we know it. Scientists control our thought not only by their choice of subjects but in their manner of description. Much scientific writing employs the passive voice, a ponderous, authority-laden style that carries an automatic sanctification of the subject under discussion. As linguist Julia Stanley points out, when B. F. Skinner writes that "The punishment of sexual behavior changes sexual behavior," he makes passive the statement "*someone* is choosing to punish and thereby change the sexual behavior of certain people."[5] Sexual behavior is punishable. Women are reinforceable. The ecosphere is manipulable. The missing questions that this linguistic structure conceals are: *by whom? for what purpose? in whose interest? under what conditions?*

Looking at scientific language, we must provide answers to these questions and delegitimatize language that masks authority and authorship. Behind every statement that something is doable are a doer and a motive.

The development of a highly technical language has always been defended by scientists as necessary for "precision." Whether this is true or not, it is important to look at the *social* function of technical language. Learning what scientists are talking about requires a long period of apprenticeship, a molding of one's consciousness to fit the information into a precise, *publicly inaccessible* mode.

Scientific knowledge must be maximally accessible and minimally obfuscated through language. Language is power in the scientific mode as elsewhere; if we want the power that comes with information to be more widely shared, we must dispense with the elite protection of unnecessary complexities (including nomenclature) and challenge them wherever we find them.

There is another pitfall that has been largely ignored, or at least underrated. It is called by all sorts of jargon such as "experimenter expectancy," but what it boils down to is that more often than not, we find what we look for. Indeed, one can prove almost any hypothesis if one gets to set the terms of the experiment: to choose appropriate conditions, ask appropriate questions, select appropriate controls. And if one does a thorough job, the conclusions will have that quality of obviousness that scientists so enjoy at the end of meticulous research. And it really *is* obvious, for it fits what we believe about the world; but the reason it fits so well is that it is founded on those very beliefs. Thus were discovered the four humors of the body, leeching to cure disease, the inheritance of acquired characteristics, and countless cast-off theories with which we engender feelings of superiority in present-day biology students. But so also were discovered the accepted theories we examine in this book—they and many others.

Most self-fulfilling theories are devised without intent to defraud, and when they are debunked, they—at worst—damage the reputations of their authors. When such theories become effective tools for oppression, however, they are social dangers. So for example, if some scientists who believe (wish?) that women's mental lives are controlled by the physical demands of their reproductive systems (or that blacks are intellectually inferior) proceed to "prove" these hypotheses by devising the necessary tests, asking the right questions, finding appropriate subjects, and then come to the obvious conclusions, sexism (or racism) becomes part of the scientific dogma.

For two centuries men have avoided the ethics of the "woman question," as they have avoided issues of racial oppression, by claiming to base the relevant political decisions on laws of nature. "It would no doubt be a great boon to the human race," men might say "if women *could* do the marvelous things men have done, like vote, own property, and get a proper education. But such was clearly not nature's design when she gave females a 'head almost too small for intellect but just big enough for love.'"[6] If women had been meant to vote, we would not have been born with a uterus.

That the division by sex of power and privilege in society is politically motivated and not based on biology is sufficiently obvious to have been said often by many different people. But in practically every generation, there arise new prophets of "biology as destiny," and each is feted as a new Galileo who must be protected against political persecution—this time from enraged women rather than from the church.

The new science of sociobiology would have us believe that women stay home with the children because their eggs are larger (hence metabolically more expensive) than their husbands' sperm and that women's "nurturing instinct" has evolved to guard these biological "investments." Though the message to women has been somewhat altered in the century since Darwin, it remains intact: we reproduce, therefore we are. And if our reproductive functions are no longer our sole destiny, they certainly remain our most sanctioned calling. Almost fifty years ago, Virginia Woolf noted Mr. John Langdon Davies' warning that "when children cease to be altogether desirable, women cease to be altogether necessary."[7] From all appearances, androcentric scientists still remain comfortable in that conviction.

What can we do to restructure and rename our scientific world? Twenty years ago a similar question was presented to playwright Lorraine Hansberry about the black revolution in this country. She said the answer is simple: one uses everything one has. One fights in the schools, at the polls, in the marketplace, in the streets. "And, in the process, [we] must have no regard whatsoever for labels and pursed lips in the light of [our] efforts. The acceptance of our present condition is the only form of extremism which discredits us before our children."[8]

And so it is for us, concerned about the impact the science of biology continues to have on our lives. We fight with science's own tools, refuting illogical and self-serving explanations, exposing unsubstantiated claims, disclaiming poorly conceived and inadequately controlled experiments. We fight by turning our talents and money to overcome what are medical problems for some women: to try to cure dysmenorrhea rather than "prove" that it makes us unfit executives, to find safe forms of contraception rather than "prove" how much safer the pill is than death by hanging. And we fight by helping women into positions of responsibility in all facets of science, within the traditional professions and outside them, as health workers, researchers, administrators, policy makers, as teachers of future generations of scientists and as their students.

II.

In our first collection, *Women Look at Biology Looking at Women*, we showed some of the ways in which purportedly scientific descriptions of women's biology have been, and remain, deeply rooted in politics. We focused primarily on the fact that our biology has been conceptualized and described, in general, by male physicians and scientists, who as individuals and as members of their social and professional groups have had powerful personal and economic interests in portraying women in ways that make it appear "natural" for us to serve societal functions that are important, and indeed essential, if their personal and professional lives are to be guarded against major upheavals.

Now, four years later, we assert in even more explicit terms the political content of the concept of *woman* in all its aspects—social, psychological and, yes, biological. "One is not born a woman; one becomes a woman," wrote Simone de Beauvoir in *The Second Sex*. In other words, "woman" is a social construct to which little girls are taught to aspire. For, inevitably, we see ourselves as others see us, and our visions are guided by the available options. These homey truths hold for women and men, but with an important difference. Little boys are perceived from birth as potential butchers, bakers, and candlestick makers; as future astronauts, machinists, garbage collectors, doctors, firemen, lawyers, policemen, mailmen, or politicians. The fact that many will one day be husbands and fathers as well has little impact on how parents and teachers perceive, encourage, and train them.

For girls, the story is quite different. Even in today's more liberated times they are perceived first and foremost as future wives and mothers. Today's perceptions permit, in addition, a limited range of relatively low-paying, white collar professions, most of them seminurturing: nurse, teacher, hostess, secretary. Each requires above all else the skill to adjust one's needs to those of others, to submerge, subdue, efface one's self. So little girls have been taught traditionally: so they are still taught—though perhaps now more through silence than words.[9]

Figure II. Illustrations from two children's books: What Girls Can Be *and* What Boys Can Be.

It is no wonder that, as Darwin noted well over a century ago, so few women have attained "higher intellectual eminence." When one considers the pressures brought to bear against them, the wonder is rather that any have attained eminence at all—that any of our foremothers survived the distractions, the threats, the warnings of inevitable failure and the vicious contempt for their fledgling successes, to see their visions through. M. Carey Thomas, founder and first President of Bryn Mawr College, gave the following account of those pressures on her early life. Her account differs from those most of her contemporaries could give only in its happier resolution:

> I was always wondering whether it could be really true, as every-
> one thought, that boys were cleverer than girls. Indeed, I cared so
> much that I never dared to ask any grown-up person the direct
> question, not even my father or mother, because I feared to hear
> the reply. I remember often praying about it, and begging God
> that if it were true, that because I was a girl I could not success-
> fully master Greek and go to college and understand things to kill
> me at once, as I could not bear to live in such an unjust world. . . .
> I can remember weeping over the account of Adam and Eve
> because it seemed to me that the curse pronounced on Eve might
> imperil girls' going to college; and to this day I can never read
> many parts of the Pauline epistles without feeling again the sinking
> of the heart with which I used to hurry over the verses referring to
> women's keeping silence in the churches and asking their husbands
> at home. . . . I can remember one endless scorching summer's day
> when sitting in a hammock under the trees with a French
> dictionary, blinded by tears more burning than the July sun. I
> translated the most indecent book I have ever read. Michelet's
> famous—were it not now forgotten, I should be able to say
> infamous—book on woman, *La femme*. I was so beside myself with
> terror lest it might prove true that I myself was so vile and patho-
> logical a thing.[10]

We are already in danger of forgetting the tangible horrors of our past: of mistaking the real tears Thomas wept over Eve's fate for metaphorical ones, of thinking that the terror engendered by Michelet's book had some other vaguer, deeper, less prosaic cause. These men and the books they wrote were real, and the toll they took on women's lives, though immeasurable, was also real.

In the face of such powerful social pressures, it is fruitless to ask what women's biology "really" is: women try to fit the model of what women can and should be that is accepted by each society in its own historical time. And so it is that after years of speculation and research there is very little we can add to what our ancestors observed millennia ago, without the aid of rats, baboons, electrodes and personal interviews. Most women can menstruate, become pregnant, and breast-feed their babies. Most men, for their part, can inseminate women, thereby contributing half the genetic material of the next generation.

To these bare bones, each society has fitted its own notions about behaviors appropriate to each sex. Anthropologists have shown that these notions can be diametrically opposed in different societies; what may be considered fit only for the goose in one, will be the sole province of the gander in another.[11] This fact, however, has hindered few from proclaiming their particular assignment of roles as natural, innate, and commendable. From the moment of birth, each of us is admitted to a social club whose membership, at least until the advent of transsexual surgery, has been considered fixed for life. The rules of this membership are often the most stringent that will ever be invoked to govern our conduct, as the Miss Peach Cartoon painfully reminds us:[12]

Copyright © by Field Syndicate.
Reprinted by permission of Mell Lazarus.

What name we will be called, what will be the color of the first article of clothing hung on our still unselfconscious bodies, what toys we will play with, what we will be taught in school (indeed whether we go to school), what books we will read, what our life's work will be, how much (if at all) we will be paid for it—no aspect of people's lives has seemed too large or too small to be subject to sexual differentiation.

No wonder, then, that we need rarely resort to physical examination to determine the sex of an individual. Society provides us with clues more readily detected, as Edwin Lewis neatly illustrates:

A four year old who had visited a family in which there was a new
baby was later asked at home whether the baby was a boy or a girl.
"I don't know," she replied. "It's so hard to tell at that age,
especially with the clothes off."[13]

The fact is, we can never see each other with our "societal clothes" off.
Rather, scientists have offered us a reversed "emperor's new clothes" by
proclaiming that they can undress the emperor. *Homo sapiens*, they tell us,
stands splendidly naked before us if only we carefully observe the behavior
of rats, monkeys, apes, and peahens, as though these animals were humans
stripped of enculturation. Scientists have compared humans across history
and continents, thinking to discard all our varying, societal "clothes" as
acquired characteristics, in a misguided attempt to delineate a consistent
remainder that they can call innate. They have tried to isolate "pure" behavior—that not subject to environmental influence—by measuring brain
waves, or observing prelanguage infants. And they have tried to cancel out
interference from "impure" behavior by drawing their subjects from what
they perceive to be identical environments.

So far each approach has proved inadequate to the herculean task—rather
like poking around the ruins of a great fire to find the match that started it
all. We cannot regulate human environments as we do the life of a laboratory
rat; we can match up quantifiable statistics, but we cannot measure the
quality of a person's world.

"The story of our lives becomes our lives," writes Adrienne Rich.[14] And
several of the articles in this new collection show that we must take this
statement literally. The ways in which we live—what we eat, how active we
are, the kinds of work we do, how the people among whom we live and work
treat us—affect our size, shape and physiological functions; in a word, our
biology. "Natural" woman (or man) is a myth because people not only shape
their environments—and have always done so—but because we also quite
literally are shaped by them. Indeed, the very notions of "the organism" and
"its environment" are fallacies that grow out of the Western tradition to set
up dichotomies. They lead us to conceptualize our biology as though people
could exist outside of society and history. Western androcentric scientists,
brought up on the nature-nurture dichotomy, continue to believe that an
intrinsic biological substratum lies "underneath it all;" and that is what
sociobiologists search for when they try to identify the biological "bases"
of our social being. But, as Leigh Star writes:

> *What we must begin to give voice to as scientists and feminists is*
> *that there is no such thing, or place, as underneath it all.* Literally,
> empirically, physiologically, anatomically, neurologically . . . the

only accurate locus for research about us who speak to each other is the changing, moving, complex web of our interactions, in light of the language, power structures, natural environments (internal and external), and beliefs that weave it in time.[15] (Leigh Star's italics)

In a very real sense, the science of human biology not only describes what exists, but by defining "human nature" and people's "normal" and "natural" attributes, it also prescribes what should be. For norms not only summarize what is; for social organisms like ourselves, they also shape it by setting up self-fulfilling prophecies of what ought to be. The social and political impact of norms becomes easy to see when norms change, as they are doing at present, for example, in sports, where it has recently become all right for women to run marathons, lift heavy weights, play rugby, climb Anapurna, and so on. But women who do such things look, act and feel very differently from the way, for example, Alice James did a hundred years ago when she saw her only socially and personally acceptable place as that of a sophisticated invalid.[16] Such athletes' biology and health is also different from those of women who work all day in electronics assembly plants, hospital operating rooms, offices, or households.

III.

In this collection, we reprint six articles from our previous book, *Women Look at Biology Looking at Women*. We have added to them six new articles, which explore some of the political dimensions of women's biology in relation to women's work and to the ways we live our personal and social lives within the constraints imposed on us by class, race and sexual orientation. Although each article stands on its own, we hope that the collection as a whole will make it easier to acknowledge the intricate interrelationships between our biology and our lives and will enable readers to be critical of facile "explanations" of our social existence in terms of hormones, brain structure, hypothetical genes, reconstructions of human evolutionary history, or some combination(s) of these. Not only need we not believe that our biology is our destiny; we must realize that in most instances it is meaningless to ask after our "real" biology, given the extent to which it is interwoven with the way we live.

The book begins with three essays—on evolution, sex and gender, and lesbianism—which show how prevailing preconceptions can warp efforts to examine scientifically the evolution of the human species and the personal development of individuals. They illustrate that scientific "facts," like all others, are generated within a fabric of societal norms; and that the societal context pushes certain realizations into the foreground, while others readily merge with the background of the unnoticed and hence remain undescribed.[17] The two articles that follow, by Lowe and by Stellman and Henifin, offer explicit examples of how women's anatomy, physiology, and state of health can be affected by the ways in which we live and work. Stellman and Henifin also show how the ideology of "woman as breeder" has been used to track women out of some higher-paying jobs that have traditionally been open only to men, because they are said to expose women's reproductive organs to environmental hazards. Yet hazards to men's reproductive processes are ignored, as are health hazards (including reproductive hazards) to which women are routinely exposed in more traditional "women's jobs." Women's reproductive capacities are thus interpreted in whatever way is convenient to employers. Disregard for women's own wishes and decisions regarding reproduction is also documented in Helen Rodriguez-Trias' account of the involuntary sterilization procedures imposed on poor and Third World women, and she describes some of the ways in which these women have organized to fight for their reproductive rights.

The three articles by Birke and Best, Grossman and Bart, and by Brack illustrate how experiences of menstruation, menopause, and pregnancy and birth can be profoundly affected by the ways in which women live, and they document the sometimes disastrous effects of unnecessary medical interventions in our natural bodily functions. These themes, and more, are touched on in Beverly Smith's annotated bibliography on Black women's health issues.

Those of us who have begun to feel complacent about the progress women have made over the past decade in gaining access to the U.S. medical profession must think hard about the issues Mary Roth Walsh raises in her article. She describes the ways in which male physicians in the late nineteenth century used their professional authority, as well as their power within professional organizations, virtually to eliminate women from the medical profession, in spite of women's earlier successes to gain access and acceptance within it.

In the final essay, Weisstein gives her view of what it is like to be young, gifted, and female in the scientific community. Many, especially among younger women, may think her story is an old and finished chapter in women's history—to which we would respond, "Watch out." There are many ways to neutralize (and neuterize) women. One is: to keep us out. Affirmative action may prevent that, though this is no time to be complacent about it. But even if it does, women can also be rendered harmless by granting us full membership in the club provided we accept its rules.

Figure III. Staff of the Department of Chemistry, Massachusetts Institute of Technology (1899-1900), including Ellen Swallow Richards (1842-1911), instructor of the first biology course at M.I.T. (M.I.T. Historical Collections)

Others, reading Weisstein, may be tempted to conclude that if she had all that trouble, it must surely somehow be her own fault. To blame the victim is one of the prime messages of our socialization; it conveniently leaves blameless those who hold power and make the rules. No: the victims are set up to lose, not because they are born losers, but because others have the power to keep them from winning. In a society that institutes elaborate rituals of equal access, this can sometimes be hard to see; indeed, one of the functions of the rituals is to obscure this fact. Yet each of us, whether we stand outside or have managed to fight our way in, must try to become conscious of the many "keep out" signs that decorate the entrance.

Enter we must, but as Weisstein stresses, we cannot afford to do so as isolated individuals who have "made it." If we do, our entry will make no difference and will be but a further exercise in denial for ourselves and for

those who succeed us in our token successes—for all the Cinderellas who will "die young / like those favored before [us], hand-picked each one / for her joyful heart."[18]
The epilogue is followed by an extensive bibliography on women's biology and health. The number of entries is more than twice what it was in the bibliography we published in 1979, at least in part owing to the large volume of excellent work being done in the areas of women's biology and health. But, as all the authors in this volume point out, unfortunately too much current work still ignores the interrelationships between women's lives and our biology and health. It is often too narrowly focused and perpetuates old myths. It is important for women to read what is being written; but we must be able to read critically. In this collection, we try not only to inform, but also to offer useful conceptual tools with which to sort and evaluate what is being written about women's biology and health.

We hope that the book will plant seeds in the minds of many women; that it will provoke many of us to ask pointed and incisive (read "unfeminine") questions and to insist on finding honest answers. Only then can we split our present, limited, empirical world wide open and expose fracture faces that we did not even know were concealed within it. Or, to conclude with words of Mary Wollstonecraft, written nearly two hundred years ago,

Let [women's] faculties have room to unfold,
and their virtues to gain strength, and then
determine where the whole sex must stand in the
intellectual scale.

October, 1981[19]

Notes

1. Charles Darwin, *The Origin of Species and The Descent of Man* (New York: Modern Library Edition), pp. 873–874.
2. *See* for examples M. Kay Martin and Barbara Voorhis, *Female of the Species* (New York: Columbia University Press, 1975); Rayna R. Reiter, ed., *Toward an Anthropology of Women* (New York: Monthly Review Press, 1975); Naomi Weisstein, "Psychology Constructs the Female," *Woman in Sexist Society*, Vivian Gornick and Barbara Moran, eds. (New York: Basic Books, 1971), pp. 207–224; and Phyllis Chesler, *Women and Madness* (Garden City, NY: Doubleday, 1972).
3. Clive Wood, *Birth Control: Now and Tomorrow* (London: Peter Davies, 1969), pp. 36–37.

4. The actual quotation is:
   "When *I* use a word," Humpty Dumpty said, in rather a scornful tone,
   "it means exactly what I choose it to mean—neither more nor less." /
   "The question is," said Alice, "whether you *can* make words mean so
   many different things." / "The question is," said Humpty Dumpty,
   "which is to be master—that's all."
   Lewis Carroll, *Through the Looking Glass*, Martin Gardner, ed. (New
   American Library) p. 269.
5. Julia P. Stanley, "Nominalized Passives" (Paper delivered at the
   Linguistic Society of America, Chapel Hill, NC, July 1972).
6. C.D. Meigs, "Lecture on Some of the Distinctive Characteristics of the
   Female" (Paper delivered at the Jefferson Medical College, Philadelphia,
   PA, 1847), p. 67; *See* Mary Roth Walsh's essay, "The Quirks of a
   Woman's Brain."
7. Virginia Woolf, *A Room of One's Own* (New York: Harcourt, Brace and
   World, 1957), p. 116.
8. Lorraine Hansberry, "To Be Young, Gifted and Black," *Her Own Words*,
   adapted by Robert Nemiroff (Englewood Cliffs, NJ: Prentice-Hall,
   1969), pp. 213-214.
9. These, incidentally, are close to the roles which that champion of
   "natural man," Jean Jacques Rousseau (1712-1788), laid out for his
   exemplar, Emile, and for Emile's perfect helpmate and companion,
   Sophie: roles that provoked Mary Wollstonecraft in 1792 to publish in
   protest *A Vindication of the Rights of Women.*
10. M. Carey Thomas. "Present Tendencies in Women's College and Uni-
    versity Education," *The Educated Woman in America*, Barbara M.
    Cross, ed. (New York: Teachers College Press, 1965), pp. 158-160.
11. *See* for example Margaret Mead, *Male and Female: A Study of the Sexes
    in a Changing World* (New York: Dell, 1949); Lila Leibowitz, *Females,
    Males, Families: A Biosocial Approach* (North Scituate, MA: Duxbury
    Press, 1978); Eleanor Burke Leacock, *Myths of Male Dominance* (New
    York: Monthly Review Press, 1981).
12. Mell Lazarus, "Miss Peach" (New York: Field Enterprises, 1970).
13. Edwin C. Lewis, *Developing Woman's Potential* (Ames, IA: Iowa State
    University Press, 1968), p. 16.
14. Adrienne Rich, *Twenty One Love Poems* (Emery, CA: Effie's Press,
    1976), No. 18.
15. Susan Leigh Star, "Sex Differences and the Dichotomization of the
    Brain: Methods, Limits and Problems in Research on Consciousness,"
    Genes and Gender II: Pitfalls in Research on Sex and Gender, Ruth
    Hubbard and Marian Lowe, eds. (New York: Gordian Press, 1979),
    p. 116.
16. The conflicts that led her to this resolution have been described by Jean
    Strouse in the recent biography *Alice James* (Boston, MA: Houghton
    Mifflin Co., 1980).

17. For an interesting discussion of the operation of unconscious fore-grounding and backgrounding in the creation and organization of knowledge, *see* the Introduction in Mary Douglas' *Implicit Meanings* (London: Routledge & Kegan Paul, 1975).
18. Olga Broumas, "Cinderella," *Beginning with O* (New Haven: Yale University Press, 1977).
19. Portions of this introduction were published in the introduction, and in the introductions to Parts I and II of our previous collection, *Women Look at Biology Looking at Women*, R. Hubbard, M. S. Henifin, and B. Fried, eds. (Cambridge, MA: Schenkman Publishing Co., 1979).

Ruth Hubbard

Have Only Men Evolved?

*"... with the dawn of scientific investigation it might have been
hoped that the prejudices resulting from lower conditions of human
society would disappear, and that in their stead would be set forth not
only facts, but deductions from facts, better suited to the dawn of an
intellectual age.... The ability, however, to collect facts, and the power to generalize
and draw conclusions from them, avail little, when brought into direct
opposition to deeply rooted prejudices."*
—Eliza Burt Gamble, *The Evolution of Woman* (1894)

Science is made by people who live at a specific time in a specific place
and whose thought patterns reflect the truths that are accepted by the
wider society. Because scientific explanations have repeatedly run
counter to the beliefs held dear by some powerful segments of the society
(organized religion, for example, has its own explanations of how
nature works), scientists are sometimes portrayed as lone heroes swimming against the social stream. Charles Darwin (1809–82) and his
theories of evolution and human descent are frequently used to illustrate
this point. But Darwinism, on the contrary, has wide areas of congruence with the social and political ideology of nineteenth-century
Britain and with Victorian precepts of morality, particularly as regards
the relationships between the sexes. And the same Victorian notions
still dominate contemporary biological thinking about sex differences
and sex roles.

Science and the Social Construction of Reality

For humans, language plays a major role in generating reality. Without words to objectify and categorize our sensations and place them in
relation to one another, we cannot evolve a tradition of what is real in
the world. Our past experience is organized through language into our

Figure IV. Reconstruction of Neanderthal "household."
(American Museum of Natural History)

history within which we have set up new verbal categories that allow us to assimilate present and future experiences. If every time we had a sensation we gave it a new name, the names would have no meaning: lacking consistency, they could not arrange our experience into reality. For words to work, they have to be used consistently and in a sufficient variety of situations so that their volume—what they contain and exclude —becomes clear to all their users.

If I ask a young child, "Are you hungry?", she must learn through experience that "yes" can produce a piece of bread, a banana, an egg, or an entire meal; whereas "yes" in answer to "Do you want orange juice?" always produces a tart, orange liquid.

However, all acts of naming happen against a backdrop of what is socially accepted as real. The question is *who* has social sanction to define the larger reality into which one's everyday experiences must fit in order that one be reckoned sane and responsible. In the past, the Church had this right, but it is less looked to today as a generator of new definitions of reality, though it is allowed to stick by its old ones even when they conflict with currently accepted realities (as in the case of miracles). The State also defines some aspects of reality and can generate what George Orwell called Newspeak in order to interpret the world for its own political purposes. But, for the most part, at present science is the most respectable legitimator of new realities.

However, what is often ignored is that science does more than merely define reality; by setting up first the definitions—for example, three-dimensional (Euclidean) space—and then specific relationships within them—for example, parallel lines never meet—it automatically renders suspect the sense experiences that contradict the definitions. If we want to be respectable inhabitants of the Euclidean world, every time we see railroad tracks meet in the distance we must "explain" how what we are seeing is consistent with the accepted definition of reality. Furthermore, through society's and our personal histories, we acquire an investment in our sense of reality that makes us eager to enlighten our children or uneducated "savages," who insist on believing that railroad tracks meet in the distance and part like curtains as they walk down them. (Here, too, we make an exception for the followers of some accepted religions, for we do not argue with equal vehemence against our funda-mentalist neighbors, if they insist on believing literally that the Red Sea parted for the Israelites, or that Jesus walked on the Sea of Galilee.)

Every theory is a self-fulfilling prophecy that orders experience into the framework it provides. Therefore, it should be no surprise that almost any theory, however absurd it may seem to some, has its sup-porters. The mythology of science holds that scientific theories lead to the truth because they operate by consensus: they can be tested by different scientists, making their own hypotheses and designing independent experiments to test them. Thus, it is said that even if one or another scientists "misinterprets" his or her observations, the need for consensus will weed out fantasies and lead to reality. But things do not work that way. Scientists do not think and work independently. Their "own" hypotheses ordinarily are formulated within a context of theory, so that their interpretations by and large are sub-sets within the prevailing orthodoxy. Agreement therefore is built into the process and need tell us little or nothing about "truth" or "reality." Of course, scientists often disagree, but their quarrels usually are about details that do not contradict fundamental beliefs, whichever way they are resolved.[1] To overturn orthodoxy is no easier in science than in philosophy, religion, economics, or any of the other disciplines through which we try to comprehend the world and the society in which we live.

The very language that translates sense perceptions into scientific reality generates that reality by lumping certain perceptions together and sorting or highlighting others. But what we notice and how we describe it depends to a great extent on our histories, roles, and expectations as individuals and as members of our society. Therefore, as we move from the relatively impersonal observations in astronomy, physics and chemistry into biology and the social sciences, our science is increasingly affected by the ways in which our personal and social experience determine what we are able or willing to perceive as real about ourselves and the organisms around us. This is not to accuse scientists of being deluded or dishonest, but merely to point out that, like other people, they find it difficult to see the social biases that are built into the very fabric of what they deem real. That is why, by and large, only children notice that the emperor is naked. But only the rare child hangs on to that insight; most of them soon learn to see the beauty and elegance of his clothes.

In trying to construct a coherent, self-consistent picture of the world, scientists come up with questions and answers that depend on their perceptions of what has been, is, will be, and can be. There is no such thing as objective, value-free science. An era's science is part of its politics, economics and sociology: it is generated by them and in turn helps to generate them. Our personal and social histories mold what we perceive to be our biology and history as organisms, just as our biology plays its part in our social behavior and perceptions. As scientists, we learn to examine the ways in which our experimental methods can bias our answers, but we are not taught to be equally wary of the biases introduced by our implicit, unstated and often unconscious beliefs about the nature of reality. To become conscious of these is more difficult than anything else we do. But difficult as it may seem, we must try to do it if our picture of the world is to be more than a reflection of various aspects of ourselves and of our social arrangements.[2]

Darwin's Evolutionary Theory

It is interesting that the idea that Darwin was swimming against the stream of accepted social dogma has prevailed, in spite of the fact that many historians have shown his thinking fitted squarely into the historical and social perspective of his time. Darwin so clearly and admittedly was drawing together strands that had been developing over long periods of time that the questions why he was the one to produce the synthesis and why it happened just then have clamored for answers. Therefore, the social origins of the Darwinian synthesis have been probed by numerous scientists and historians.

A belief that all living forms are related and that there also are deep connections between the living and non-living has existed through much of recorded human history. Through the animism of tribal cultures that endows everyone and everything with a common spirit; through more elaborate expressions of the unity of living forms in some Far Eastern and Native American belief systems; and through Aristotelian notions of connectedness runs the theme of one web of life that includes humans among its many strands. The Judaeo-Christian world view has been exceptional—and I would say flawed—in setting man (and I mean the male of the species) apart from the rest of nature by making him the namer and ruler of all life. The biblical myth of the creation gave rise to the separate and unchanging species which that second Adam, Linnaeus (1707-78), later named and classified. But even Linnaeus— though he began by accepting the belief that all existing species had been created by Jehovah during that one week long ago ("Nulla species nova")—had his doubts about their immutability by the time he had identified more than four thousand of them: some species appeared to be closely related, others seemed clearly transitional. Yet as Eiseley has pointed out, it is important to realize that:

> Until the scientific idea of 'species' acquired form and distinctness there could be no dogma of 'special' creation in the modern sense. This form and distinctness it did not possess until the naturalists of the seventeenth century began to substitute exactness of definition for the previous vague characterizations of the objects of nature.[3]

And he continues:

> ... it was Linnaeus with his proclamation that species were absolutely fixed since the beginning who intensified the theological trend. ... Science, in its desire for classification and order, ... found itself satisfactorily allied with a Christian dogma whose refinements it had contributed to produce.

Did species exist before they were invented by scientists with their predilection for classification and naming? And did the new science, by concentrating on differences which could be used to tell things apart, devalue the similarities that tie them together? Certainly the Linnaean system succeeded in congealing into a relatively static form what had been a more fluid and graded world that allowed for change and hence for a measure of historicity.

The hundred years that separate Linnaeus from Darwin saw the development of historical geology by Lyell (1797-1875) and an incipient effort to fit the increasing number of fossils that were being uncovered into the earth's newly discovered history. By the time Darwin came along, it was clear to many people that the earth and its creatures had histories. There were fossil series of snails; some fossils were known to be very old, yet looked for all the world like present-day forms; others had no like descendants and had become extinct. Lamarck (1744-1829), who like Linnaeus began by believing in the fixity of species, by 1800 had formulated a theory of evolution that involved a slow historical process, which he assumed to have taken a very, very long time.

Possibly one reason the theory of evolution arose in Western, rather than Eastern, science was that the descriptions of fossil and living forms showing so many close relationships made the orthodox biblical view of the special creation of each and every species untenable; and the question, how living forms merged into one another, pressed for an answer. The Eastern philosophies that accepted connectedness and relatedness as givens did not need to confront this question with the same urgency. In other words, where evidences of evolutionary change did not raise fundamental contradictions and questions, evolutionary theory did not need to be invented to reconcile and answer them. However one, and perhaps the most, important difference between Western evolutionary thinking and Eastern ideas of organismic unity lies in the materialistic and historical elements, which are the earmark of Western evolutionism as formulated by Darwin.

Though most of the elements of Darwinian evolutionary theory existed for at least hundred years before Darwin, he knit them into a consistent theory that was in line with the mainstream thinking of his time. Irvine writes:

The similar fortunes of liberalism and natural selection are significant. Darwin's matter was as English as his method. Terrestrial history turned out to be strangely like Victorian history writ large. Bertrand Russell and others have remarked that Darwin's theory was mainly 'an extension to the animal and vegetable world of laissez faire economics.' As a matter of fact, the economic conceptions of utility, pressure of population, marginal fertility, barriers in restraint of trade, the division of labor, progress and adjustment by competition, and the spread of technological improvements can all be paralleled in *The Origin of Species*. But so, alas, can some of the doctrines of English political conservatism. In revealing the importance of time and the hereditary past, in emphasizing the persistence of vestigial structures, the minuteness of variations and the slowness of evolution, Darwin was adding Hooker and Burke to Bentham and Adam Smith. The constitution of the universe exhibited many of the virtues of the English constitution.[4]

One of the first to comment on this congruence was Karl Marx (1818-83) who wrote to Friedrich Engels (1820-95) in 1862, three years after the publication of *The Origin of Species*:

It is remarkable how Darwin recognizes among beasts and plants his English society with its division of labour, competition, opening up of new markets, 'inventions,' and the Malthusian 'struggle for existence.' It is Hobbes's 'bellum omnium contra omnes,' [war of all against all] and one is reminded of Hegel's *Phenomenology,* where civil society is described as a 'spiritual animal kingdom,' while in Darwin the animal kingdom figures as civil society.[5]

A similar passage appears in a letter by Engels:·

The whole Darwinist teaching of the struggle for existence is simply a transference from society to living nature of Hobbes's doctrine of 'bellum omnium contra omnes' and of the bourgeois-economic doctrine of competition together with Malthus's theory of population. When this conjurer's trick has been performed . . . the same theories are transferred back again from organic nature into history and now it is claimed that their validity as eternal laws of human society has been proved.[5]

The very fact that essentially the same mechanism of evolution through natural selection was postulated independently and at about the same time by two English naturalists, Darwin and Alfred Russel Wallace (1823-1913), shows that the basic ideas were in the air—which is not to deny that it took genius to give them logical and convincing form. Darwin's theory of *The Origin of Species by Means of Natural Selection,* published in 1859, accepted the fact of evolution and undertook to explain how it could have come about. He had amassed large quantities of data to show that historical change had taken place, both from the fossil record and from his observations as a naturalist on the Beagle. He pondered why some forms had become extinct and others had survived to generate new and different forms. The watchword of evolution seemed to be: be fruitful and modify, one that bore a striking resemblance to the ways of animal and plant breeders. Darwin corresponded with many breeders and himself began to breed pigeons. He was impressed by the way in which breeders, through careful selection, could use even minor variations to elicit major differences, and was searching for the analog in nature to the breeders' techniques of selecting favorable variants. A prepared mind therefore encountered Malthus's *Essay on the Principles of Population* (1798). In his *Autobiography,* Darwin writes:

> In October 1838, that is, fifteen months after I had begun my systematic enquiry, I happened to read for amusement Malthus on *Population,* and being well prepared to appreciate the struggle for existence which everywhere goes on from long-continued observation of the habits of animals and plants, it at once struck me that under these circumstances favourable variations would tend to be preserved and unfavourable ones to be destroyed. The result of this would be the formation of new species. Here, then, I had at last got a theory by which to work.[6]

Incidentally, Wallace also acknowledged being led to his theory by reading Malthus. Wrote Wallace:

> The most interesting coincidence in the matter, I think, is, that I, *as well as Darwin,* was led to the theory itself through Malthus. . . . It suddenly flashed upon me that all animals are necessarily thus kept down—'the struggle for existence'—while *variations,* on which I was always thinking, must necessarily often be *beneficial,* and would then cause those varieties to increase while the injurious variations diminished.[7] (Wallace's italics)

Both, therefore, saw in Malthus's struggle for existence the working of a natural law which effected what Herbert Spencer had called the "survival of the fittest."

The three principal ingredients of Darwin's theory of evolution are: endless variation, natural selection from among the variants, and the resulting survival of the fittest. Given the looseness of many of his arguments—he credited himself with being an expert wriggler—it is surprising that his explanation has found such wide acceptance. One reason probably lies in the fact that Darwin's theory was historical and materialistic, characteristics that are esteemed as virtues; another, perhaps in its intrinsic optimism—its notion of progressive development of species, one from another—which fit well into the meritocratic ideology encouraged by the early successes of British mercantilism, industrial capitalism and imperialism.

But not only did Darwin's interpretation of the history of life on earth fit in well with the social doctrines of nineteenth-century liberalism and individualism. It was used in turn to support them by rendering them aspects of natural law. Herbert Spencer is usually credited with having brought Darwinism into social theory. The body of ideas came to be known as social Darwinism and gained wide acceptance in Britain and the United States in the latter part of the nineteenth and on into the twentieth century. For example, John D. Rockefeller proclaimed in a Sunday school address:

> The growth of a large business is merely the survival of the fittest . . . . The American Beauty rose can be produced in the splendor and fragrance which bring cheer to its beholder only by sacrificing the early buds which grow up around it. This is not an evil tendency in business. It is merely the working-out of a law of nature and a law of God.[8]

The circle was therefore complete: Darwin consciously borrowed from social theorists such as Malthus and Spencer some of the basic concepts of evolutionary theory. Spencer and others promptly used Darwinism to reinforce these very social theories and in the process bestowed upon them the force of natural law.[9]

Sexual Selection

It is essential to expand the foregoing analysis of the mutual influences of Darwinism and nineteenth-century social doctrine by looking critically at the Victorian picture Darwin painted of the relations between th sexes, and of the roles that males and females play in the evolution of animals and humans. For although the ethnocentric bias of Darwinism is widely acknowledged, its blatant sexism—or more correctly, androcentrism (male-centeredness)—is rarely mentioned, presumably because it has not been noticed by Darwin scholars, who have mostly been men. Already in the nineteenth century, indeed within Darwin's life time, feminists such as Antoinette Brown Blackwell and Eliza Burt Gamble called attention to the obvious male bias pervading his arguments.[10,11] But these women did not have Darwin's or Spencer's professional status or scientific experience; nor indeed could they, given their limited opportunities for education, travel and participation in the affairs of the world. Their books were hardly acknowledged or discussed by professionals, and they have been, till now, merely ignored and excluded from the record. However, it is important to expose Darwin's androcentrism, and not only for historical reasons, but because it remains an integral and unquestioned part of contemporary biological theories.

Early in *The Origin of Species,* Darwin defines sexual selection as one mechanism by which evolution operates. The Victorian and androcentric biases are obvious:

> This form of selection depends, not on a struggle for existence in relation to other organic beings or to external conditions, but on a struggle of individuals of one sex, generally males, for the possession of the other sex.[12]

And,

> Generally, the most vigorous males, those which are best fitted for their places in nature, will leave most progeny. But in many cases, victory depends not so much on general vigor, as on having special weapons confined to the male sex.

The Victorian picture of the active male and the passive female becomes even more explicit later in the same paragraph:

> the males of certain hymenopterous insects [bees, wasps, ants] have been frequently seen by that inimitable observer, M. Fabre, fighting for a particular female who sits by, an apparently unconcerned beholder of the struggle, and then retires with the conqueror.

Darwin's anthropomorphizing continues, as it develops that many male birds "perform strange antics before the females, which, standing by as spectators, at last choose the most attractive partner." However, he worries that whereas this might be a reasonable way to explain the behavior of peahens and female birds of paradise whose consorts anyone can admire, "it is doubtful whether [the tuft of hair on the breast of the wild turkey-cock] can be ornamental in the eyes of the female bird." Hence Darwin ends this brief discussion by saying that he "would not wish to attribute all sexual differences to this agency."

Some might argue in defense of Darwin that bees (or birds, or what have you) do act that way. But the very language Darwin uses to describe these behaviors disqualifies him as an "objective" observer. His animals are cast into roles from a Victorian script. And whereas no one can claim to have solved the important methodological question of how to disembarrass oneself of one's anthropocentric and cultural biases when observing animal behavior, surely one must begin by trying.

After the publication of *The Origin of Species,* Darwin continued to think about sexual selection, and in 1871, he published *The Descent of Man and Selection in Relation to Sex,* a book in which he describes in much more detail how sexual selection operates in the evolution of animals and humans.

In the aftermath of the outcry *The Descent* raised among fundamentalists, much has been made of the fact that Darwin threatened the special place Man was assigned by the Bible and treated him as though he was just another kind of animal. But he did nothing of the sort. The Darwinian synthesis did not end anthropocentrism or androcentrism in biology. On the contrary, Darwin made them part of biology by presenting as "facts of nature" interpretations of animal behavior that reflect the social and moral outlook of his time.

In a sense, anthropocentrism is implicit in the fact that we humans have named, catalogued, and categorized the world around us, including ourselves. Whether we stress our upright stance, our opposable thumbs, our brain, or our language, to ourselves we are creatures apart and very different from all others. But the scientific view of ourselves is also profoundly androcentric. *The Descent of Man* is quite literally *his* journey. Elaine Morgan rightly says:

> It's just as hard for man to break the habit of thinking of himself as central to the species as it was to break the habit of thinking of himself as central to the universe. He sees himself quite unconsciously as the main line of evolution, with a female satellite revolving around him as the moon revolves around the earth. This not only causes him to overlook valuable clues to our ancestry, but sometimes leads him into making statements that are arrant and demonstrable nonsense . . . . Most of the books

forget about [females] for most of the time. They drag her on stage rather suddenly for the obligatory chapter on Sex and Reproduction, and then say: 'All right, love, you can go now,' while they get on with the real meaty stuff about the Mighty Hunter with his lovely new weapons and his lovely new straight legs racing across the Pleistocene plains. Any modifications of her morphology are taken to be imitations of the Hunter's evolution, or else designed solely for his delectation.[13]

To expose the Victorian roots of post-Darwinian thinking about human evolution, we must start by looking at Darwin's ideas about sexual selection in *The Descent,* where he begins the chapter entitled "Principles of Sexual Selection" by setting the stage for the active, pursuing male:

With animals which have their sexes separated, the males necessarily differ from the females in their organs of reproduction; and these are the primary sexual characters. But the sexes differ in what Hunter has called secondary sexual characters, which are not directly connected with the act of reproduction; for instance, the male possesses certain organs of sense or locomotion, of which the female is quite destitute, or has them more highly-developed, in order that he may readily find or reach her; or again the male has special organs of prehension for holding her securely.[14]

Moreover, we soon learn:

in order that the males should seek efficiently, it would be necessary that they should be endowed with strong passions; and the acquirement of such passions would naturally follow from the more eager leaving a larger number of offspring than the less eager.[15]

But Darwin is worried because among some animals, males and females do not appear to be all that different:

a double process of selection has been carried on; that the males have selected the more attractive females, and the latter the more attractive males . . . . But from what we know of the habits of animals, this view is hardly probable, for the male is generally eager to pair with any female.[16]

Make no mistake, wherever you look among animals, eagerly promiscuous males are pursuing females, who peer from behind languidly drooping eyelids to discern the strongest and handsomest. Does it not sound like the wishfulfillment dream of a proper Victorian gentleman? This is not the place to discuss Darwin's long treatise in detail. Therefore, let this brief look at animals suffice as background for his section on Sexual Selection in Relation to Man. Again we can start on the first page: "Man is more courageous, pugnacious and energetic than woman, and has more inventive genius."[17] Among "savages," fierce, bold men are constantly battling each other for the possession of women and this has affected the secondary sexual characteristics of both. Darwin grants that there is some disagreement whether there are "inherent differences" between men and women, but suggests that by analogy with lower animals it is "at least probable." In fact, "Woman seems to differ from man in mental disposition, chiefly in her greater tenderness and less selfishness,"[18] for:

Man is the rival of other men; he delights in competition, and this leads to ambition which passes too easily into selfishness. These latter qualities seem to be his natural and unfortunate birthright.

This might make it seem as though women are better than men after all, but not so:

The chief distinction in the intellectual powers of the two sexes is shown by man's attaining to a higher eminence, in whatever he takes up, than can women—whether requiring deep thought, reason, or imagination, or merely the use of the senses and hands. If two lists were made of the most eminent men and women in poetry, painting, sculpture, music (inclusive both of composition and performance), history, science, and philosophy, with half-a-dozen names under each subject, the two lists would not bear comparison. We may also infer . . . that if men are capable of a decided pre-eminence over women in many subjects, the average of mental power in man must be above that of woman. . . . [Men have had] to defend their females, as well as their young, from enemies of all kinds, and to hunt for their joint subsistence. But to avoid enemies or to attack them with success, to capture wild animals, and to fashion weapons, requires the aid of the higher mental faculties, namely, observation, reason, invention, or imagination. These various faculties will thus have been continually put to the test and selected during manhood.[19]

"Thus," the discussion ends, "man has ultimately become superior to woman" and it is a good thing that men pass on their characteristics to their daughters as well as to their sons, "otherwise it is probable that man would have become as superior in mental endowment to woman, as the peacock is in ornamental plumage to the peahen."

So here it is in a nutshell: men's mental and physical qualities were constantly improved through competition for women and hunting, while women's minds would have become vestigial if it were not for the fortunate circumstance that in each generation daughters inherit brains from their fathers.

Another example of Darwin's acceptance of the conventional mores of his time is his interpretation of the evolution of marriage and monogamy:

> ... it seems probable that the habit of marriage, in any strict sense of the word, has been gradually developed; and that almost promiscuous or very loose intercourse was once very common throughout the world. Nevertheless, from the strength of the feeling of jealousy all through the animal kingdom, as well as from the analogy of lower animals ... I cannot believe that absolutely promiscuous intercourse prevailed in times past....[20]

Note the moralistic tone; and how does Darwin know that strong feelings of jealousy exist "all through the animal kingdom?" For comparison, it is interesting to look at Engels, who working largely from the same early anthropological sources as Darwin, had this to say:

> As our whole presentation has shown, the progress which manifests itself in these successive forms [from group marriage to pairing marriage to what he refers to as "monogamy supplemented by adultery and prostitution"] is connected with the peculiarity that women, but not men, are increasingly deprived of the sexual freedom of group marriage. In fact, for men group marriage actually still exists even to this day. What for the woman is a crime entailing grave legal and social consequences is considered honorable in a man or, at the worse, a slight moral blemish which he cheerfully bears .... Monogamy arose from the concentration of considerable wealth in the hands of a single individual—a man—and from the need to bequeath this wealth to the children of that man and of no other. For this purpose, the monogamy of the woman was required, not that of the man, so this monogamy of the woman did not in any way interfere with open or concealed polygamy on the part of the man.[21]

Clearly, Engels did not accept the Victorian code of behavior as our natural biological heritage.

Sociobiology: A New Scientific Sexism

The theory of sexual selection went into a decline during the first half of this century, as efforts to verify some of Darwin's examples showed that many of the features he had thought were related to success in mating could not be legitimately regarded in that way. But it has lately regained its respectability, and contemporary discussions of reproductive fitness often cite examples of sexual selection.[22] Therefore, before we go on to discuss human evolution, it is helpful to look at contemporary views of sexual selection and sex roles among animals (and even plants).

Let us start with a lowly alga that one might think impossible to stereotype by sex. Wolfgang Wickler, an ethologist at the University of Munich, writes in his book on sexual behavior patterns (a topic which Konrad Lorenz tells us in the Introduction is crucial in deciding which sexual behaviors to consider healthy and which diseased):

> Even among very simple organisms such as algae, which have threadlike rows of cells one behind the other, one can observe that during copulation the cells of one thread act as males with regard to the cells of a second thread, but as females with regard to the cells of a third thread. The mark of male behavior is that the cell actively crawls or swims over to the other; the female cell remains passive.[23]

The circle is simple to construct: one starts with the Victorian stereotype of the active male and the passive female, then looks at animals, algae, bacteria, people, and calls all passive behavior feminine, active or goal-oriented behavior masculine. And it works! The Victorian stereotype is biologically determined: even algae behave that way.

But let us see what Wickler has to say about Rocky Mountain Bighorn sheep, in which the sexes cannot be distinguished on sight. He finds it "curious":

> that between the extremes of rams over eight years old and lambs less than a year old one finds every possible transition in age, but no other differences whatever; the bodily form, the structure of the horns, and the color of the coat are the same for both sexes.

Now note: ". . . the typical female behavior is absent from this pattern." Typical of what? Obviously not of Bighorn sheep. In fact we are told that "even the males often cannot recognize a female," indeed, "the females are only of interest to the males during rutting season." How does he know that the males do *not* recognize the females? Maybe these sheep are so weird that most of the time they relate to a female as though she were just another sheep, and whistle at her (my free translation of "taking an interest") only when it is a question of mating. But let us get at last to how the *females* behave. That is astonishing, for it turns out:

> that *both* sexes play two roles, either that of the male or that of the young male. Outside the rutting season the females behave like young males, during the rutting season like aggressive older males. (Wickler's italics)

In fact:

> There is a line of development leading from the lamb to the high ranking ram, and the female animals (♀) behave exactly as though they were in fact males (♂) whose development was retarded . . . . We can say that the only fully developed mountain sheep are the powerful rams. . . .

At last the androcentric paradigm is out in the open: females are always measured against the standard of the male. Sometimes they are like young males, sometimes like older ones; but never do they reach what Wickler calls "the final stage of fully mature physical structure and behavior possible to this species." That, in his view, is reserved for the rams.

Wickler bases this discussion on observations by Valerius Geist, whose book, *Mountain Sheep,* contains many examples of how androcentric biases can color observations as well as interpretations and restrict the imagination to stereotypes. One of the most interesting is the following:

> Matched rams, usually strangers, begin to treat each other like females and clash until one acts like a female. This is the loser in the fight. The rams confront each other with displays, kick each other, threat jump, and clash till one turns and accepts the kicks, displays, and occasional mounts of the larger without aggressive displays. The loser is not chased away. The point of the fight is not to kill, maim, or even drive the rival off, but to treat him like a female.[24]

This description would be quite different if the interaction were interpreted as something other than a fight, say as a homosexual encounter, a game, or a ritual dance. The fact is that it contains none of the elements that we commonly associate with fighting. Yet because Geist casts it into the imagery of heterosexuality and aggression, it becomes perplexing.

There would be no reason to discuss these examples if their treatments of sex differences or of male/female behavior were exceptional. But they are in the mainstream of contemporary sociobiology, ethology, and evolutionary biology.

A book that has become a standard reference is George Williams's *Sex and Evolution*.[25] It abounds in blatantly biased statements that describe as "careful" and "enlightened" research reports that support the androcentric paradigm, and as questionable or erroneous those that contradict it. Masculinity and femininity are discussed with reference to the behavior of pipefish and seahorses; and cichlids and catfish are judged downright abnormal because both sexes guard the young. For present purposes it is sufficient to discuss a few points that are raised in the chapter entitled "Why Are Males Masculine and Females Feminine and, Occasionally, Vice-Versa?"

The very title gives one pause, for if the words masculine and feminine do not mean of, or pertaining, respectively, to males and females, what *do* they mean—particularly in a scientific context? So let us read.

On the first page we find:

Males of the more familiar higher animals take less of an interest in the young. In courtship they take a more active role, are less discriminating in choice of mates, more inclined toward promiscuity and polygamy, and more contentious among themselves.

We are back with Darwin. The data are flimsy as ever, but doesn't it sound like a description of the families on your block?

The important question is who are these "more familiar higher animals?" Is their behavior typical, or are we familiar with them because, for over a century, androcentric biologists have paid disproportionate attention to animals whose behavior resembles those human social traits that they would like to interpret as biologically determined and hence out of our control?

Williams' generalization quoted above gives rise to the paradox that becomes his chief theoretical problem:

Why, if each individual is maximizing its own genetic survival should the female be less anxious to have her eggs fertilized than a male is to fertilize them, and why should the young be of greater interest to one than to the other?

Let me translate this sentence for the benefit of those unfamiliar with current evolutionary theory. The first point is that an individual's *fitness* is measured by the number of her or his offspring that survive to reproductive age. The phrase, "the survival of the fittest," therefore signifies the fact that evolutionary history is the sum of the stories of those who leave the greatest numbers of descendants. What is meant by each individual "maximizing its own genetic survival" is that every one tries to leave as many viable offspring as possible. (Note the implication of conscious intent. Such intent is not exhibited by the increasing number of humans who intentionally *limit* the numbers of their offspring. Nor is one, of course, justified in ascribing it to other animals.)

One might therefore think that in animals in which each parent contributes half of each offspring's genes, females and males would exert themselves equally to maximize the number of offspring. However, we know that according to the patriarchal paradigm, males are active in courtship, whereas females wait passively. This is what Williams means by females being "less anxious" to procreate than males. And of course we also know that "normally" females have a disproportionate share in the care of their young.

So why these asymmetries? The explanation: "The *essential* difference between the sexes is that females produce large immobile gametes and males produce small mobile ones" (my italics). This is what determines their "different optimal strategies." So if you have wondered why men are promiscuous and women faithfully stay home and care for the babies, the reason is that males "can quickly replace wasted gametes and be ready for another mate," whereas females "can not so readily replace a mass of yolky eggs or find a substitute father for an expected litter." Therefore females must "show a much greater degree of caution" in the choice of a mate than males.

E. O. Wilson says the same thing somewhat differently:

> One gamete, the egg, is relatively very large and sessile; the other, the sperm, is small and motile. . . . The egg possesses the yolk required to launch the embryo into an advanced state of development. Because it represents a considerable energetic investment on the part of the mother the embryo is often sequestered and protected, and sometimes its care is extended into the postnatal period. *This is the reason why* parental care is *normally* provided by the female. . . .[26] (my italics)

Though these descriptions fit only some of the animal species that reproduce sexually, and are rapidly ceasing to fit human domestic arrangements in many portions of the globe,[27] they do fit the patriarchal model of the household. Clearly, androcentric biology is busy as ever trying to provide biological "reasons" for a particular set of human social arrangements.

The ethnocentrism of this individualistic, capitalistic model of evolutionary biology and sociobiology with its emphasis on competition and "investments," is discussed by Sahlins in his monograph, *The Use and Abuse of Biology*.[5] He gives many examples from other cultures to show how these theories reflect a narrow bias that disqualifies them from masquerading as descriptions of universals in biology. But, like other male critics, Sahlins fails to notice the obvious androcentrism.

About thirty years ago, Ruth Herschberger wrote a delightfully funny book called *Adam's Rib*,[28] in which she spoofed the then current androcentric myths regarding sex differences. When it was reissued in 1970, the book was not out of date. In the chapter entitled "Society Writes Biology," she juxtaposes the then (and now) current patriarchal scenario of the dauntless voyage of the active, agile sperm toward the passively receptive, sessile egg to an improvised "matriarchal" account. In it the large, competent egg plays the central role and we can feel only pity for the many millions of miniscule, fragile sperm most of which are too feeble to make it to fertilization.

This brings me to a question that always puzzles me when I read about the female's larger energetic investment in her egg than the male's in his sperm: there is an enormous disproportion in the *numbers* of eggs and sperms that participate in the act of fertilization. Does it really take more "energy" to generate the one or relatively few eggs than the large excess of sperms required to achieve fertilization? In humans the disproportion is enormous. In her life time, an average woman produces about four hundred eggs, of which in present-day Western countries, she will "invest" only in about 2.2.[29] Meanwhile the average man generates several billions of sperms to secure those same 2.2 investments!

Needless to say, I have no idea how much "energy" is involved in producing, equipping and ejaculating a sperm cell along with the other necessary components of the ejaculum that enable it to fertilize an egg, nor how much is involved in releasing an egg from the ovary, reabsorbing it in the oviduct if unfertilized (a partial dividend on the investment), or incubating 2.2 of them to birth. But neither do those who propound the existence and importance of women's disproportionate energetic investments. Furthermore, I attach no significance to these questions, since I do not believe that the details of our economic and social arrangements reflect our evolutionary history. I am only trying to show how feeble is the "evidence" that is being put forward to argue the evolutionary basis (hence *naturalness*) of woman's role as home-maker.

The recent resurrection of the theory of sexual selection and the ascription of asymmetry to the "parental investments" of males and females are probably not unrelated to the rebirth of the women's movement. We should remember that Darwin's theory of sexual selection was put forward in the midst of the first wave of feminism.[30] It seems that when women threaten to enter as equals into the world of affairs, androcentric scientists rally to point out that our *natural* place is in the home.

The Evolution of Man

Darwin's sexual stereotypes are doing well also in the contemporary literature on human evolution. This is a field in which facts are few and specimens are separated often by hundreds of thousands of years, so that maximum leeway exists for investigator bias. Almost all the investigators have been men; it should therefore come as no surprise that what has emerged is the familiar picture of Man the Toolmaker.

Figure V. Discussion of the Piltdown skull.
(American Museum of Natural History)

This extends so far that when skull fragments estimated to be 250,000 years old turned up among the stone tools in the gravel beds of the Thames at Swanscombe and paleontologists decided that they are probably those of a female, we read that "The Swanscombe woman, or her husband, was a maker of hand axes...."[31] (Imagine the reverse: The Swanscombe man, or his wife, was a maker of axes....) The implication is that if there were tools, the Swanscombe *woman* could not have made them. But we now know that even apes make tools. Why not women?

Actually, the idea that the making and use of tools were the main driving forces in evolution has been modified since paleontological finds and field observations have shown that apes both use and fashion tools. Now the emphasis is on the human use of tools as weapons for hunting. This brings us to the myth of Man the Hunter, who had to invent not only tools, but also the social organization that allowed him to hunt big animals. He also had to roam great distances and learn to cope with many and varied circumstances. We are told that this entire constellation of factors stimulated the astonishing and relatively rapid development of his brain that came to distinguish Man from his ape cousins. For example, Kenneth Oakley writes:

Men who made tools of the standard type ... must have been capable of forming in their minds images of the ends to which they laboured. Human culture in all its diversity is the outcome of this capacity for conceptual thinking, but the leading factors in its development are tradition coupled with invention. The primitive hunter made an implement in a particular fashion largely because as a child he watched his father at work or because he copied the work of a hunter in a neighbouring tribe. The standard hand-axe was not conceived by any one individual *ab initio,* but was the result of exceptional individuals in successive generations not only copying but occasionally improving on the work of their predecessors. As a result of the co-operative hunting, migrations and rudimentary forms of barter, the traditions of different groups of primitive hunters sometimes became blended.[32]

It seems a remarkable feat of clairvoyance to see in such detail what happened some 250,000 years in pre-history, complete with the little boy and his little stone chipping set just like daddy's big one.

It is hard to know what reality lurks behind the reconstructions of Man Evolving. Since the time when we and the apes diverged some fifteen million years ago, the main features of human evolution that one can read from the paleontological finds are the upright stance, reduction in the size of the teeth, and increase in brain size. But finds

Figure VI. Reconstructions of the "progression of prehistoric man," including (second from left) the Piltdown hoax.
(American Museum of Natural History)

are few and far between both in space and in time until we reach the Neanderthals some 70,000 to 40,000 years ago—a jaw or skull, teeth, pelvic bones, and often only fragments of them.[33] From such bits of evidence as these come the pictures and statues we have all seen of that line of increasingly straight and upright, and decreasingly hairy and ape-like men marching in single file behind *Homo sapiens*, carrying their clubs, stones, or axes; or that other one of a group of beetle-browed and bearded hunters bending over the large slain animal they have brought into camp, while over on the side long-haired, broad-bottomed females nurse infants at their pendulous breasts.

Impelled, I suppose, by recent feminist critiques of the evolution of Man the Hunter, a few male anthropologists have begun to take note of Woman the Gatherer, and the stereotyping goes on as before. For example Howells, who acknowledges these criticisms as just, none-theless assumes "the classic division of labor between the sexes" and states as fact that stone age men roamed great distances "on behalf of the whole economic group, while the women were restricted to within the radius of a fraction of a day's walk from camp." Needless to say, he does not *know* any of this.

One can equally well assume that the responsibilities for providing food and nurturing young were widely dispersed through the group

Figure VII. Reconstruction of Cro-Magnon "mammoth hunters."
(American Museum of Natural History)

that needed to cooperate and devise many and varied strategies for survival. Nor is it obvious why tasks needed to have been differentiated by sex. It makes sense that the gatherers would have known how to hunt the animals they came across; that the hunters gathered when there was nothing to catch, and that men and women did some of each, though both of them probably did a great deal more gathering than hunting. After all, the important thing was to get the day's food, not to define sex roles. Bearing and tending the young have not necessitated a sedentary way of life among nomadic peoples right to the present, and both gathering and hunting probably required movement over large areas in order to find sufficient food. Hewing close to home probably accompanied the transition to cultivation, which introduced the necessity to stay put for planting, though of course not longer than required to harvest. Without fertilizers and crop rotation, frequent moves were probably essential parts of early farming.

Being sedentary ourselves, we tend to assume that our foreparents heaved a great sigh of relief when they invented agriculture and could at last stop roaming. But there is no reason to believe this. Hunter/gatherers and other people who move with their food still exist. And what has been called the agricultural "revolution" probably took considerably longer than all of recorded history. During this time, presumably some people settled down while others remained nomadic, and some did some of each, depending on place and season.

We have developed a fantastically limited and stereotypic picture of ways of life that evolved over many tens of thousands of years, and no doubt varied in lots of ways that we do not even imagine. It is true that by historic times, which are virtually now in the scale of our evolutionary history, there were agricultural settlements, including a few towns that numbered hundreds and even thousands of inhabitants. By that time labor was to some extent divided by sex, though anthropologists have shown that right to the present, the division can be different in different places. There are economic and social reasons for the various delineations of sex roles. We presume too much when we try to read them in the scant record of our distant prehistoric past.

Nor are we going to learn them by observing our nearest living relatives, among the apes and monkeys, as some biologists and anthropologists are trying to do. For one thing, different species of primates vary widely in the extent to which the sexes differ in both their anatomy and their social behavior, so that one can find examples of almost any kind of behavior one is looking for by picking the appropriate animal. For another, most scientists find it convenient to forget that present-day apes and monkeys have had as long an evolutionary history as we have had, since the time we and they went our separate ways many millions of years ago. There is no theoretical reason why their behavior should tell us more about our ancestry than our behavior tells us about theirs. It is only anthropocentrism that can lead someone to magine that "A possible preadaptation to human ranging for food is the behavior of the large apes, whose groups move more freely and widely compared to gibbons and monkeys, and whose social units are looser."[34] But just as in the androcentric paradigm men evolved while women cheered from the bleachers, so in the anthropocentric one, humans evolved while the apes watched from the trees. This view leaves out not only the fact that the apes have been evolving away from us for as long a time as we from them, but that certain aspects of their evolution may have been a response to our own. So, for example, the evolution of human hunting habits may have put a serious crimp into the evolution of the great apes and forced them to stay in the trees or to hurry back into them.

The current literature on human evolution says very little about the role of language, and sometimes even associates the evolution of language with tool use and hunting—two purportedly "masculine" characteristics. But this is very unlikely because the evolution of language probably went with biological changes, such as occurred in the structure of the face, larynx, and brain, all slow processes. Tool use and hunting, on the other hand, are cultural characteristics that can evolve much more quickly. It is likely that the more elaborate use of tools, and the social arrangements that go with hunting and gathering, developed in part as a consequence of the expanded human repertory of capacities and needs that derive from our ability to communicate through language.

It is likely that the evolution of speech has been one of the most powerful forces directing our biological, cultural, and social evolution, and it is surprising that its significance has largely been ignored by biologists. But, of course, it does not fit into the androcentric paradigm. No one has ever claimed that women can not talk; so if men are the vanguard of evolution, humans must have evolved through the stereotypically male behaviors of competition, tool use, and hunting.

How to Learn Our History? Some Feminist Strategies

How *did* we evolve? Most people now believe that we became who we are by a historical process, but, clearly, we do not know its course, and must use more imagination than fact to reconstruct it. The mythology of science asserts that with many different scientists all asking their own questions and evaluating the answers independently, whatever personal bias creeps into their individual answers is cancelled out when the large picture is put together. This might conceivably be so if scientists were women and men from all sorts of different cultural and social backgrounds who came to science with very different ideologies and interests. But since, in fact, they have been predominantly university-trained white males from privileged social backgrounds, the bias has been narrow and the product often reveals more about the investigator than about the subject being researched.

Since women have not figured in the paradigm of evolution, we need to rethink our evolutionary history. There are various ways to do this:

(1) We can construct one or several estrocentric (female-centered) theories. This is Elaine Morgan's approach in her account of *The Descent of Woman* and Evelyn Reed's in *Woman's Evolution*.[35] Except as a way of parodying the male myths, I find it unsatisfactory because it locks the authors into many of the same unwarranted suppositions that underlie those very myths. For example, both accept the view that our behavior is biologically determined, that what we do is a result of what we were or did millions of years ago. This assumption is unwarranted given the enormous range of human adaptability and the rapid rate of human social and cultural evolution. Of course, there is a place for myth-making and I dream of a long poem that sings women's origins

and tells how we felt and what we did; but I do not think that carefully constructed "scientific" mirror images do much to counter the male myths. Present-day women do not know what prehistoric hunter/gatherer women were up to any more than a male paleontologist like Kenneth Oakley knows what the little toolmaker learned from his dad.

(2) Women can sift carefully the few available facts by paring away the mythology and getting as close to the raw data as possible. And we can try to see what, if any, picture emerges that could lead us to questions that perhaps have *not* been asked and that should, and could, be answered. One problem with this approach is that many of the data no longer exist. Every excavation removes the objects from their locale and all we have left is the researchers' descriptions of what they saw. Since we are concerned about unconscious biases, that is worrisome.

(3) Rather than invent our own myths, we can concentrate, as a beginning, on exposing and analyzing the male myths that hide our overwhelming ignorance, "for when a subject is highly controversial—and any question about sex is that—one cannot hope to tell the truth."[36] Women anthropologists have begun to do this. New books are being written, such as *The Female of the Species*[37] and *Toward an Anthropology of Women,*[38] books that expose the Victorian stereotype that runs through the literature of human evolution, and pull together relevant anthropological studies. More important, women who recognize an androcentric myth when they see one and who are able to think beyond it, must do the necessary work in the field, in the laboratories, and in the libraries, and come up with ways of seeing the facts and of interpreting them.

None of this is easy, because women scientists tend to hail from the same socially privileged families and be educated in the same elite universities as our male colleagues. But since we are marginal to the mainstream, we may find it easier than they to watch ourselves push the bus in which we are riding.

As we rethink our history, our social roles, and our options, it is important that we be ever wary of the wide areas of congruence between what are obviously ethno- and androcentric assumptions and what we have been taught are the scientifically proven facts of our biology. Darwin was right when he wrote that "False facts are highly injurious to the progress of science, for they often endure long . . . ."[39] Androcentric science is full of "false facts" that have endured all too long and that serve the interests of those who interpret as women's biological heritage the sexual and social stereotypes we reject. To see our alternatives is essential if we are to acquire the space in which to explore who we are, where we have come from, and where we want to go.

Notes

I want to thank Gar Allen, Rita Arditti, Steve Gould and my colleagues in the editorial group that has prepared this book for their helpful criticisms of an earlier version of this manuscript.

1. For a discussion of this process, see Thomas S. Kuhn, *The Structure of Scientific Revolutions*, 2nd ed. (University of Chicago Press, 1970).
2. Berger and Luckmann have characterized this process as "trying to push a bus in which one is riding." [Peter Berger and Thomas Luckmann, *The Social Construction of Reality* (Garden City: Doubleday & Co., 1966) p. 12.]. I would say that, worse yet, it is like trying to look out of the rear window to *watch* oneself push the bus in which one rides.
3. Loren Eiseley, *Darwin's Century* (Garden City: Doubleday & Co., Anchor Books Edition, 1961), p. 24.
4. William Irvine, *Apes, Angels, and Victorians* (New York: McGraw-Hill, 1972), p. 98.
5. Quoted in Marshall Sahlins, *The Use and Abuse of Biology* (Ann Arbor: University of Michigan Press, 1976), pp. 101–102.
6. Francis Darwin, ed., *The Autobiography of Charles Darwin* (New York: Dover Publications, 1958), pp. 42–43.
7. *Ibid.*, pp. 200–201.
8. Richard Hofstadter, *Social Darwinism in American Thought* (Boston: Beacon Press, 1955), p. 45.
9. Though not himself a publicist for social Darwinism like Spencer, there can be no doubt that Darwin accepted its ideology. For example, near the end of *The Descent of Man* he writes: "There should be open competition for all men; and the most able should not be prevented by laws or customs from succeeding best and rearing the largest number of offspring." Marvin Harris has argued that Darwinism, in fact, should be known as biological Spencerism, rather than Spencerism as social Darwinism. For a discussion of the issue, *pro* and *con*, see Marvin Harris, *The Rise of Anthropological Theory: A History of Theories of Culture* (New York: Thomas Y. Crowell, 1968), Ch. 5: Spencerism; and responses by Derek Freeman and others in *Current Anthropology* 15 (1974), 211–237.
10. Antoinette Brown Blackwell, *The Sexes Throughout Nature* (New York: G. P. Putnam's Sons, 1875; reprinted Westport, Conn.: Hyperion Press, 1978). Excerpts in which Blackwell argues against Darwin and Spencer have been reprinted in Alice S. Rossi, ed., *The Feminist Papers* (New York: Bantam Books, 1974), pp. 356–377.
11. Eliza Burt Gamble, *The Evolution of Woman: An Inquiry into the Dogma of her Inferiority to Man* (New York: G. P. Putnam's Sons, 1894).

12. Charles Darwin, *The Origin of Species and the Descent of Man* (New York: Modern Library Edition), p. 69.
13. Elaine Morgan, *The Descent of Woman* (New York: Bantam Books, 1973), pp. 3–4.
14. Darwin, *Origin of Species*, p. 567.
15. *Ibid.*, p. 580.
16. *Ibid.*, p. 582.
17. *Ibid.*, p. 867.
18. *Ibid.*, p. 873.
19. *Ibid.*, pp. 873–874.
20. *Ibid.*, p. 895.
21. Frederick Engels, *The Origin of the Family, Private Property and the State*, E. B. Leacock, ed. (New York: International Publishers, 1972), p. 138.
22. One of the most explicit contemporary examples of this literature is E. O. Wilson's *Sociobiology: The New Synthesis* (Cambridge: Harvard University Press, Belknap Press, 1975); *see* especially chapters 1, 14–16 and 27.
23. Wolfgang Wickler, *The Sexual Code: The Social Behavior of Animals and Men* (Garden City: Doubleday, Anchor Books, 1973), p. 23.
24. Valerius Geist, *Mountain Sheep* (Chicago: University of Chicago Press, 1971) p. 190.
25. George C. Williams, *Sex and Evolution* (Princeton: Princeton University Press, 1975).
26. Edward O. Wilson, *Sociobiology: The New Synthesis* (Cambridge: Harvard University Press, Belknap Press, 1975), pp. 316–317. Wilson and others claim that the growth of a mammalian fetus inside its mother's womb represents an energetic "investment" on her part, but it is not clear to me why they believe that. Presumably the mother eats and metabolizes, and some of the food she eats goes into building the growing embryo. Why does that represent an investment of *her* energies? I can see that the embryo of an undernourished woman perhaps requires such an investment—in which case what one would have to do is see that the mother gets enough to eat. But what "energy" does a properly nourished woman "invest" in her embryo (or, indeed, in her egg)? It would seem that the notion of pregnancy as "investment" derives from the interpretation of pregnancy as a debilitating disease that Datha Brack discusses in her essay in this collection.
27. For example, at present in the United States, 24 per cent of households are headed by women and 46 per cent of women work outside the home. The fraction of women who work away from home while raising children is considerably larger in several European countries and in China.

28. Ruth Herschberger, *Adam's Rib* (1948; reprinted ed., New York: Harper and Row, 1970).

29. Furthermore, a woman's eggs are laid down while she is an embryo, hence at the expense of her mother's "metabolic investment." This raises the question whether grandmothers devote more time to grandchildren they have by their daughters than to those they have by their sons. I hope sociobiologists will look into this.

30. Nineteenth-century feminism is often dated from the publication in 1792 of Mary Wollstonecraft's (1759–1797) *A Vindication of the Rights of Woman;* it continued right through Darwin's century. Darwin was well into his work at the time of the Seneca Falls Declaration (1848), which begins with the interesting words:

    When, in the course of human events, it becomes necessary for one portion of the family of man to assume among the people of the earth a position different from that which they have hitherto occupied, but one to which the *laws of nature and of nature's God* entitle them ... (my italics).

    And John Stuart Mill (1806–1873) published his essay on *The Subjection of Women* in 1869, ten years after Darwin's *Origin of Species* and two years before the *Descent of Man and Selection in Relation to Sex.*

31. William Howells, *Evolution of the Genus* Homo (Reading: Addison-Wesley Publishing Co., 1973), p. 88.

32. Kenneth P. Oakley, *Man the Toolmaker* (London: British Museum, 1972), p. 81.

33. There are also occasional more perfect skeletons, such as that of *Homo erectus* at Choukoutien, commonly known as Peking Man, who was in fact a woman.

34. Howells, p. 133.

35. Evelyn Reed, *Woman's Evolution* (New York: Pathfinder Press, 1975).

36. Virginia Woolf, *A Room of One's Own* (1945; reprinted ed., Penguin Books, 1970), p. 6.

37. M. Kay Martin and Barbara Voorhis, *Female of the Species* (New York: Columbia University Press, 1975).

38. Rayna R. Reiter, ed., *Toward an Anthropology of Women* (New York: Monthly Review Press, 1975).

39. Darwin, *Origin of Species,* p. 909.

[Ruth Hubbard *is Professor of Biology at Harvard University where she teaches courses dealing with the interaction of science and society. In her scientific research she has shown how the visual pigment molecules in the retina of the eye change when they absorb light and has tried to understand how their changes initiate the signals that travel to the brain. In recent years she has been thinking, writing, and lecturing about how the assumptions scientists make about the world influence their work and how the society in which they live influences their assumptions. She is particularly interested in the way gender—the fact of one's having grown up female or male— affects these questions. She is also interested in health care, particularly as it relates to women. She has written numerous articles and reviews on all these subjects.*]

Barbara Fried

Boys Will Be Boys Will Be Boys: The Language of Sex and Gender

Our first record of the use of the word "tomboy" is in 1553 as a term of censure for "a rude, boisterous or forward boy." Twenty-five years later, the censure is redirected to a "girl who behaves like a spirited or boisterous boy," and there it has remained ever since.

While the specific attributes of "tomboyism" have seen some modifications over the past four centuries to conform to our changing notions of impropriety, the function of the label remains the same: to separate from their "normal" sisters those girls who display behavior considered fitting only for boys. It is a word particularly laden with society's fears and disapproval. It is not a word one would expect to find employed as a scientific measure of behavior.

But employed it is. In their celebrated book, *Man and Woman, Boy and Girl,* published in 1972, John Money and Anke Ehrhardt once more enlisted its services to designate deviance: this time of girls who have been exposed, while still in their mothers' wombs, to a greater than usual amount of "male" hormone, an androgen (andros = male, Greek). The list of tomboyish characteristics they compile is more methodical, less poetical, than those of their literary counterparts who found their way into the *Oxford English Dictionary.* It overlooks some of the activities previously condemned so stoutly, while including some which the Elizabethans never would have dreamed of. But "tomboy" conveys still the original distress at discovering girls whose behavior violates the bounds of what is deemed "natural."

Words like tomboy, with all their peculiar distresses and presumptions, have set the terms for every "scientific" exploration of the relationship between sex and gender identity.[1] And it is no longer possible to think we can study that relationship without also studying the role language plays in establishing it—language, that is, in its broadest sense, meaning our collective memory as well as the specific enlistment of that memory in oral and written communication.

Language is the means by which we abstract an experience from its

Figure VIII. Women overcoming gravity.
(Photo: E. Wald)

immediate physical context, and thereby attach to the experience a meaning independent of that context. With the acquisition of language, a child's experience is moved from a purely *subjective* plane to a partially *objective* one. But both these words have implications opposite from those commonly understood. The process of objectification, rather than removing experience from the realm of personal prejudice, injects that prejudice into it for the first time. The child, one would suppose, formerly had no preconceptions through which to understand experience, except perhaps a predisposition to distinguish between pleasure and pain, and a predilection for the former. But the memory newly acquired in language will henceforth prejudice the child's perception of each new experience by setting up the increasingly complex framework in which s/he assigns it meaning.

Throughout our adult life, our accumulated language thus mediates between us and our sensory experience, and then mediates among all our personal syntheses to produce communication. The problems with which we are confronted in any "scientific" study are: To what extent does that medium become the message? And to what extent are the properties we perceive in what we study more accurately attributed to the properties of our own perception? This problem is usually ignored by scientists, who tend to regard language as a neutral medium that can push subjective experience into an objective realm without tampering with its meaning. My particular concern is the extent to which this blithe disregard for the complex interrelationships of language, thought, and reality has underlain (and hence undermined) the work done thus far on sex and gender differentiation.

It is accepted fact that "gender identity is fundamentally established in more or less the same period as native language—in the first two years of life" (*MWBG,* 164). This simultaneous advent of language and gender identity is not coincidental. Language does not simply communicate the link between one's sex and one's gender identity; it *constitutes* that link. The first time a man watched a woman drive a car into a lamp post and observed, "Ah yes, women drivers," he did not merely voice a preexisting correspondence between female genitals and poor driving skills. He created that correspondence. And whether at the time it was, in fact, statistically accurate rapidly becomes unimportant: once such a correspondence is planted in people's minds, they will be predisposed to notice all the examples that bear it out (and a random distribution of driving skills guarantees abundant examples of women with poor ones). And the more examples observed, and hence the more firmly planted the correspondence, the more likely it is to be borne out by statistics as women's behavior changes to conform to their newly acquired identity as automotive incompetents. When the process has gone on long enough, the correspondence is absorbed into our consciousness as an inchoate law of human nature, permanently codified and transmitted to our children in simple units of meaning like "women drivers."

This simple example should make it obvious how complex a task it necessarily must be to isolate the genetic from the environmental sources of sexually differentiated behavior. But scientists have approached the relationship of sex to gender not as a complex system but as a simple dichotomy—whatever isn't sex is gender, and vice versa. In doing so, they have disregarded those mechanisms in biological and cultural evolution that have produced the current state of sexual dimorphism we see before us as "immutable fact." They have disregarded as well those mechanisms in us that influence the way we perceive those "facts." By ignoring these complexities, much of their work founders on the fundamental fallacy of using a language that has become sexually separate and unequal to determine the parameters of sexual separateness and inequality in society—a society whose development is channeled by the same (separate and unequal) language.

The work that John Money and his school have done on sex and gender, which is presented for the non-specialist in the book *Man & Woman, Boy & Girl,* has been accepted by many as the authoritative work in the field. Their methodology can be called into serious question on grounds that are outside the scope of this essay;[2] the grounds to be considered here are the roles language plays in 1) structuring the "facts" that have been studied; 2) structuring the way those facts are perceived; and 3) controlling how those perceptions are communicated to the reader.

How Language Shapes Reality

The first problem, the role of language in constructing the reality that is being studied, concerns the fallacy of animal analogy (or why mounting behavior in *rats* tells us absolutely nothing about the sexual life of *humans*). Most "sex and gender" research is designed to isolate genetically-controlled sexually dimorphic behavior from that which is culturally induced.[3] There are two ways that behavioral scientists have traditionally tried to achieve this isolation. The first is through exhaustive cross-cultural studies, assuming that whatever common elements of behavior emerge in widely disparate societies must be genetic in origin.[4] The second is to study a "pure" population, one that has not yet defined itself differentially by gender, which usually means pre-verbal infants. Money and Ehrhardt use neither approach, aside from a brief cross-cultural foray into sex-dimorphic erotic behavior. While it is true that both approaches present problems of their own,[5] the alternative methodology that the authors adopt is even more questionable. They have taken as their subjects the rare cases of spontaneously occurring hermaphroditism or sexual ambiguity—persons born with external genital characteristics that do not accord with their chromosomal sex. They have matched these subjects with "normal" controls who share all but those characteristics (that is, they accord in economic class, family situation, IQ, etc.), presuming that whatever *behavioral* differences are observed between the two groups are attributable to the differing *biological* traits.

The following is an example of their methodology in action. Two of the groups they studied were composed of girls who were exposed to excess androgen during their mothers' pregnancies. The first group (ten girls) experienced the increased androgen level as an unknown side effect of the drug progestin, which had been given to their pregnant mothers. Ironically, progestins are "female" hormones, administered to women who were threatened by miscarriages to allow them to carry their pregnancies to term. However, once in the body, progestins were in some instances metabolized to androgens that masculinized *in utero* the external genitals of the girl babies.[6]

The second group (fifteen girls) were afflicted with adrenogenital syndrome, a defect in the adrenal gland that makes it produce greater than normal amounts of androgen. This can now be treated after birth, but not while the infants are still in the womb. Such girls are born with abnormally large clitorises, so large that they are sometimes mistaken for boys.

The fifteen "andrenogenital" girls whom Money and Ehrhardt studied had all been treated soon after birth by reducing their androgen levels to normal and correcting surgically their enlarged clitorises.[7] The ten progestin-affected girls underwent corrective surgery where necessary, and required no further treatment since the influence of progestin on their hormone levels ceased at birth.

Therefore, although all twenty-five girls had been "fetally androgenized," their biographies were considered otherwise normal: they had two X chromosomes (like all females), their genitals after early surgery were of normal female appearance, their androgen levels after birth were normal, and all of them were raised as females. The authors compared these girls with a normal "control" group matched by age, IQ, socio-economic backround and race. They assumed that whatever behavioral differences were found in the "fetally androgenized" girls could be attributed to a single divergent element in their biographies: that the excess of "prenatal androgens may have left a presumptive effect on the brain, and hence on subsequent behavior" (*MWBG*, 98). If Money and Ehrhardt were studying rats, their presumption perhaps might be reasonable, but with humans it is not. A high fetal androgen level is *not* the only divergent element in the biographies of these girls. Every one of the families knew that their daughters had been born "abnormal" because they were prenatally androgenized; all the girls were exposed to continuous and extensive treatment by doctors and psychologists; in addition to immediate surgical correction for genitals that were masculine in appearance, some of the girls required vaginal surgery in their teens (surgery Money and Ehrhardt dismiss as minor [97]). And all the girls with adrenogenital syndrome required regular treatment from birth to keep their androgen levels down.

Clearly, the families as well as the girls themselves were aware of their anomalous sexual histories. But the authors never question what effect this knowledge may have had on the parents' and the girls' expectations of their behavior, and hence on that behavior itself. Yet common sense would suggest that the effect could have been great enough to account for what the authors have labeled as the girls' "tomboyish" behavior without contriving to explain it by some hypothetical, ill-defined process in which their brains were ."masculinized" *in utero*.[8]

The same disregard of human consciousness as a variable underlies the authors' extensive discussion of dimorphic erotic behavior in humans. They maintain that the sexual fantasies of both women and men universally center on the man's active desire for the woman. This requires that a woman respond to erotic images by "identifying herself with the female to whom men respond [with the result that] she herself becomes the sexual object," while men respond by objectifying the female (252).[9] Even if Money and Ehrhardt *had* documented the universal occurrence of these two different patterns of response to erotic imagery, *which they do not*,[10] they disregard the fact that our language, through every public medium (TV, movies, books, newspapers, etc.), devotes more energy to teaching girls and boys "appropriate" erotic responses than to any other single pursuit. Even if one *were* to prove those patterns to be quite general, what would one prove beyond the fact that one of society's blitzkriegs had had its intended effect? But Money and Ehrhardt would like us to believe they have proved far more:

> Of course, it is possible for a woman as well as a man to respond to the visual stimulus of a lover or potential partner in sex . . . .

Her imagery of arousal will, however, tend to be different, as though geared to the premise that he, the man, will be coitally useless to her, except that the stimulus of her receptive body is capable of erecting his penis. Her arousal fantasy tends to build itself around the sentiment and romance of his reacting to her and wanting her—of wanting to hold, caress, and kiss her. If he wants her enough, then his penis will erect and want her too, and perhaps not only once, but again and again for a lasting love affair, and even an entire lifetime. [251]

One can hardly imagine a more concise and illuminating picture of what the male fantasizes to be a woman's inner life. This may be fascinating biography or socio-cultural history, but is it science? And again:

Depictions of sexual intercourse, especially in a movie, are erotically stimulating to women, as well as men, but the same basic difference of identification versus objectification applies once again. The woman viewer is likely to build her erotic excitement into a fantasy of enlarging her repertory of sexual skills, learning something from the female in the movie, with the intention of utilizing it on the next available occasion with her lover or sexual partner. The man viewer builds up a level of erotic excitement, imagining that the woman on the screen is having intercourse with him on the spot. It would not, in fact, be difficult for him to copulate with any live surrogate for the female in the movie. [252]

Certainly there are many women not inclined to dispute the truth of at least the last sentence. But proving something's existence—even (if the authors had) its wide-spread existence—is not the same thing as proving it is innate.

Through the same faulty methodology, Money and Ehrhardt identify an innate sense of territoriality in males (although they themselves demonstrate in their preceding paragraph how such dimorphic traits as territoriality could arise from divergent childhood experiences):

Territoriality is less prominent in the human male than in various lower species, but some signs of it are evident. Boys rather than girls are youthful explorers, fort-builders, and scouts, and boys are the ones who form gangs or troops that set up territories, dare rivals to trespass, and attack them if they do. [182]

Aside from the fact that what the authors are offering up as biological fact, by an implied analogy to lower animals, is nothing more than undocumented cliché concerning male and female roles in this particular society, any girl who has had the not uncommon experience of being ejected from the fort or banned from the neighborhood troop could tell Money and Ehrhardt that there is nothing very mysterious or genetic about the lack of territoriality she subsequently displays.

But the authors reach that conclusion through the same spurious logic that maintains that Blacks have "natural rhythm." The fallacy goes: "If a particular pattern of behavior recurs frequently enough, we can conclude that it is innate." It is possible to attack this doctrine on a conservative level: that is, by conserving its basic premise but questioning its application. We would then ask, how frequent is "frequently enough?" For example, if research showed that in all but two known human societies women cry more often than men, are we to conclude that the difference is genetic? Or must we conclude on the strength of those two exceptions that the difference is socially acquired? And if the research turns up no exceptions, are we sure that it has considered all available evidence, and without observer bias?[11]

But it is also possible to attack the doctrine on a more radical basis. Even if diligent observations turn up no exceptions to the rule, we may have proved only that the influence of sex-role stereotyping on human history has been more universal than we suspected. Such a possibility should not surprise us, because of the powerful nature of language. It not only shapes our understanding of our past and present; it engenders our future as well, by establishing the limits of our futuristic vision. This is the profound meaning of oppression, that it not only circumscribes our experience, but it also denies us an intuition of that circumscription. It thus guarantees its own perpetuation, because lacking that intuition, we will only wander out of our prescribed limits individually and by accident, when we can be picked off and packed away as deviants one by one. So where the bias is most deep, we should not be surprised to find it universal—but that does not make it any less a bias or any more a biological fact.

This process of implying but never proving a biological basis for an observed behavioral trait underlies much of the research that has been done on sex and gender. Erik Erikson's well-known study on sexual differences in spatial orientation is another example of this phenomenon. Erikson gave 300 children the task of constructing a scene with toys on a table. After studying the resulting configurations, he concluded that girls and boys differ in their utilization of space: girls constructed around what he called "inner space," whereas boys built into "outer space," a difference he predictably ties to their respective anatomies.[12]

Ann Oakley provides an excellent critique of Erikson's study, one that can serve as a model for attacking all similarly constructed experiments:

> Although he does not actually say the differences are innate, the intimate association between them and the body ground-plan means that he is giving a general biological explanation. The words he underlines in the description of the girls' constructions refer to the female's relegation to interior, domestic, sedentary occupations in our society. In the boys' constructions, the underlined words refer to activity, aggression, self-projection, exteriority, and male pursuits. When faced with this result, the question he asks is, 'Does the anatomy of male and female suggest a difference along these lines?' The question he fails to ask is 'How do boys and girls translate this anatomical difference into such detailed and specific differences of activity and interest?' An even more pertinent question is 'Why do these differences so exactly parallel the roles society defines for male and female?' The children in Erikson's sample had had ample time to observe these differences between male and female roles, and ample time to incorporate them into their own play. Aged between ten and twelve, they were far too old to be used for studying the possibility of innate sex differences.

And we can add to Oakley's list of fallacies, one more to be drawn from her earlier description of Erikson's methodology:

> To test whether the qualities he believed he had detected had an objective existence that could be recognized by other observers, he asked other people to sort photographs of the toy scenes into male and female piles, using his criteria: the correlation between his own evaluation and those of the other observers was in fact statistically significant.[13]

How Language Shapes Our Perceptions of Reality

Erikson's supposed corroboration leads to our second area of concern, the role of language in structuring our perceptions of reality, or the fallacy of the "objective eye witness." "Surely," Erikson seems to argue, "if 150 girls display the same 'female' spatial orientation that mimics their womb, and if all my colleagues detect the same quality of 'femaleness' in their constructions, then there can be no doubt that we have discovered an innate, eternal principle of femaleness." Right?

Wrong. Forty million Frenchmen can be wrong, and they usually are when it comes to their observations on French women. And we should not be surprised to find that they are all wrong in exactly the same way, since they have all "inherited" the same set of cultural prejudices that determine how they see what they look at. Money and Ehrhardt do not even trouble to call in "disinterested" parties to corroborate their perceptions of their subjects' behavior. They simply arrogate to themselves the role of objective eye witness and present their observations as incontrovertible fact. Thus they offer the following "documentation" of tomboyish behavior in the group of fetally androgenized girls:

> The discrepancy between the diagnostic groups and their controls reached statistical significance on the criterion of baby care in the adrenogenital syndrome only. Some girls in this group disliked handling babies and believed they would be awkward and clumsy. By contrast, many of the control girls rated high in enthusiasm for little children; they adored them and took every opportunity to get in close contact with them.

> All control girls were sure that they wanted to have pregnancies and be the mothers of little babies when they grew up, whereas one third of the fetally androgenized girls with the adrenogenital syndrome said they would prefer not to have children. The remainder, as well as the ten girls with a history of fetal progestin, did not reject the idea of having children, but they were rather perfunctory and matter-of-fact in their anticipation of motherhood, and lacking the enthusiasm of the control girls.

> When queried about the priority of a nondomestic career versus marriage and being a housewife in the future, the majority of fetally androgenized girls subordinated marriage to career, or else wanted an occupational career other than housewife concurrent with being married, and regarded occupational and marital status as equally important. Among the control girls, the emphasis was in favor of marriage over non-marital career. For the majority of these girls, marriage was the most important goal of their future. [101-102]

I have already suggested what might have occurred in the girls' environment to make them more interested in a career than in marriage. (What were their parents' expectations of them? What effect did their own knowledge of their anomalous sexual status have on the directions of their interest?) At issue now is how the authors measured their data. On what criteria were the control girls "rated high in enthusiasm for little children?" How was it determined that "they adored them and took every opportunity to get in close contact with them?" How do we know that the diagnostic group was "lacking the enthusiasm of the control girls" at the prospect of motherhood?

These questions are never answered, and arise again in their discussion of androgen-insensitive children. The subjects in this group were all genetic males. But due to a biochemical inability to utilize the androgen it produced, the fetus failed to differentiate male genitals. They were all assigned and raised as females, undergoing, with the aid of administered hormones, normal female puberty (except for the absence of menstruation, because of the absence of female gonads). Money and Ehrhardt present the following data in an attempt to prove that their medical histories did not prevent the subjects from developing what the authors regard as "normal" female gender identity:

> With respect to marriage and maternalism, the girls and women
> with the androgen-insensitivity syndrome showed a high incidence
> of preference for being a wife with no outside job (80 percent);
> of enjoying homecraft (70 percent); of having dreams and
> fantasies of raising a family (100 percent); of having played
> primarily with dolls and other girls' toys (80 percent); of having
> a positive and genuine interest in infant care, even though they
> had to forfeit the care of the newborn (60 percent); and of high
> or average affectionateness, self-rated (80 percent). Two of the
> married women each had adopted two children, and they proved
> to be good mothers with a good sense of motherhood. [111]

How did these women demonstrate to Money and Ehrhardt their "preference for being a wife with no outside job?" (Had they looked for jobs and failed to find them? Had they never looked at all, having resigned themselves to the unlikelihood of finding them, since they lacked necessary credentials? Had they never wanted a job because they feared even more than the "normal" woman snide remarks about being the one who wore the pants in the family?) What precisely is "homecraft" (washing dishes? needlepoint?) and how do we measure their enjoyment of it? (Because they do it with a smile? Or just because they do it at all?) How do we determine that their interest in infant care is either positive or genuine? And what exactly is a "good mother" or a "good sense of motherhood," and how does one prove to be the first and have the second?

Money and Ehrhardt do not provide answers to these questions; they do not even ask them. They offer us only a self-proclaimed objective eyewitness presenting hearsay evidence. Where in this is any outside point of reference? At least Erikson had a briefcase full of photographs we could pore over while we argued endlessly about their origin and their meaning. But Money and Ehrhardt give us nothing more concrete than *their* impressions of their subjects' self-impressions. When they conclude their description of the androgen-insensitive females (i.e., the genetic boys who were turned into social girls) by saying "the majority (90 per cent) of androgen-insensitive women rated themselves as fully content with the female role" (112), all they have given us is two networks of prejudice talking with each other in a common language. What questions do we ask to elicit that answer (e.g., "Are you fully content with the female role?")? And how do we evaluate their answers (e.g., "Yes I am fully content with the female role.")? Money and Ehrhardt accept this statement as proof-positive for the existence of such contentment in the diagnostic group, and by extension in "normal" females as well. Any amateur psychiatrist would be more sophisticated in the use s/he made of such "data."

It is hard to imagine, in America 1972, a scientist gathering together a group of any ten women and receiving without skepticism the information that nine out of ten were "fully content with the female role." In a society noted for its discontent with almost everything, should we not wonder at 90 per cent of any sample rating themselves "fully content" with anything? And what precisely is the "female role?" Is it the role of females in an impoverished Appalachian village? The role of females in a hunter/gatherer tribe in Africa? Of eighty-year-old widows in nursing homes? Money and Ehrhardt do not ever say, though we can probably assume (from their obvious orientation) that they intend something approximating the stereotype, to borrow their own phrase, of "shoddy popular psychology and the news media" (10): a white middle-class suburban housewife with some leisure time, spending money, and 1.9 small children. Possibly, we can even assume that the same was intended by those ten women. But whatever their individual intentions might be, the authors disregard the fact that the information is outrageously culture-bound, as is almost all of the other "scientific data" presented.

I alluded in the beginning to the authors' use of the word "tomboy" to describe the deviance displayed by the fetally androgenized females from their "normal" counterparts. Here is their up-to-date definition for that irrepressible word:

1. The ratio of athletic to sedentary energy expenditure is weighted in favor of vigorous activity, especially outdoors. Tomboyish girls like to join with boys in outdoors sports, specifically ball games . . . . Tomboyish girls prefer the toys that boys usually play with . . . .
2. Self-adornment is spurned in favor of functionalism and utility in clothing, hairstyle, jewelry, and cosmetics. Tomboyish girls generally prefer slacks and shorts to frills and furbelows, though they do not have an aversion to dressing up for special occasions. Their cosmetic of choice is perfume.
3. Rehearsal of maternalism in childhood dollplay is negligible. Dolls are relegated to permanent storage. Later in childhood, there is no great enthusiasm for baby-sitting or any caretaker activities with small children . . . .
4. Romance and marriage are given a second place to achievement and career . . . . The tomboyish girl reaches the boyfriend stage in adolescence later than most of her compeers. Priority assigned to career is typically based on high achievement in school and on high IQ. . . . [10]

It is possible that if one were to study exhaustively the role of women throughout history, one would discover certain common experiences in their lives (which is a long way still from discovering the reasons for commonalities). But it is likely that the list would *not* include "cosmetics," "frills and furbelows" (instead of "slacks and shorts"), "babysitting," "boyfriends," "priority of marriage over career," lower IQ scores, and all other such arch-American (vintage 1972) variations on femaleness.

One cannot study a fetally androgenized woman's choice of cosmetics as one would study the mounting behavior of fetally androgenized rats— a fact one futilely hopes would be obvious to scientists by now. A twenty-year-old woman's choice of cosmetics has been so thoroughly influenced by verbal and non-verbal communications from her environment since that time twenty-one years earlier when she received excess androgen *in utero,* that the only remote connection between the two events that one could postulate would be: Her high androgen level *in utero* caused her, *in some completely undetermined way,* to develop after birth some degree of identification with the male half of the population, which she *chose* to express by rejecting some of her culture's traditional signs of femininity. Given the dubiousness of the hypothesis, it seems strange that one would choose it to explain her deviant behavior, in preference to what we *know* was a factor in her development—that she and her family were aware of her abnormal condition, which provoked extensive medical treatment, including a clitoridectomy. But this is precisely the choice that Money and Ehrhardt make:

> The most likely hypothesis to explain the various features of *tomboyism* in fetally *masculinized* genetic females is that their tomboyism is a sequel to a *masculinizing* effect on the fetal brain. This *masculinization* may apply specifically to pathways, most probably in the limbic system or paleocortex, that mediate dominance assertion ... *Masculinization* of the fetal brain may also apply to the inhibition of pathways that should eventually subserve *maternal* behavior. More correctly, one might say partial inhibition of these pathways, for normal males are capable of *paternalism,* much of which is identical with *maternalism,* both being manifestations of parentalism or caretaking. [103] [emphasis mine]

Their conclusion brings us to the third role of language in the study of sex and gender: how the language we use controls what we communicate.

How Language Shapes Our Descriptions of Reality

The authors offer us above an evaluation of the extent of sexually differentiated behavior through a language itself thoroughly differentiated according to sex. Having accepted in the first conception (and title) of the book the dualities of man & woman, boy & girl, in which our language has codified our belief in a sex-dichotomous world, they then proceed to "prove" the existence of that dichotomy by fortifying it with all the cognates and derivatives of those two words that our language has spawned: male-female, masculine-feminine, masculinization-femininization, maternal-paternal. Is it not clear that their "most likely hypothesis" is nothing more than a tautology? In the act of naming the disease "tomboyism," they already impute to it a cause—namely, a woman imitating a man's behavior. They then name the possessors of the disease "fetally *masculinized* genetic females," thus doubly reinforcing the cause—of course, the "tomboy" is a "masculinized" female. Finally, they formally name the cause "a *masculinizing* effect on the fetal brain." Proof is executed: A equals A equals A.

They then digress briefly to a discussion of possible biological mechanisms involved in the presumed "masculinization," and return once more to the tried and truistic circular proof with the statement "Masculinization of the fetal brain may apply also to the inhibition of pathways that should eventually subserve maternal behavior. More correctly, one might say partial inhibition of these pathways, for *normal males* are capable of *paternalism,* much of which is identical with *maternalism,* both being manifestations of parentalism or caretaking" (emphasis mine).[14] We are thus presented with the peculiar task of measuring the extent of paternalism in fathers and maternalism in mothers to determine the degree of sexual differentiation in an activity, "parentalism," which the authors simultaneously admit *not* to be differentiated by sex.

The same fallacy underlies their statement elsewhere in the book that in the selection of playtoys, "boys tended to be more uniformly masculine in their preferences than girls were feminine" (181). And it underlies their entire discussion of dimorphic erotic behavior, in statements such as "Another extreme [in love patterns] is that of the Don Juan (in females the nymphomaniac) psychopathic type" (189). A neutral word could have been chosen to describe the same types of behavior in men and women, but once more it was not. So here, as in all the other examples, by choosing to assign different names to a characteristic when it occurs in men or in women, the authors guarantee that they will discover that characteristic to be differentiated by sex.

Money and Ehrhardt are careful to point out the fallacy of dichotomizing gender identity and gender role, as in the definition of "one type of male homosexual as 'having a masculine identity juxtaposed against a feminine role, that is a man who gives a masculine impression except that he relates erotically to a male, not a female'" (146). But they embrace the deeper fallacy of dichotomizing between masculine and feminine, as can be seen in their "correction" of the first mistake:

> Actually, such a person has an identity/role that is partially
> masculine, partially feminine. The issue is one of proportion:
> more masculine than feminine. Masculinity of identity manifests
> itself in his vocational and domestic role. Femininity of identity
> appears in his role as an erotic partner; it may be great or slight
> in degree, and it is present regardless of whether, like a woman,
> he receives a man's penis or, also like a woman, he has a man giving
> him an orgasm. It goes without saying that the ratio of masculinity
> to femininity varies among individuals . . . . [146]

No—it only goes with saying, and saying, and saying that there exist such polar concepts as masculine and feminine which, placed in varying ratios to each other, form the bedrock of our personality. It only goes with saying that a man who chooses to have sex with another man has a feminine identity as an erotic partner. We could just as easily name it spiritual, or bestial, or classically Greek, or anything else we like, and thereby alter our interpretations of the experience to conform to our preconception of it. But the masculine-feminine duality in behavior is the reigning construct of Western civilization. So, finding it difficult to fit male homosexuality into their conception of the "masculine" role, Money and Ehrhardt must therefore assign it to the "feminine" role— a strange solution, one would think.

The fundamental problem with accepting *a priori* the sexual duality as the primary construct of reality is that all of our discussions about sex and gender must then take place *within* this construct. All that the recent work on the relationship of sex to gender has done is move us from a consideration of human activity in two categories (male and female) to one in two pairs of categories (male, masculine; female, feminine). And we are left to spend our time squabbling over whether each trait displayed by a man is more rightly attributed to his maleness (sex) or his masculinity (gender).

Money and Ehrhardt make it clear that they accept the duality of gender identity as innate and indispensable when they draw their analogy to bilingualism. A child exposed to two different languages from birth is believed to assimilate both more quickly if they are spoken in two non-overlapping environments (e.g., one by father, the other by mother; or one at home, the other at school). In the same manner, the authors believe that a child will most easily acquire a healthy gender identity when exposed to two clearly demarcated gender roles in the parents:

> The traditional content of the masculine and feminine roles is
> of less importance than the clarity and lack of ambiguity with
> which the tradition is transmitted to a child. . . . When the models
> of gender identification and complementation have unambiguous
> boundaries, then a child is able to assimilate both schemas,
> the same way that a bilingual child assimilates two languages,
> the users of which are clearly demarcated and non-overlapping.
> The analogy with bilingualism is closer than it seems, if one consid-
> ers those cases of the children of immigrants who learn to listen and
> to talk in the language of the new country, but only to listen in the
> language of the old country. The parental language is enveloped
> with shame and indignity. . . . The two gender schemas are, in
> the development of the ordinary child, similarly coded as positive
> and negative in the brain. The positive one is cleared for everyday
> use. The negative one is a template of what not to do and say,
> and also of what to expect from members of the opposite
> sex. [164-5]

The ideal is for a child to have parents who consistently reciprocate one another in their dealings with that child. Then a five-year-old daughter is able to go through the stage of rehearsing flirtatious coquetry with her father, while the mother appropriately gives reciprocal directives as to where the limits of rivalry lie; conversely for boys. [186]

The authors' sense of the ideal and the appropriate in child-rearing leaves much to be desired. But while we may repudiate with ease the particular partisan interpretation which they have given to the sexual duality, we must still consider the existence of that duality in some form. So the question remains, is society's proclivity to view the world in a dualistic way supported by biological fact?[15]

There is good reason to suggest that the division of humanity into two *distinct* groups is at least in part a distortion of our biology. The widely believed conclusion Money and Ehrhardt reach through their comparative studies of hermaphrodites is that some sexually dimorphic behavioral traits have a genetic origin, most likely hormonally controlled. As has already been discussed, they never clearly state which particular traits they consider to be differentiated by sex (though they seem to fall into the traditionally accepted duality of male equals active and female equals passive). They do not ever attempt to verify the universal occurrence of these differences, with the exception of a brief and inconclusive study of erotic behavior in five primitive societies (see footnote 9). And they do not ever explain the mechanism by which they presume hormones to influence behavior in a predictable, universal way.[16] But even supposing all of these problems solved, sex and gender researchers approach the study of "sex" hormones with one more fallacy unexamined, as is manifested in the fact that estrogen is commonly referred to as the "female" hormone, and androgen as the "male" hormone. It is general knowledge that both estrogen and androgen are present in all males and females. It is perhaps not generally known how close, on average, the production of the two is in both sexes throughout most of our lives.

The only uniformly wide divergence is the elevated estrogen level in women during ovulation. The average level of androgen is somewhat higher in males than in females; the average level of estrogen (excluding the dramatic increase at the time of ovulation) is slightly higher in females than in males, until the age of fifty-five.[17] But average differences are rather small if we consider the ratio of the two numbers, and what effect those relative differences might have on individual behavior is purely speculative. The relationship of hormones to behavior is not well understood; and further, one would expect their impact on an individual to depend not on the absolute amount of hormones present, but on their relationship to the size of the individual and to the internal chemical environment of which they are only a part. Furthermore, within those average differences, there are wide variations that make it *not* uncommon to find women with higher androgen levels than those of the "average" man, and men with higher estrogen levels than those of the "average" (non-ovulating) woman. And finally, perhaps most importantly, androgen and estrogen are interconvertible in our normal body processes (see footnote 6).

And yet even Ann Oakley, in an otherwise excellent book, persists in calling androgen the "male," and estrogen the "female," hormone, leading to such verbal gymnastics as:

> [After age ten,] *in both sexes,* the production of *both male and female* hormones increases. With the approach of puberty, the increase in the production of *male* hormones *in both sexes* becomes pronounced. . . . The increase in the *female* hormone is much greater for *girls* than *boys* . . . but estrogen production in boys does increase at puberty. . . . While *men and women* produce *both male and female* hormones, the relative amount and proportions vary a great deal between individuals and one cannot establish biological maleness or femaleness from the hormone count alone. [24-26] [emphasis mine]

Wouldn't it make more sense, in studying the effects of "sex" hormones on behavior and physical development, to adopt the suggestion of Roger Williams that we classify humans into nine categories, representing all the possible combinations of low, medium, or high androgen and estrogen levels? There is far more at stake than semantics, because in recognizing the inaccuracy of the old terminology, we would recognize as well the inaccuracy of all of the sex-based generalizations that have followed from it. Even Williams' suggestion would not be precise terminology, because in talking about androgen and estrogen levels, we are discussing a phenomenon which is not dimorphic, tetramorphic or nonamorphic. It is di-spectral with wide overlaps, and so the number of points on each spectrum into which we choose to group humanity to study the effects of each hormone is arbitrary. But the larger the number of points we consider, the less distortion will be involved in fitting each person into one of the groups. And if we at least begin by dislodging androgen and estrogen from their inaccurate one-to-one correspondence to men and women respectively, we will have gone a long way toward clearing the air for honest scientific inquiry into their actual effects on behavior.

In the same way, the common division of somatotypes (body physiques) into female and male, paralleling the popular image of the "tall dark and handsome man" and the "lithe little woman," distorts reality. Most "secondary sex characteristics" are in fact *not* distributed dimorphically according to sex, but once again spectrally and with wide overlaps between women and men. Height, weight, musculature, body hair, breast size, pitch of voice are all characteristics that vary widely within each sex. While we find more men than women clustered at the tall end of the height spectrum, many women are there as well. And while the averages for the male and female populations diverge in all of these characteristics, it is important to realize that society is likely to have contributed even to this divergence 1) by establishing, over time, dimorphic criteria for sexual desirability; 2) by encouraging different diets and activities for women and men; 3) by encouraging people to mask their "deviant" characteristics (we can, for instance, guess that a large number of women have facial hair from the number of people who make their livings removing it); and no doubt by other mechanisms as well.

If in the end we limit our discussion of genetically prescribed dimorphic behavior, as Money and Ehrhardt finally do, to saying that "women menstruate, gestate and lactate and men don't." (*MWBG*, 163), we are at least approaching a truth (providing we amend that statement to read that *most* women menstruate, and *can* gestate and lactate for part of their adult lives if they so choose). What significance this difference has depends upon how much of her adult life a woman in fact spends gestating and lactating, and how much mental and physical effort is invested in each. And the more those decisions come to be personal choices for *each* woman, the less appropriate it is to generalize about their significance for *all* women.

Societally prescribed behavior, however, is quite another matter. Gender is by definition a dualistic concept, since the word is merely the symbol for our belief in a dualistic world. That belief has played a role so fundamental in ordering human experience that, in general, the lives of women and men in this and every other known society are undeniably different. Any attempt to study those lives must therefore be dual as well (with the proviso that if we analyze gender in its *public* terms of femininity and masculinity, we will never probe beneath its *public* image). But it is critical that discussions of sex and of gender be kept scrupulously unconfused.

Money and Ehrhardt's work is a self-proclaimed effort to achieve that clarity by "scientifically" determining the parameters of each. They have for the most part succeeded in confounding the truth even further. Indeed it may be an impossible task at this time. We know that cultural forces fuelling a sexually separate and unequal society are strong; one must question why scientists are so eager to ferret out whatever weak (and variable) genetic factors may be feeding the fire as well. Sex and gender researchers' predisposition to try, at a time when American women are once again coming to recognize the extraordinary impact of society on our destiny, seems a suspect effort to repin that destiny once more to our biology.[18]

Notes

1. I am using the definitions of "sex" and "gender" used by Money and Ehrhardt in their book, *Man & Woman, Boy & Girl* (Baltimore: Johns Hopkins University Press, 1972). Subsequent references will appear in the text with the abbreviation, *MWBG*. "Sex" should be understood to refer only to the information transmitted to the individual genetically through the sex chromosomes. "Gender identity" is defined as "the sameness, unity, and persistence of one's individuality as male, female, or ambivalent, in greater or lesser degree, especially as it is experienced in self-awareness and behavior; gender identity is the private experience of gender role, and gender role is the public expression of gender identity" (p. 4). I think these definitions are themselves a distortion of the phenomena they purport to describe, for reasons that I hope will become clear in the course of this article.

2. For example, they attempt to isolate the various genetic causes of sexually dimorphic behavior by studying spontaneously occurring examples of human hermaphroditism. The groups they are studying are hence very small, ranging from a low of one, to a high of twenty-three members (pp. 98, 104, 106). Much of their evidence must therefore be considered inconclusive. Furthermore, it is questionable what a study of psychosexual differentiation under extremely stressful abnormal conditions tells us about the course of normal psychosexual development.

3. While Money and Ehrhardt in particular pay lip service to modern genetic theory that frowns upon the "antiquated dichotomy" of genetics and environment (p. 1), in fact, the greater part of the book is devoted to determining the definitive line dividing the two.

4. *See* for example Margaret Mead's classic, *Male and Female* (New York: Dell Publishing Co., 1968).

5. The problems inherent in cross-cultural studies will be discussed later in the paper. As for infants, they are a difficult population to study, since they are relatively immobile, with a limited range of activities in which to display differentiation of any sort, sexual or otherwise. The statistics collected are further limited in their usefulness by the fact that they have already been affected by sexually dimorphic sensory stimuli administered by parents and others. For an excellent discussion of the methodological problems in infant studies, *see* Hugh Fairweather, "Sex Differences in Cognition," *Cognition*, 4 (1976), pp. 231–80.

6. It is important to understand that both females and males secrete what we call "male" and "female" hormones—"androgens" and "estrogens" —which are regularly interconverted by our normal body processes. For an excellent discussion of the mechanisms of interconversion, *see* Ruth Bleier, "Myths of the Biological Inferiority of Women: An Exploration of the Sociology of Biological research," *University of Michigan Papers in Women's Studies*, 2, No. 2 (1976), pp. 39-63. Bleier draws her data on hormone interconversion from two studies: K. H. Ryan et al., "Estrogen Formation in the Brain," *American Journal of Obstetrics and Gynecology*, 114 (1972), pp. 454-60; and Judith Weisz and Carol Gibbs, "Conversion of Testostereone and Androstenedione to Estrogens *in vitro* by the Brain of Female Rats," *Endocrinology*, 94 (1973), pp. 616-20.

7. As Patricia Farnes and Ruth Hubbard have noted, that surgery in fact routinely involved removal of the entire clitoris and its nerve supply— in other words, a total clitoridectomy (*see* Letter to the Editor, *Ms. Magazine*, April 1981, pp. 9-10; Letter to the Editor, *Science for the People*, March/April 1981, p. 2.) While that fact is never explicitly acknowledged in *Man & Woman, Boy & Girl*, it can easily be detected from the photographs of the "reconstructed" (that is, clitoridectomized) female genitals exhibited on pp. 168, 170 and 172. It is also described explicitly in the relevant medical literature (*see*, for example, Money, Hampson and Hampson, *Bull. Johns Hopkins Hospital 97*: 284-300, 1955). In addition to the ethical problems raised by this procedure, it casts further doubt on the accuracy of the authors' findings (*see* footnote 8).

8. The authors also never question what effect having undergone a clitoridectomy (*see* footnote 7) might have had on the girls' behavior, particularly as regards such "findings" as that "[r]omance and marriage are given second place to achievement and career," and that the girls reach "the boyfriend stage in adolescence later than most of [their] compeers." (p. 10). Indeed, the authors' failure even to acknowledge that one effect of the "treatment" was to perform a clitoridectomy on their subjects must cast serious doubt on the thoroughness and reliability of their findings and deductions elsewhere.

9. For a spoof of this view, see R. Herschberger's *Adam's Rib* (New York: Harper & Row, 1970), especially chapter 8, "Society Writes Biology."

10. Here, as elsewhere in the book, Money and Ehrhardt do not make explicit the intended scope of their thesis, though we are encouraged to view it as universal, through their implication that the difference is grounded in the mechanics of sexual intercourse. But whether their subject is America or the world, the only "proof" the authors offer for their assertion is, "There is good empirical support for this imagistic difference between the sexes, even in the absence of experimental design. The reader can test this hypothesis on the basis of his or her own experience" (p. 252).

11. This particular line of attack is useful in responding to some of the cross-cultural data Money and Ehrhardt present on "gender dimorphic traditions in sexual partnerships." They describe five different patterns of sexual partnership in primitive societies in order to show that although there is great variation between societies in the particular patterns prescribed, the patterns are invariably dimorphic (p. 145) Yet by their own indirect admission, their information is grossly incomplete. The researchers neglected to bring along a female investigator as part of their team, and since "talk about sexual activity . . . is taboo between the sexes" (p. 130), they obtained no information directly from the female members of each society. But rather than acknowledging the doubt this oversight must cast on the findings, Money and Ehrhardt treat that lack of information as information: accompanying their length discussions of prescribed and ritualized male homosexuality in these societies, they announce that "there is no evidence of a female homoscxual relationship" (p. 133), and this in spite of the fact that they have admitted that such evidence would *not* have been available to the all-male team sent to collect it. One is of course forced to wonder if they would have been so ready to accept as definitive information any comparable *lack* of information about the male half of the population.

12. Erik Erikson, "Inner and Outer Space: Reflections on Womanhood," *The Woman In America*, R. J. Lifton, ed., (Boston: Houghton Mifflin, 1965), pp. 1–26.

13. Ann Oakley, *Sex, Gender and Society*, (London: Harper & Row, 1972), pp. 96–97, 83. For another excellent critique of Erikson *see* Kate Millett, *Sexual Politics* (New York: Doubleday & Co., 1969), pp. 210–220.

14. It is important to stress that all of this is entirely speculative. No neural pathways are known to exist that *do* subserve maternal behavior, much less "pathways that should eventually subserve" such behavior.

15. It has been a major contention of feminist theory (see, for example, de Beauvoir's *Second Sex*, Millet's *Sexual Politics*, Daly's *Beyond God the Father*) that men would find it in their self-interest to maintain this dualistic world-view with or without a biological basis for it. I think this is true, and serves in part to explain why John Money is predisposed to believe the duality of gender to be both innate and commendable, before he has proved it to be either. But that is a topic too large to be pursued further here.

16. This is a particularly distressing aspect of the book—that its argument is put forth in terms obvious enough that it is absorbed, but indirect and understated enough that one is not encouraged to confront it. No doubt, if pushed to defend some of the more spurious assumptions behind their argument, the authors would retreat safely behind the protection of having said the connection is only "presumed." But the fact remains that readers are allowed, indeed encouraged, to accept the connection as proved. And as the book progresses' through a predictable "carelessness," the presumed connection ceases to be treated

as hypothesis, and comes to be regarded as fact. Thus, when discussing the adrenogenital syndrome, in which excess androgen is introduced into the female fetus, the authors say, "masculinization of the body is typically a source of mortification to the woman afflicted. The hormonal source of body masculinization does not also masculinize her mind" (p. 213). While ostensibly disproving one suSPect fact, Money and Ehrhardt plant a far more outrageous one as fact in everyone's mind: i.e., "We still don't know exactly what masculinization of the brain is, but we now know it exists, though it is, by the way, not triggered by the adrenogenital syndrome."

17. Oakley, *Sex, Gender and Society*, pp. 24, 27.

18. In this light, it is not surprising that his research has led John Money to become one of the most outspoken proponents of transsexual operations, surely one of society's most reactionary institutional responses to the problems created by its sexual split. Considering the way in which doctors have dealt with homosexuality, some people have been surprised at the ease with which the medical establishment has granted transsexuality the status of normality. It should not be surprising at all: the establishment may have conceded a battle or two along the way, but it has won another war. Under the guise of the "revolutionary" insight that is possible medically to cross the sex barrier, the proponents of transsexual surgery relink that barrier in the strongest possible terms to our biology. They affirm with all the sanctity of their profession behind them that there exists an *innate* difference between the *mental* lives of women and men that is so clear that it makes sense to speak of a "woman trapped inside a man's body." If there is any doubt about the reactionary implications of that belief, one need only look at the accounts male-to-female transsexuals have given of their own mental lives, many of which are grotesque parodies of our most degrading stereotypes of feminity. (*See* Janice Raymond's article, "Transsexualism: The Ultimate Homage to Sex-Role Power," *Chrysalis*, 3 [1977].) While transsexual operations may offer some sort of relief to a privileged few suffering an acute case of this society's sexual madness, they offer no comfort to those millions who find they are women in the right body, "trapped inside" the wrong society.

[Barbara Fried *has been active in the women's movement for many years, including as an owner and manager of Bread and Roses, Cambridge's first Women's Restaurant; and producer of the 1977-78 Women of Words Series and other cultural events designed to bring women artists and scientists to a larger feminist audience. Her previous published works include a critical study of William Faulkner entitled* The Spider in the Cup *(Harvard University Press). She is currently a student at Harvard Law School, where she is one of the editors of the* Harvard Women's Law Journal.]

Figure IX. Lesbian Family, Maine, 1979.
(Photo: Catherine Allport)

Lynda I. A. Birke

From Sin to Sickness: Hormonal Theories of Lesbianism

This chapter is concerned with scientific and medical views of lesbianism. Many theories have attempted to identify the "cause" of lesbianism, ranging from those that implicate faulty. families to those suggesting the failure of some biological function. Whatever differing explanations each of these theories may in the end reach, most start from the common assumption that to be a lesbian is to be abnormal. The first part of this chapter looks at some notions of abnormality, particularly as they inform this assumption that lesbians are biologically different from other women. The second part considers the various forms of "treatment" that have been meted out to lesbians and gay men, and that result directly from considering homosexuals as deviants.[1]

From Evil to Sickness: The Rise of the Medical View

Lesbians are, and have always been, social outcasts. For centuries, this treatment was actively encouraged by the Christian Church,[2] which viewed lesbian activities as an offense before the eyes of God, like many other sexual acts that did not potentially result in procreation. Lesbianism is mentioned somewhat obscurely in some penitentials of the early Christian Church as meriting several years' penitence, with subsequent offenses requiring more drastic punishment.[3]

The dominance of these religious views began to lessen, however, during the eighteenth century and eventually gave way to a form of medical orthodoxy. Where once the Church had condemned homosexuality as unnatural in the sight of God, now the ascendant medical profession was condemning homosexuality as physiologically deviant and thus requiring medical attention.

Foucault has argued in his *History of Sexuality* that attitudes toward sexuality began to change during the eighteenth and nineteenth centuries as a consequence of a dramatic increase in discourse concerning sex and sexuality, which led, among other things, to a greater awareness of the many different forms of sexual expression.[4] A whole new range of definitions and categories emerged from this renewed debate, including that of "the" homosexual woman or man. While traditional religious doctrine had regarded homosexuality as "a potentiality in all sinful nature, unless severely . . . punished,"[5] emerging concepts treated the homosexual as a specific kind of *person*, clearly differentiated from everyone else.[6] This was an important change, for once the possibility was allowed that such a person might be suffering from some kind of defective anatomy or physiology, it became possible, and indeed appropriate, to think of medical treatments for the condition—a reaction still echoed in much writing about lesbians and gay men today, and one that has considerable implications for the way lesbians are regarded.

The multiplication of categories of sexuality to which Foucault refers led to discussions in medical journals of many aspects of non-procreative sex. Doctors now took over where the Church had left off, and warnings about the dire consequences of masturbation, oral sex, or even "excessive" marital intercourse abounded in the medical literature.[7] In this sense, lesbianism was not very different from other nonreproductive sex; all of it was to be avoided, and was considered likely to lead to derangement, insanity, or even death.

Some feminists have conjectured that lesbianism was more severely punished than other forms of "deviant" sexuality, not least because it implied that women were less available to men. In fact, there is little evidence that this has ever been so, perhaps because lesbianism was treated either as something of a joke or as highly improbable.[8] That lesbianism is severely condemned is not at issue; but evidence to date fails to suggest that lesbians have been punished *more* than any other woman who failed to fit the patriarchal, heterosexual standard. Any such woman is likely to be thought of as abnormal, and treated or punished accordingly.

As homosexuality has come to be increasingly viewed as a sickness, speculations as to how it develops have become more frequent. An enormous variety of theories ascribe it to biology and try to explain its development. Even those reformers who tried around the turn of the the twentieth century to increase bourgeois society's acceptance of homosexuals endeavoured to do so by arguing that homosexuality was biologically determined and that homosexuals therefore should not be faulted for it. This, for example, was the position of Havelock Ellis, whose views on sex were progressive for his time. Ellis, writing just as the effects of hormones were beginning to be documented,[9] suggested that hormones might be instrumental in causing the development of homosexuality. Since homosexuals thus were victims of aberrant endocrine systems, he urged that they be pitied rather than condemned.

The notion of lesbians and gay men as somehow sick or evil has, of course, persisted, and is with us still. Gay people are condemned because they might seduce or corrupt children, especially if they are professionally involved with them as teachers or nurses. Such condemnation conveniently ignores the fact that most sexual assaults on children are committed by *hetero-sexuals*, and frequently by the father or guardian of the child.[10] The idea that gay people might "infect" children with their sickness is often raised, especially in connection with gay parents. "Can lesbians make normal mothers?" asked the British press when it discovered that lesbians were having children. And that question is often implicit in research purporting to compare the lesbian mother with her heterosexual counterpart.[11] Threading through such research is the unlikely assumption that the lesbian mother's sexual preference profoundly changes the way she raises her children.

Researchers tend to assume, *a priori*, that homosexuality is abnormal, and therefore find it necessary to seek explanations of it. Heterosexuality, on the other hand, is assumed to be so normal that it requires no explanation. Freud[12] was one of the few who pointed out that any useful theory of the development of human sexual orientations has to account not only for those people who prefer their own sex, but also for those who grow up to seek partners *exclusively* of the opposite sex. But few other researchers on sexuality have shared this conviction that both types of preference merit explanation, and have instead accepted the unspoken assumption that the only valid expressions of human love and sexuality are heterosexual. No matter that this is not true in many societies in which homosexual behavior is actively encouraged or condoned,[13] the assumption that it is deviant runs through the scientific and medical literature, and is rarely questioned. Once a group of people has been defined as deviant, of course, it is reasonable to ask how they became so: is it a function of an aberrant biology, or did they have a bad start in life?

Since Havelock Ellis wrote his *Sexual Inversion* at the end of the last century, a range of hypotheses has been proposed to account for homosexuality, all starting from the assumption of abnormality. In some, the homosexual's family is blamed; for example, gay men are said to have "over-intimate" mothers[14] — a notion that is sufficiently well known to have engendered considerable guilt and anxiety in a great many mothers. Yet, as one author has asked, what is wrong with such a mother, unless she happens to be the mother of someone you have judged beforehand to be deviant?[15]

Many feminist scholars have pointed out that the way in which language is used helps to perpetuate without question society's implicit values. This is particularly evident in the literature on biology and gender, as Fried points out in her article in this collection. Within the literature on homosexuality, the assumption of abnormality is reinforced through language. For example, in a study on the electrical brain response (electroencephalogram, E.E.G.) of gay and heterosexual men, when the authors noted some small differences between the two groups, they concluded that the gay men had a "degree of cerebral immaturity;"[16] this despite the fact that interpreting the E.E.G. is notoriously difficult, so that it is debatable whether it tells us much at all about healthy brain function.[17] Even more blatant, in another study that *failed* to demonstrate consistent hormonal differences between lesbians and heterosexual women, the researchers concluded that the lesbians did not exhibit a "consistent pattern of hormonal abnormality."[18] By such use of language scientists can reinforce by verbal trickery the prevailing prejudice that lesbianism *is* the product of abnormal hormone levels, even as they report data that disprove it.

A second assumption threading through the scientific literature on homosexuality is that it is perfectly reasonable to generalize from studies of other species to humans. The animal of choice for studies of the relationship of hormones to sexual behavior, as we shall see later, is the much-exploited rat. Despite our knowledge of the enormous variety of human sexual expressions throughout the world, rat sex is taken as the biological model.[19] And even the rat becomes impoverished in the process, since the only ratty behaviors that are considered relevant are "mounting" (defined as "male," although females mount quite frequently too), and "lordosis," in which the female presents by (usually) sticking her rear in the air. Ruth Bleier considers this implicit equation of human eroticism with rat posture a typical case of the simplistic thinking that pervades much of the research in this area and comments:

> [Such simplistic thinking] posits the existence in women and men
> of a set of stereotypical sexual behaviors that are reflexly
> unleashed by certain hormonal states, and implies that people
> [especially homosexuals] assume stereotyped sexual postures,
> equivalent to mounting or lordosis, that can be used as a measure
> of human sexuality.[20]

A related, usually unquestioned assumption is that hormones affect behavior, and not the other-way around. Yet there is clear evidence that in a variety of animals behavior can influence hormone levels, and behavioral effects on hormones are also evident in humans. Stress, for example, can affect levels of steroid hormones,[21] as most women are aware who have noted how susceptible their menstrual cycles are to stress. Yet, although there is plenty of evidence of behavioral effects on hormone secretion, this fact is hardly ever mentioned in the literature on homosexuality. For example, several papers that report that a group of lesbians has more of a particular hormone than has a group of heterosexual women,[22] then go on to assume that the hormonal differences in some way produce the behavioral differences. Occasionally, authors note that some aspect of behavior *might* affect the hormone level, but they rarely pursue this further. In this context, it is important to recognize that the term behavior must not be taken to signify only sexual behavior. Obviously, the life of a lesbian, hence her "behavior," can differ in many ways from that of a heterosexual woman, since the latter is not forced into a ghetto existence, and does not need to deny part of her life in order to keep her job, her house, or her children; yet any of these differences in life situations might affect hormone levels.

There is another way of interpreting hormonal differences, should they exist. Lesbians are not solely women who relate sexually to other women; they are also women who do not, in general, relate sexually to men. And this too could affect hormone levels. Not surprisingly, these kinds of interpretations are not found in the pages of scientific/medical journals where, instead, the hormone levels and behavior of heterosexual women are taken as the standard against which lesbians are judged—and found wanting.

Another assumption that permeates this research is that lesbians are somehow more masculine than heterosexual women, gay men more feminine than heterosexual men. This assumption is just one example of the general tendency to describe sex and gender in terms of the polarized categories of two opposing and non-overlapping groups: sexual orientation is seen as homosexuality *versus* heterosexuality;[23] gender identity as feminine *versus* masculine, and so forth.

The primary assumption in the literature on sexual preference is that since homosexuality is defined as abnormal, gay people must *also* be abnormal according to one of the other polarized categories. Lesbians, for example, are thought to have a more "masculine" gender identity. Then, having made that association, observers are more likely to notice only individuals who tend to conform to the preconceived model. Thus women who are considered to be "unfeminine" in appearance often are simply assumed to be lesbians, especially if they do not seem to have a man in their lives; while "feminine" women, like "masculine" men, are categorized as heterosexual without question.

The result is that most hormone research regarding homosexuality begins with a firmly entrenched model of homosexuals as "masculine" females and "effeminate" males, and simply seeks to explain this "fact" by looking for an excess of hormones defined beforehand as "male" in lesbians, and for a corresponding deficit in gay men.

Hormones and Homosexuality

During the 1940s, shortly after the steroid hormones commonly known as "sex hormones" (for example, estrogens and androgens) had first been synthesized, there was considerable interest in their possible relationship with homosexuality. The hunt was on for hormonal differences between lesbians and heterosexual women, between gay and straight men. Despite considerable interest and attempts to use hormones to "treat" homosexuality,[24] this research effort failed to produce any significant results.

There have been a few similar studies since then, many of them using extremely dubious methods. For example, studies rarely match the groups of lesbians and heterosexual women for important social variables such as age, number of children, or class.[25] The "methods" sections of these papers make interesting reading. For instance, in one report, which claimed to have found raised levels of androgens (the so-called "male hormones") in lesbians (which would, of course, accord with the stereotype of lesbians as exceptionally masculine), the number of women tested was four—two lesbian couples, hardly the stuff of sophisticated statistics.[26] Despite these ludicrously low numbers, as well as a variety of other methodological flaws, this report was published in the highly prestigious scientific journal *Nature*.

Of the studies which are less questionable on methodological grounds, the majority exhibits no consistent hormonal differences between lesbians and heterosexual women.[27] Given the bias of scientific journals in favor of publishing *positive* findings, one would expect that many studies showing no difference may have gone unpublished. Hence it seems safe to conclude that no consistent hormonal differences have been shown to exist between adult lesbian and straight women.

Before leaving the problem of methodology, I want to raise another point. Few of the papers cited here address the problem of definition. "Homosexual" is simply a blanket word applied to anyone who has sexual relations with a member of the same biological sex. Thus, some studies include within the homosexual sample individuals who may also be having heterosexual sex. One study, for example, included lesbian women who at the time were taking oral contraceptives.[28] Alternatively, the heterosexual sample may include data from bisexuals,[29] or transsexuals may be included within the homosexual sample.[30] There are many reasons why transsexuals should not be thought of as similar to homosexuals,[31] not the least of which is that transsexuals commonly seek hormone treatments.

Since the search for hormonal correlates of homosexuality in adults has failed to produce any significant findings, researchers have turned instead to the embryo—and specifically to the ways in which hormones might influence the brain during embryonic life. The idea that brain development might be in some way defective in homosexuals is not new. One nineteenth-century reformer, Ulrich, pleaded for tolerance toward homosexuals on the grounds that they were products of an anomalous embryonic development.[32] And when the question of legislation against lesbianism was being discussed by the

British Parliament in 1921, it was suggested that lesbians were victims of an "abnormality of the brain." [33]

Prenatal steroid hormones are now known to be crucial for normal sexual differentiation in mammals. The embryonic testes secrete androgens, which enter the body's cells. There the major androgen, testosterone, is converted to an estrogen, which in turn is responsible (paradoxically) for the so-called masculinizing effect of testosterone on the hypothalamus of the brain. Among other things, this part of the brain sends hormonal messages to the pituitary gland. In this way it controls, in turn, the cyclic production of hormones that are responsible for the periodic release of fertilizable eggs from the ovaries.

In laboratory rats—the animals on which much of this work has been done—these hormones also seem to affect later behavior that scientists have stereotypically associated with mating. The rat is used for many of these studies because it is born while its nervous system is still very immature and hence can be affected by administering hormones to newborns. For example, if a newborn female rat is given androgens, she may be more likely than her untreated sisters to show behaviors that are defined as "male." For instance, she may mount other individuals more and be less inclined to raise her rump into the receptive lordosis position. Normal female rats, as I have indicated, also mount others when in heat, as well as presenting their rumps in lordosis. The hormone treatment merely has shifted the balance and made mounting more likely and lordosis less likely. By contrast, a male rat's development can be pushed in the opposite direction by castrating it at birth, which removes its major source of androgens. Because it is thus possible to manipulate particular *postures* of the laboratory rat by early exposure to hormones, the suggestion has been made that what makes a homosexual person behave as she or he does is an aberrant pattern of hormone levels during comparably early stages of development, which for people means before birth.

One of many difficulties with interpreting this literature is that a large number of terms is used as though they are scientific and neutral, when in fact they are heavily ideological. I have already referred to the problems inherent in using the terms *gender identity, sex role*, and so on. The hormone literature further relies heavily on concepts such as "masculinization" and "defeminization" as a way of explaining how early exposure to these hormones can produce behavioral changes in later life (at least in rats). These concepts are, to say the least, problematic. "Masculinization," for example, refers to tipping the balance in favor of behavior patterns that *have already been defined as male*. The point is that, for example, mounting is clearly something that males of most mammalian species have to do during copulation. However, as I have noted, mounting is also something that females of many species normally do at various times. For both sexes, it may have a social as well as a sexual significance. Similarly, once lordosis has been defined as a *female* response, it becomes "natural" for scientists to speak of "defeminized" females, meaning those who as adults show lordosis less frequently following exposure to hormones in early life.

In the literature on humans, "masculinization" is simply assumed to be a basis for lesbianism.[34] Consider the following, quoted from Meyer-Bahlberg's 1978 paper on "Sex Hormones and Female Homosexuality." Having made reference to the standard rat studies, he continues:

> Studies on lesbian women have provided scattered supportive evidence for a shift toward generalized masculinity. . . . [One study] found that 70% of the lesbian women but only 16% of the heterosexual women had been tomboys in childhood. . . . To establish a generalized shift toward masculine behavior in lesbian women requires comparisons with both male and female controls. For instance, the argument that lesbian sexuality is femalelike in comparison to the sexuality of male homosexuals . . .—on account of such indicators as number of partners or need for affective rather than purely sexual relationships—bypasses the observation that, compared to female heterosexual controls, lesbian women show a greater number of partners and of short-lived relationships. This again may be interpreted as a shift toward the masculine behaviour spectrum.[35]

No it may not. It is more reasonably interpreted as a valiant attempt on the part of the author to fit lesbianism into his biodeterministic model. There may be a multitude of reasons why lesbians, on average, have more sexual relationships than heterosexuals—their way of life, the social approval and support given heterosexuality through marriage, and so on.

"Masculinization" and "defeminization" are, then, terms that are based upon an *assumed* distinction between types of behavior, with one behavior defined as the exclusive prerogative of males and the other of females. The absurdity of this distinction is revealed by the suggestion that some "defeminization" can occur in normal females—of course, female rats.[36] Thus, by giving pups drugs that oppose androgen activity, these authors claim to have produced a superfeminine rat—one that sticks her rump in the air more readily than usual. Stereotyping is reinforced also by the fact that, when sex hormones are studied, usually only their effects on the "passive" and "receptive" posture of females are considered. The fact that the females actively control whether intercourse will take place[37] does not often enter into the definitions of femininity and masculinity on which the research is based.

In the light of these various implicit and explicit assumptions about the behavior of females, it is not surprising that a tendency to show malelike behavior (for example, more mounting in rats) is called "lesbian." Among the absurdities of the designation is that it is usually applied only to the rat that does the mounting. What about the female whom she mounts?

One study, however, not only denotes behavior more typical of the other sex as equivalent to "homosexuality;" the author, Gunter Dörner, also suggests that the frequency of sexual encounters by rats with members of the same sex is increased by early hormone treatments. That is, the female that has received extra doses of androgens more frequently has "sexual" encounters with other females than with males.[38] At first sight, this might suggest that he has indeed succeeded in creating something of a lesbian rat. On closer consideration, however, it does not. Females given androgens, as noted above, are more likely to mount than are other females. But most rats, unless they are females during the few hours of heat, normally repel another rat trying to mount them. It follows that the only rats that will allow the hormone-treated females to mount will be other females who are at that point receptive. A similar sort of explanation holds for castrated males who, indeed, are more likely than untreated males to show lordosis to the mounts of other males. Since, on average, males are more likely to try to mount than are females, it follows that a castrated male is likely to be mounted by other males. All that the hormonal manipulation has done is to increase the relative probabilities of mounting and lordosis, which in turn makes what are defined as "homosexual encounters" more likely.

These kinds of experiments are as flawed as the earlier work that looked for hormonal differences between lesbians and heterosexual women. In both, by accepting the idea that the lesbian is socially deviant, the researcher can infer that her biology is deviant, and hence that it *causes* behavior. The obvious correlation of that inference is that, with appropriate hormonal treatment, homosexuality can be cured. These suggestions are simply false. They are also dangerous.

To date, it has not proved possible to alter an animal's—or a person's—sexual orientation by biological means, despite occasional claims to the contrary. When we do hear that a form of treatment has been "successful" (as is sometimes claimed for aversion therapy, which I discuss below), what is usually meant is that the person's sex drive has been diminished, not that her or his sexual orientation has been permanently altered from homo- to heterosexual.

There is often an implicit assumption that human behavior is to a large extent determined by prenatal or adult hormones, thus denying that significant contributions are made by the individual's interaction with her or his complex environment. At present, there is no evidence that hormones as such determine human behavior; so it seems somewhat premature to posit that they determine sexual behavior. Human sexuality is complex, and is a product of a great many interrelated factors, whose interactions we understand little, if at all. Insofar as it is biological fact that I have a woman's anatomy, and that fact is likely to affect the way I experience my sexuality, biology is of course relevant to human sexuality. But that, on present evedence, is all one can posit with any certainty about biological "determination" of sexuality. I would say that my sexuality is no more determined by my estrogens than is, say, the fact that I am a scientist.

Many women are now coming out as lesbians within the Women's Movement, which may present some problems for people who cannot give up the idea of hormonal determination. How can you offer biological theories to explain the life story of the woman who was once married and has children, but who now thinks of herself as lesbian? Lesbian-feminists usually consider their lesbianism to be a matter of choice; they tend to see it as an extension of the concern and affection for other women that feminism teaches us to value. But rather than face that contradiction squarely, the medical literature instead dodges it by proclaiming that when a woman *chooses* lesbian relationships, that is "pseudohomosexuality,"[39] thus making a distinction between lesbians in the Women's Movement, and "real" lesbians.

Sickness or Evil? Prevailing Forms of Punishment

As I pointed out at the beginning of this chapter, the Church for a long time taught that people indulging in non-procreative sex of any kind deserved punishment. Homosexual activity, like practicing contraception, was held to be a sin. Although views of homosexuality have changed in ensuing centuries, the punitive response still prevails, either directly in the Church's version of homosexuality as evil, or indirectly in the medical version of homosexuality as disease requiring "treatment."

Since "treatment" is based on the supposition that homosexuality is abnormal, the usual object is to eliminate the abnormal behavior. Among the various forms that the treatment can take are ridicule, hormone treatments, surgery, and aversion therapy. I will deal with each of these in turn.

A. Ridicule: The Psychotherapeutic Approach. As a result of trying to live in a world that condemns their very existence, lesbians sometimes end up seeking, or being referred to, psychotherapy. Once in psychotherapy, they are subjected to whatever biases their therapist might hold. Sometimes he or she may be reasonably sympathetic to gays; more often, he or she is not. Phyllis Chesler describes the attitudes of some male therapists toward lesbians:[40]

"How could you? Women are dirty, they smell." A strange comment, one would think, from a heterosexual male.

One of the clearest indications of the attitudes of many therapists is gained from some of the things they write for other members of their profession. Here are a few samples:

> . . .the homosexual relation is an immature one, an arrest of normal sexual development at an adolescent stage.[41]

> Basically, all homosexuals are alike, looking for love where there can be no love, and looking for sexual satisfaction where there can be no lasting satisfaction.[42]

> One must guard against the facile assumption of many homosexual apologists that what is biologically natural is necessarily desirable or permissible. Murder does not become any more tolerable because it can be viewed as the expression of a natural impulse.[43]

> Most lesbian women are content to keep their homosexual inclinations hidden from general view and it is only the most psychopathic among them who make a show of their abnormality.[44]

> . . .one vagina plus another vagina still equals zero.[45]

> Those lesbians who protest that, for them, this kind of relationship is better than any possible intimacy with a man do not know what they are missing . . . to be a woman loved by a man and who has children by him is the first and most important aim of feminine existence.[46]

That these statements only indicate the writer's prejudice, rather than anything about homosexuality, should be clear. It is also clear that many of these statements could equally well apply to a great many relationships that are respectably heterosexual.

B. Hormone Treatments: Possible Prevention? One type of treatment that has been used, although predominantly on men, is the administration of a hormone to diminish the sex drive. This is sometimes used on homosexuals imprisoned for sexual offenses. These drugs are commonly antiandrogens—synthetic chemicals that work by counteracting the effects of the man's own androgens. The treatment does nothing to alter the sexual orientation, only the sex drive.

Use of such drugs raises a number of important ethical questions. Inmates of psychiatric institutions are sometimes *expected* by doctors to be willing to try the drugs if they can thereby avoid a long stay. In prisons, sometimes the drugs are offered as a precondition to obtaining parole. As long as homosexuality remains stigmatized, there continues to be a risk of such pressure being exerted.

That potentially behavior-controlling drugs are used extensively to control people whose behavior is defined as deviant has now been well documented.[47] Lesbians are somewhat less at risk than gay men from this kind of abuse, if only because they are less likely to end up in prison since the law usually denies their existence. But they can, and sometimes do, end up in psychiatric institutions, whose programs of "therapy" often include behavior-modifying drugs of all types. Hormones to diminish sex drive are only one part of the armory of chemicals that can be used to control.

Copyright © by Jo Nesbitt

Another type of "treatment" has recently been suggested. As I have indicated, research on other mammals suggests that prenatal steroid hormones (that is, androgens) affect differentiation of the hypothalamic portion of the brain. From his work with rats, Dörner has proposed that the brains of homosexuals (of either sex) respond differently to these hormones than do those of "normal" people, and he attributes this difference to as yet hypothetical prenatal effects of the hormones.[48] Based on the work that I described above, which led Dörner to designate as *homosexual behavior* the tendency of male rats who had been castrated at birth to show lordosis, he has now hypothesized that it might be feasible to "prevent homosexuality" by administering appropriate hormones to pregnant women. He has been working hard to develop suitable procedures, which begin with withdrawing samples of the fluid that surrounds the embryo in order to determine its sex and its hormone levels. This work is repugnant from the point of view not only of the unborn children who are needlessly exposed to possibly dangerous steroid hormones,[49] but also of the women who become his guinea pigs.

Dörner has also been trying to prove that the brains of homosexuals are somehow different from those of heterosexuals. So far, he has published results only on men, but he suggests that lesbians' brains are "abnormal" too, a proposition for which he has turned up no evidence.

At the end of his 1976 book, Dörner expresses doubts whether one ought to tamper with homosexual proclivities, given that so many (male) artists of merit were homosexual! However, he overcomes his qualms by noting that "[those] with inborn sexual deviations are suffering from psychosexual pressure," which produces a high rate of suicide. He does not try to prove this assumed causal link, nor to consider the possibility that the pressure originates from a society intolerant of people who do not conform to its rigid notions of acceptable behavior.

C. The Surgeon's Knife. Genital surgery on women, in the form of clitoridectomy and a variety of other sexual abuses, has long been performed to keep female sexuality under control. While genital surgery is used less frequently now, at least in the West, as a means of controlling homosexuality, psychosurgery has sometimes been employed in recent years. Dörner, amid his equations of people and rats, discusses the possibility of psychosurgery for homosexual men. Studies of hormone-treated rats have indicated to him that injuring the hypothalamus may reduce their capacity to "behave homosexually." From this, he concludes that:

> If our data can be confirmed in other species, a destruction of the "female mating center" [i.e. the ventromedial hypothalamus] . . . could be taken into consideration for homosexual men.[50]

"Could be taken into consideration" is dishonest since this has *already* been done, for "studies" have been conducted in which the hypothalamus of gay men was destroyed.[51] This may be the most extreme current example of the forms of social control that can result from defining homosexuality as sickness. But whereas such treatment is relatively uncommon, it should be remembered that lesbians and gay men have been subjected to a range of nonsurgical but nonetheless brutal treatments to "cure" them of their deviant behavior, beginning with the very act of labeling homosexuals as sick or deviant. Relatively common among non-surgical forms of violence against homosexuals is aversion therapy.

D. Shocking Tactics: Aversion Therapy. The use of aversion therapy is based upon the assumption that it is possible to condition a "homosexual" into a preference for the opposite sex and a revulsion for one's own sex. The approach is basically a behaviorist one and, as such, an antithesis to the biological models that I have discussed up to this point. The assumption is that, by a process of earlier conditioning, the person has attached her or his sexual impulses to the "wrong" stimulus. One suggestion has been that the person may have become conditioned by masturbating to fantasies of these "wrong" stimuli, and slowly built up the association. Masturbation to the "right" stimuli can form part of what is then called "therapy."[52] Apart from all the heterosexist assumptions underlying this form of therapy, the treatments ostensibly used to eliminate the deviant behavior are particularly punitive.

Aversion therapy is carried out by such abhorrent methods as the use of drugs to make the person vomit or become very dehydrated, or the use of electric shocks. Patients are usually men. At the first sign of sexual arousal to "deviant" stimuli (for example, pictures of naked men), the victim is given a painful electric shock.[53] Following this, he may be given hormone injections while being shown pin-ups of "attractive young women" (more heterosexist assumptions: the reference is never made simply to pictures of women; they are invariably of the "pin-up" variety). In the various reports of such procedures, we are never told of the "patient's" subsequent fate. Indeed, follow-up studies of these "patients" are rarely done. The attitude of many of the therapists using this kind of "therapy" is well illustrated by a report of treatment of a transvestite male:

> At one session, by a particularly happy [*sic*] chance, one of his favorite pictures fell into the vomit in the basin so that the patient had to see it every time he puked.[54]

But although the literature is primarily concerned with men, women do not escape entirely. I have met many lesbians who have been given electric shock "treatment" to "cure" them. In some cases, this has been through the genitals. In one case shocks were given through the brain to "correct" a woman's bisexual behavior, and in this case, it "succeeded." The woman is now exclusively lesbian.

These "treatments" are meted out by the "caring professions:" therapists, social workers, and doctors. It is immaterial which one of them actually administers the shocks, because the victim is often powerless in the face of their collusion. They show even less concern for the former lover who is totally forgotten in the enthusiastic reports of the "success" of electric-shock treatments. Jane Rule points out: "It is left to other kinds of therapists to deal with the rejected lover, spattered with all that healthy and righteous vomit, shaken by grief and rage,"[55] and having to continue her life without her lover.

It is clear that all these "treatments" are inherently punishments. Lesbians have, throughout history, been punished in one way or another for our failure to conform. That punishment reached its height in the Nazi gas chambers, in which thousands of lesbians and gay men were murdered, but it is with us still. Gay people, and especially lesbians, are continually harassed, thrown out of jobs, deprived of their homes and of their children, and are victims of physical assault. The punishments are everywhere, but it is only as medical "treatment" that such actions are elevated to the status of "scientific" procedures.

What Can We Do?

The advent of the Women's and Gay Liberation Movements since the 1960s has made people less afraid to admit that they are gay. These social changes have been the first step toward confronting prejudice. But prejudice against lesbians continues to exist at many levels in our society, and I have documented only a small fraction of it here. If feminists are to fight back, to attempt to change this massive and deep-rooted prejudice, then we too have to work at many levels. From forms of consciousness raising, to street demonstrations, to abstract theorizing: we need it all.

"Scientific" arguments are used increasingly to justify a range of power relations in our society, whether these are relations of class, race, sex, or sexual preference. The arguments used to "prove" that lesbians are abnormal, maladjusted, or hormonally deficient are just another example. We clearly need feminist voices to be raised long and loud in opposition to this pseudo-science. Even if, as so often happens, our words are censored or ignored, we can begin to effect change, however small, just by speaking up. For example, in 1979 I gave a talk in which I criticized the hormonal theories of homosexuality at a conference on "Psychosexual Medicine" in Liverpool, England. The audience was composed mainly of medical doctors. I was told, and still have the letter to prove it, that my paper would subsequently be published in the *British Journal of Sexual Medicine*. I need hardly add that it was not, although the *Journal* continues to publish papers that support a biological theory of homosexuality even when the scientific methodology is atrociously bad.[56] All this is depressing, though scarcely surprising. I mention it here because I discovered later that, although most of the audience did not want to hear my criticisms of their pet prejudices, a few listened and began to question. If just a few listened, then I believe it was worthwhile.

At the same time as we criticize the poorly thought out and executed work, though, we have to try to move forward, to rethink existing assumptions and concepts within biology and medicine. I am not sure what people mean when they write of creating feminist biology, because I am not sure that we yet understand what the scientifically important questions are, nor in what social and intellectual framework we should be asking them. We know what is wrong with the current theories that limit and devalue us, but we need to work hard at developing our understanding of the kinds of analysis we would like to move toward. This is beginning to happen, and is an urgent task. It is only by counterposing more liberating visions of biology to what we now have that we can begin to change existing dogma. And only when we have changed that dogma can lesbians begin to be free of the oppression and punishments that have for centuries been our lot.

Notes

1. I am aware that many lesbians, especially in the Women's Movement, dislike being subsumed under the label "homosexual," as this ignores the reality of our oppression as *women*, which is generally more pervasive than our oppression as "homosexuals." However, I use the term here as the medical literature that I describe uses it, to describe women and men. Thus, the biological models that are produced are commonly phrased to cover both sexes. That the theories are applied to both sexes does not, however, mean that the punishments that I also describe are equal; lesbians are as likely to be "treated" for their assumed rejection of the feminine role as they are for their "abnormal" sexuality.
2. V. L. Bullough, *Sexual Variance in Society and History* (Chicago, IL: University of Chicago Press, 1976).
3. *Ibid.*, especially p. 361.
4. M. Foucault, *The History of Sexuality, Vol. 1: An Introduction* (London: Allen Lane, 1979).
5. J. Weeks, "Movements of Affirmation: Sexual Meanings and Homosexual Identities" (Paper presented at the Annual Meeting of the British Sociological Association, University of Sussex, England, 1978).
6. M. McIntosh, "The Homosexual Role," *Social Problems* 16 (1968), 182–192; and J. Weeks, *Coming Out: Homosexual Politics in Britain from the Nineteenth Century to the Present* (London: Quartet Books, 1977).
7. V. L. Bullough, "Homosexuality and the Medical Model," *Journal of Homosexuality* 1 (1974), 99–110.

From Sin to Sickness: Hormonal Theories of Lesbianism 87

8. C. Wolff, *Love Between Women* (London: Duckworth, 1971).
9. M. Borrell, "Organotherapy, British Physiology, and the discovery of the internal secretions," *Journal of the History of Biology* 9 (1976), 235–268.
10. *See* for example S. Butler, *Conspiracy of Silence: The Trauma of Incest* (San Francisco, CA: New Glide Publications, 1978).
11. E. g., B. M. Mucklow and G. K. Phelan, "Lesbian and Traditional Mothers' Responses to Adult Response to Child Behavior and Self Concept," *Psychological Reports* 44, 3 Part 1, (1979), 880–882.
12. S. Freud, *Three Essays on the Theory of Sexuality* (1905) (London: Pelican Freud Library, 1973).
13. C. S. Ford and F. A. Beach, *Patterns of Sexual Behavior* (London: Eyre and Spotiswoode, 1952). These authors cite data suggesting that male homosexuality is acceptable in some form or another in approximately half the human societies for which data were then available. Such comparisons are, however, difficult to assess, since the meaning attached to homosexuality within a given society may be very different from that given by anthropologists studying the society. There are few data on lesbianism cross-culturally principally, it appears, because researchers have neglected to take the most rudimentary steps necessary to discover them. *See* for example J. Money and A. Ehrhardt, *Man and Woman, Boy and Girl* (Baltimore: Johns Hopkins University Press, 1972), Chapter 7, as discussed in B. Fried's article in this collection.
14. C. W. Socarides *et al.*, "Homosexuality in the Male." *International Journal of Psychiatry* 11 (1973), 461–479. But see reply by:
15. G. C. Davison, "Homosexuality–the Ethical Challenge," *Journal of Consulting and Clinical Psychology* 44 (1976), 157–162.
16. R. Papatheophilour, S. James, and A. Orwin, "Electroencephalographic Findings in Treatment-seeking Homosexuals Compared with Heterosexuals: A Comparative Study," *British Journal of Psychiatry* 127 (1975), 63–66.
17. The chief diagnostic use of the E.E.G. is, of course, in determining seizure foci in cases of epilepsy. For this purpose, it has considerable value. Nonetheless, what meaning to attach to the various wave forms of the E.E.G. beyond this is difficult to estimate. It is certainly impossible to use it to detect "cerebral immaturity."
18. P. D. Griffiths *et al.*, "Homosexual Women: An Endocrine and Psychological Study," *Journal of Endocrinology* 63 (1974), 549–556.
19. The impulse to draw sweeping generalizations from rat data to human sexuality seems to be commonplace. Two particular examples are provided by: M. McCulloch, "Biological Origins of Homosexual Behavior in Man," *British Journal of Sexual Medicine* 7 (1980) 36–41; and G. Dörner, *Hormones and Brain Differentiation* (Amsterdam: Elsevier, 1976).

20. R. Bleier, "Social and Political Bias in Science: An Examination of Animal Studies and Their Generalizations to Human Behaviors and Evolution," *Genes and Gender II: Pitfalls in Research on Sex and Gender*, R. Hubbard and M. Lowe, eds. (New York: Gordian Press, 1979), p. 54.
21. *See* for example R. M. Rose, P. Bourne, and R. Poe, "Androgen Responses to Stress," *Psychosomatic Medicine* 31 (1969), 418–436; and L. E. Kreuz, R. M. Rose and J. R. Jennings, "Suppression of Plasma Testosterone Levels and Psychological Stress," *Archives of General Psychiatry* 62 (1972), 479–482.
22. For example, J. A. Loraine *et al.*, "Patterns of Hormonal Secretion in Male and Female Homosexuals," *Nature* (London) 234 (1970), 552–554. Also N. K. Gartrell, D. L. Loriaux, and T. N. Chase, "Plasma Testosterone in Homosexual and Heterosexual Women," *American Journal of Psychiatry* 134 (1977), 1117–1119. For a review of similar studies, see H. L. Meyer-Bahlberg, "Sex Hormones and Female Homosexuality: A Critical Examination," *Archives of Sexual Behavior* 8 (1978), 101–119.
23. While many people think of themselves as "bisexual," they usually tend to see this in terms of the prevailing dichotomy, and hence tend to believe that their sexuality contains elements of *both* homosexuality and heterosexuality. See the verbatim reports in C. Wolff, *Bisexuality: A Study* (London: Quartet Books, 1979).
24. G. I. Swyer, for instance, referred to earlier attempts to give homosexuals hormones as a cure in his paper "Homosexuality: The Endocrine Aspects," *Practitioner* 172 (1954), 374–377.
25. *See* for example F. E. Kenyon, "Studies in Female Homosexuality IV and V," *British Journal of Psychiatry* 114 (1968), 1337–1350; and P. D. Griffiths *et al.*, "Homosexual Women: An Endocrine and Psychological Study," *Journal of Endocrinology* 63 (1974), 549–556.
26. Loraine *et al.*, *Patterns of Hormonal Secretion*.
27. *See* for example A. J. Eisenger *et al.*, "Female Homosexuality," *Nature* (London) 238 (1972), 106; and Griffiths *et al.*, *Homosexual Women*.
28. Griffiths *et al.*, *Homosexual Women*.
29. Dörner, *Hormones and Brain Differentiation*.
30. Meyer-Bahlberg, *Sex Hormones*.
31. *See* J. G. Raymond, *The Transsexual Empire* (London: The Women's Press, 1979).
32. Weeks, *Coming Out*.
33. *Ibid.*, pp. 126–127.
34. *See* for example McCulloch, *Biological Origins*; Meyer-Bahlberg, *Sex Hormones*.
35. Meyer-Bahlberg, *Sex Hormones*, p. 102.
36. L. G. Clemens and B. A. Gladue, "Feminine Sexual Behavior in Rats Enhanced by Prenatal Inhibition of Androgen Aromatization," *Hormones and Behavior* 11 (1978), 190, 201.

37. M. McClintock and N. T. Adler, "The Role of the Female During Copulation in Wild and Domesticated Norway Rats (*Rattus Norvegicus*)," *Behavior* 68 (1978), 67–96.

38. *See* for example Dörner, *Hormones and Brain Differentation* and G. Dörner, "Hormones and Sexual Differentation of the Brain," *Sex, Hormones and Behavior*, Ciba Foundation Symposium 62 (1979), 81–112. Herbert similarly makes an equation between "male-like" behavior in rhesus monkeys, and lesbianism, as though the two were synonymous. *See* J. Herbert, "Neurohormonal Integration of Sexual Behavior in Female Primates," *Biological Determinants of Sexual Behavior*, J. B. Hutchison (ed.), (Chichester: Wiley, 1978).

39. Z. DeFries, "Pseudohomosexuality in Feminist Students," *American Journal of Psychiatry* 133 (1976), 400–404.

40. P. Chesler, *Women and Madness* (New York: Avon, 1972).

41. A. L. Winner, 1947, cited by F. E. Kenyon, "Female Homosexuality—A Review," *Understanding Homosexuality: Its Biological and Psychological Bases*, J. A. Loraine, ed., (Lancaster, England: Medical and Technical Publishing, 1974).

42. D. J. West, *Homosexuality* (London: Duckworth, 1955).

43. *Ibid*.

44. A. Munor and W. McCulloch, *Psychiatry for Social Workers* (Oxford: Pergamon, 1969).

45. D. Reuben, *Everything You Always Wanted to Know About Sex—But Were Afraid to Ask* (London and New York: W. H. Allen 1970), p. 215.

46. A. Storre, *Sexual Deviation* (Harmondsworth: Penguin, 1964).

47. *See* for example S. L. Chorover, *From Genesis to Genocide*, (Cambridge, MA: M.I.T. Press, 1980).

48. Dörner, *Hormones and Brain Differentiation*.

49. We should be very wary of treating fetuses with hormones, bearing in mind the awful story of diethylstilbesterol (D.E.S.). Following treatment of women with this hormone in pregnancy, to prevent miscarriage, their daughters were found to be likely to contract a normally extremely rare form of vaginal cancer. *See* K. Weiss, "Vaginal Cancer: An Iatrogenic Disease," *International Journal of Health Services* 5 (1975), 235–252.

50. Dörner, *Hormones and Brain Differentiation*, p. 140.

51. F. D. Roeder, D. Müller, H. Orthner, "Weitere Erfahrungen mit der Stereotaktischen Behandlung Sexueller Perversionen," *Journal Neuro-Visc. Relat.* 10 (1971), 317–324. *See* also review by E. Schorsch and G. Schmidt, "Hypothalamotomie Bei Sexuellen Abweichungen," *Der Nervenarzt* 50 (1979), 689–699, which takes a more critical view.

52. *See* for example J. W. Blitch and S. N. Haynes, "Multiple Behavioral Techniques in a Case of Female Homosexuality," *Journal of Behavior Therapy and Experimental Psychiatry* 3, (1972), 319–322.

53. For example, R. Matthews, "Behavior Therapy for Sex Offenders," *British Journal of Sexual Medicine* 8 (1980), 12–17.

54. M. P. Feldman, "Aversion Therapy for Sexual Deviation: A Critical Review," *Psychological Bulletin* 65 (1966), 65 *et seq.*
55. J. Rule, *Lesbian Images*, (London: Peter Davies, 1976).
56. Two examples that might interest readers are the paper by McCulloch that I have cited above, and S. Crown, "Sex and Feminism" (*sic*), *British Journal of Sexual Medicine* 7 (1980), 42–43.

[*Lynda Birke is a research Fellow in the Biology Department at the Open University in Milton Keynes near London. She is involved with the Open University's Women's Studies course and teaches and does research on animal behavior. Her current research interests include the effects of hormones on behavioral development. She was a member of the Brighton Women and Science Group, which collectively edited* Alice Through the Microscope: The Power of Science over Women's Lives *(London: Virago, 1980). She has been actively involved in feminist and lesbian groups.*]

Figure X. Lifting
(Photo: D. Wald)

Marian Lowe

Social Bodies: The Interaction of Culture and Women's Biology

The stereotypes we accept about sex roles have far-reaching effects. Ideas about appropriate behavior for women and men act as powerful constraints on behavior and often become self-fulfilling prophecies. The effects of sex-role stereotyping can go even further, however; not only behavior but also biology can be modified by the constraints imposed by rigidly defined sex roles. In part, we literally may be shaped by our social roles. The observation that this is so is in sharp contrast to the assumption that has been made over and over again during the last 150 years that sex roles are themselves the products of biologically determined differences in women's and men's natures.

Current theories based on this assumption, such as sociobiology, do not claim that sex differences in behavior are completely determined by biology, but speak of "propensities" or "predispositions" for certain behaviors depending on one's sex.[1] While somewhat less rigidly deterministic than their nineteenth-century counterparts, such modern theories continue to assume that innate biological differences lurk "underneath it all." As various critics have pointed out, there are a number of difficulties with these theories,[2] perhaps the most fundamental being that a great many aspects of biological functioning—including body size and strength, hormone levels, and, possibly, brain development—can be changed considerably by changes in the environment.

When considering the origins of sex differences in behavior, it is important to acknowledge the extent to which, in contemporary American culture (and in most other present-day societies) women and men grow up and live in radically different social environments. The kinds of work we do, our physical environments, the way we spend our leisure time, our relationships with other people, and many other aspects of life differ greatly for women and men. At present, some women's environments are undergoing rapid change. In many cases, these changes expose women to work and life experiences that are much more like those traditionally seen as appropriate for men. This is true for women entering such male-dominated fields as medicine, law, science, coal mining or heavy construction; and also 'for women who become involved in athletics. As women's environments and activities become more like men's, both in our society and in others, the extent of many sex differences decreases: spatial visual ability, some measures of aggression, and a number of other behavioral traits have become more alike in women and men[3] and, as we will see, so have a number of biological traits such as size and strength. We do not know how much of the remaining sex differences may be due to inborn differences and how much to environmental differences. Certainly, there is no reason to suppose that whatever differences in behavior and biology remain must be due only to innate factors.

Sex Differences in Strength and Physical Performance

A great many cultures, including our own, assume that there are fairly substantial innate sex differences in body form that explain why men generally show superior performance in tasks reflecting strength or athletic ability. Such assumptions have powerful personal and social consequences. Different attitudes toward our bodies constitute a basic part of our self-identification as feminine or masculine. The assumption that women cannot do anything that requires strength powerfully reinforces social roles; at the same time, it also produces some of the actual strength differences. For example, stereotypes about sex differences in strength have affected not only women's participation in various sports and physical activities, but also the kinds of exercise and jobs thought suitable for them.

The United States wage labor force is highly segregated by sex, and this economically important division of work into women's and men's jobs is in part rationalized by ideas about innate differences in strength between the sexes. Even though many women work at jobs where a great deal of strength is needed (such as nursing, which involves the lifting of sick and sometimes obese adults), still the image of the fragile female is used to keep women out of various higher-paying occupations, such as construction work or heavy industry. This is only one example of the fact that a smaller, frailer female frame is seen as not just *consistent* with women's social role and lack of social power, but as one of the *reasons* for it.[4] However, the idea that physical differences in strength can be used to explain differences in social position is clearly erroneous. Most "male" occupations, particularly those with prestige and power such as corporate executive, politician, doctor, or army officer, do not require physical strength. In general, jobs that involve physical labor, such as digging ditches or construction work, tend not to be highly valued in our society. (The "jobs" held by a few highly paid male athletes are among the few exceptions to this rule.) In spite of this reality, men's presumed greater innate physical strength has come to function as a symbol of greater male social power.[5]

In reality, sex differences in strength have had little to do with causing the current sexual division of labor and indeed, the origin of these differences is open to question. There is growing evidence that differences in physical strength could come as much from differences in life experience as from innate factors. The relative contributions of environment and biology to sex differences in strength and other physical abilities, just as to sex differences in behavior, turn out to be impossible to pry apart. After all, not only may women's and men's biology be different, but sex differences in social roles result in major differences in the ways in which we use our bodies. From earliest childhood, the activities that are encouraged for girls and boys are different. As little girls, we have probably all been admonished to "act like a lady;" and to move, speak, and think like a "lady" is not without biological effects on our shape, strength and other abilities. Studies have indicated that, beginning at birth, parents exercise the limbs of their baby boys more frequently and vigorously than they do their girls'.[6] Later on, girls and boys are channeled into different kinds of play, with the result that many girls are involved in more sedentary activities. In adult life, the earlier patterns tend to continue, although we are now seeing some changes.

In examining possible origins of sex differences in size and strength, it is helpful to look first at some of the sex differences that exist at present in the United States and other Western societies and then to look at the changes that occur when women's activities become more like men's. In discussing these differences, it is important to keep in mind that we are always comparing *average* values. All traits vary widely for each sex, so that the difference between groups is usually smaller than are the variations within each group. Extensive overlaps between women and men are found in the distributions of most traits except for the anatomical and physiological differences directly involved in reproduction. With this in mind, let us consider what the norms are for some physical traits. On average, American men are about 5 inches taller than American women.[7] Men are heavier and more muscular. Women have a higher proportion of body fat: the average for women is 23%; the average for men, 15%.[8] Body fat also is distributed somewhat differently: for women, more of it is found around the hips and breasts. Women have somewhat smaller, less dense bones than men. This, combined with the higher proportion of fat, means that a woman, on the average, weighs less than a man of the same height. Aside from lighter bones, the only established skeletal differences seem to be women's slightly wider pelvises and perhaps slightly shorter legs relative to the length of the trunk. There is no evidence for differences in elbows or shoulder joints, notwithstanding the explanation usually given as to why girls cannot throw a football or baseball "properly."

Another difference is in sweat production.[9] The Victorian adage, "A horse sweats, a man perspires, but a lady only glows," seems to have some truth in it. Women's body temperature must be two or three degrees higher than men's before we begin to sweat, though experts disagree as to whether this means that women are more or less heat-tolerant than men. A number of other physiological differences between the sexes may also affect relative abilities to perform physical activities. These include relative numbers of red blood cells, energy utilization, and capacity to carry oxygen from the lungs to the tissues.[10]

Differences in performance also clearly exist, but they do not correspond exactly with the stereotypes.[11] In measures of strength, it appears that, overall, women are about two-thirds as strong as men. However, the size of the difference changes for different muscle groups. It is greatest for arm, chest and shoulder muscles, and least for the leg muscles. Part of the difference is due to the fact that men are larger. When strength is compared for people of equal weight, the difference becomes smaller with women about three-quarters as strong as men of the same weight. When allowance is made for sex differences in body fat, and hence for differences in muscle mass for equal weights, the differences in strength become even smaller: leg strength per unit of lean body weight is actually slightly greater for women than for men.[12]

In competitions requiring power and speed, both of which depend on strength, women's records consistently lag behind men's. Trained men perform better in events such as sprints, medium-distance running or swimming, discus throwing, shot putting, and in the various jumping events. In some of these, the difference in performance appears to be due primarily to sex differences in size. For the high jump and long jump, for instance, performances are essentially the same for women and men when corrected for body weight. In others, even on an equal weight basis, women's performances are lower. However, in athletic events that are primarily tests of endurance, such as supermarathons (fifty- or 100-mile runs) or long-distance swimming, women are beginning to outperform men.[13] One theory is that, in endurance events, fat as well as carbohydrate (glycogen) becomes an important source of energy, and that women may be better at utilizing their stored fat. Women distance runners do not seem to experience the phenomenon known as "hitting the wall"—the point at which all the muscle glycogen is used up, which seems to occur at around twenty miles for male marathoners.[14]

Clearly there are significant sex differences in body shape, physiology and functioning. The question of interest here is to what extent they have been created and hence can be modified by environmental factors and social changes.

Changes in Muscular Development and Strength

A recent *Science News* survey of current research on sex differences in strength suggests that stereotypes about inherent female and male body types may need to be revised:

> Several researchers have concluded that much of this difference in strength is the result of society's encouraging the average man to be more active than the average woman. They feel that the social influences are so great that inherent physiological differences in strength cannot yet be estimated.[15]

The work of J. H. Wilmore has been particularly important in demonstrating that women's lack of strength in the upper body comes at least in part from disuse rather than totally from biology.[16] In several studies in which non-athletic women and men were tested during training with weights, women's strength was found to increase faster than men's, and women's greatest gains were made in the muscles of the upper body. This much greater relative improvement in women's arm and shoulder strength was attributed to the fact that, in daily life, American women already use their lower bodies in ways similar to men, but make much less use of the upper body. Results similar to those of Wilmore have been found during training programs for women cadets at West Point.[17] Furthermore, when we look at differences

Figure XI. Flexing
(Photo: D. Wald)

between *individuals* in our society, we find that women who use their arms and shoulders a good deal in their work are significantly stronger than average.[18] In some other societies, "women's work" regularly includes carrying heavy loads or hauling water. It would be interesting to see whether in these societies sex differences in upper-body strength are much the same as sex differences in leg strength.

Increased use and development of muscles does not mean that women will necessarily develop the same *type* of musculature as men. Although strength increases, women in general do not develop the bulging muscles usually associated with men. Typically, women can increase their strength by 50 to 75 percent without any increase in muscle bulk. This difference in responses to training is probably due to sex differences in levels of testosterone, which seems to be involved in the development of large muscle mass.

The relationship between muscle mass, exercise, and strength is unknown at present. What is clear is that women need not have as large muscles as men in order to be equally strong.[19] The recent research on women is not the only indication of this. It has been known for a long time that lighter male athletes possess proportionally greater muscle power than do heavier ones. Wilmore and others have postulated, on the basis of their work with women, that women may have the same potential for strength as men of comparable size even though muscular development is different.

Changes in strength are not the only ones that occur when women become physically more active. Although an average difference in percentage and distribution of body fat persists, women lose relatively more body fat during training than do men.[20] (Estrogens and body fat mutually affect each other, but the relationship is complex, with levels of estrogen affecting amounts and distribution of body fat, and body fat, in turn, affecting estrogen levels.) There are clear differences between women athletes and the general population of women. The percentage of body fat, on the average, is lower for the athletes and, for them, it is similar to that for non-athletic men. Some female long-distance runners show values as low as 6 or 7 percent. Sex differences in other physiological measures, such as oxygen uptake or energy utilization, also show a decrease during physical training for women.[21] Changes in sex differences in heat adaptation and sweating are not well established but, since it is well known that adaptation to heat improves with training, it has been suggested that, as girls become more active and sweat more when they are young, this "sex difference" may also decrease.[22]

Somewhat more speculative is the possible effect of exercise on bone growth. Bones are known to become stronger when used. In right-handed women, for example, the bones of the right hand are denser and more substantial than those of the left. Calcium is metabolized best when the muscles that surround the bone are being used, thus stimulating blood flow to the area. The pull of the muscles on bone may also have an effect. It is possible that similar activity patterns in girls and boys might result in a smaller sex difference in bone formation. There has been speculation that heavy training, particularly prior to puberty, might influence the structural development of the female pelvis[23] — a possibility that some people view with alarm. There may also be some effects on height, a possibility which I will discuss later.

Changes in athletic records show the same trends toward equalization. The number of women athletes has increased significantly in the last few decades, and training programs for women have been taken more seriously, although often only after legal challenges to high-school and university sports programs under Title VII of the Civil Rights Act. During this time, in many athletic events women's records have been increasing faster than men's. The greatest relative gains have come in swimming, where almost all the top women competitors routinely break men's records of the recent past. In long distance running the female/male gap has also decreased substantially.

In general, the physical performance of American women athletes differs from that of the average populations of non-athletic women far more than that of men athletes does from non-athletic men.[24] Furthermore, women athletes are somewhat closer in physical attributes to men athletes than the average woman is to the average man.

Cross-cultural studies also illustrate the point that physical development is under a great deal of cultural control, offering further proof that the origins of sex differences in musculature are not simple. In most contemporary societies, there is a tendency for men to perform the activities that require a great deal of exertion. Balinese society is an exception, since there neither women nor men traditionally do much heavy work, with the result that both are slender and show minimal sex differences in body build. European visitors have complained that they cannot tell the women and men apart, even from the front. However, when Balinese men are hired to work on the docks loading ships for Europeans, they develop the heavy musculature we consider typical of males.[25] There are other examples, as well, of societies in which sex differences in body type appear to be much smaller than in our culture.

Height and Environmental Factors

Height is another anatomical feature that is affected by culture in complex ways which illustrate the difficulties of separating nature and nurture. While height appears to have considerable hereditary components, for genetically similar populations reared in different environments average height can change substantially.[26] Immigrant families to the United States show dramatic increases in height from one generation to the next. Furthermore, although sex differences in average height appear to exist in all cultures, the magnitude can vary considerably. Differences ranging from less than two inches to more than eight have been observed;[27] and within a given society, sex differences in height may change with time. For example, over the last eighty years in Japan the difference has decreased by more than half an inch.[28]

A number of factors other than heredity have been shown to influence height. There is clear evidence of a direct effect of change in diet on height.[29] Other factors have been pointed to as well: incidence of diseases, amount of sunshine, physical and emotional stress, amount of physical activity.

Diet is often given as the sole cause of changes in the height of immigrants to the U.S. However, changes in diet are often accompanied by other changes which may also affect height. Much of the evidence linking diet with stature is based on the positive correlation between height and socioeconomic class, but clearly, many differences other than diet distinguish classes. Hard physical labor and increased incidence of diseases, as well as a bad diet, are often the lot of those at the bottom of the social hierarchy. Infant care practices that differ by class may also affect growth. Thus, class differences in stature, which are well-documented in nineteenth-century Britain and in a number of other countries, are probably due to a number of circumstances of which diet is only one.

In one study, two anthropologists have used cross-cultural data to try to assess the effect of physical stress in infancy.[30] They examined the effects on average male height of stressful infant-rearing practices, such as customs that involved either piercing or molding some part of the body. Average adult male height was found to be over two inches greater in the stressed groups. The usual variables thought to affect height—genetic factors and diet—were controlled for in this work. Thus, the difference would appear to be due primarily to cultural variables. It could be due to differences in stress, or it might be due to some other cultural variables that correlate with this particular type of treatment. Corroborating such a cultural interpretation is another study cited by these authors, though on a very small sample, in which it was found that handling premature infants was correlated with accelerated growth.

Cultural effects on height are probably mediated by hormones, for various hormones affect height, including the pituitary growth hormone and some sex hormones; and the levels of these hormones are influenced by a variety of cultural factors, including stress and exercise.[31] For example, the levels of growth hormone seem to respond to even minor changes in the environment. In normal individuals, levels of growth hormone vary in relation to food intake and activity and there is a significant response to exercise. However, the degree to which growth is affected by the normal variations in hormone levels that result from different environments and behaviors is largely unknown. Some studies have claimed that psychologically stressful situations not only can inhibit growth hormone secretion in children but that this can be sufficient to affect growth.[32] For example, one study showed that children who suffered severe emotional deprivation in their homelife were severely retarded in their growth and had abnormal growth hormone responses. When they were removed to new and better environments, they began to grow rapidly.[33]

Differences in upbringing, then, for a variety of reasons, may affect growth and it is possible that these differences influence sex differences in height. There are many ways in which girls and boys are treated differently, and some of the height differences between the sexes may be a reflection of early environmental differences.

Sex differences in height also could be due in part to the consistent sex differences that exist between girls and boys in activity and in the kind of sports and play that are encouraged. A long-term growth study on competitive young female swimmers has shown that their growth, as a group, was accelerated above established norms and that the acceleration was greater during their training years.[34]

Correlations between activity levels and height may be attributable to a number of hormonal effects. As we have seen, activity may directly affect growth hormone production. It may also affect "sex hormone" levels, which, in turn, can affect height. The production of estrogens by females at puberty, for example, retards long bone growth, thus leading on the average to smaller females, but since estrogen levels vary with activity, final adult height may be affected by physical activity during adolescence. Estrogen levels are also affected by the degree of body fat, itself indirectly affected by activity levels.

In some cultures, not only average height but also sex differences in height may be influenced by diet. Among peoples where dietary deficiencies affect height, the effects may be different for women and men since in marginal situations boys sometimes get a more adequate diet than do girls. There is also the possibility that even where the diet is adequate nutritionally, *differences* in diet may cause differences in height, and in many societies there is a significant difference in the foods eaten by girls and boys.

I do not mean to suggest that if females and males were raised in the same way there would be no average difference in strength or height. At this point, we have no way of knowing what would happen under these circumstances. However, the variation in sex differences in height cross-culturally, historically, and currently in different social classes indicates that, just as with strength, these differences are at least in part the result of environmental factors. How significant these factors are or how they might change from one environment to another remains to be established.

Intellectual Abilities and the Brain

Sex differences in mental capacities are often said to fit women and men for their different social roles. In fact, one of the most tenacious ideas of the last 100 years has been the notion that inherited mental abilities are a major factor in determining social position. People—usually those in positions of privilege—have, for example, repeatedly ascribed the disadvantaged positions of women relative to men, or of Blacks relative to whites, to differences in intellectual abilities or other mental traits. Widespread assumptions of innate sex, class, and ethnic differences in mental characteristics have had a major influence on efforts to measure mental abilities and have encouraged attempts to find a biological basis for differences in these abilities. At times, scientists have concentrated on trying to show that differences in abilities exist between different groups. At other times, they have been more interested in finding some measurable physical differences in the brains of different groups which might give rise to differences in mental performance. The effect of the underlying assumptions is clear in both cases. Either way, scientists attempt to find tools to measure the differences that are *assumed* to be there, rather than to examine whether or not the differences actually exist. Results that do not come out as expected are blamed on faulty methodology or bad data and rejected. Since the criterion for acceptability of a given test of mental abilities (such as the IQ test) is usually not based in some theory of intelligence (since none exists) but rather in the ability to provide expected results, it is often not at all clear what is being measured. This means that, far more than for the body characteristics discussed in the previous section, for mental characteristics it is often difficult to say what differences actually exist between groups or what any measured differences might reveal about the way people function in society.[35]

In order to understand the problems of current research in this area it is helpful to take a brief look at past work. Craniology, the measurement of skull or brain size, was an important scientific field during the nineteenth century, when much work on mental abilities and the brain was being done.[36] Craniologists assumed that "intelligence" was directly related to brain size and therefore set about measuring the sizes of brains, or rather, of skulls. The results at first bore out expectations: whites appeared to have bigger brains than Native Americans, Asian Americans or Blacks. In fact, the measurements were manipulated to obtain these results since, on average, there are no racial differences in brain size. The force of the underlying assumptions was so great that the expected results were obtained, however, and in at least one case, the misinterpretation of the measurements appears to have been unconscious.[37] Needless to say, it was also clear to craniologists that if brain size were correlated with intelligence, men must have bigger brains than women. In this case, no manipulation of data was required to reach the expected result, since on average women do have smaller brains than men. It became apparent, however, that here body size had to be taken into account since, if sheer size of the brain were to be the measure of mental capacity, elephants and whales should be the most intelligent creatures. This dilemma turned out to be downfall of craniology: there was no obvious and generally agreed-upon way of handling the measurements so as to come out with proper anatomical reasons for men's assumed intellectual superiority over women. By some measures, such as the ratio of brain surface to body surface, women appeared to be inferior intellectually to men, while by others, such as ratio of brain weight to body weight, women came out ahead.

This problem eventually led to the disappearance of craniology as a serious scientific field. It was replaced by IQ testing, which undertook the measurement of mental abilities rather than of physical differences.[38] IQ tests were originally developed as a way to predict performance in school. They have now not only come to be seen as measures of intelligence, but intelligence has come to be defined in terms of whatever abilities are measured on an IQ test. For the last fifty years these tests, and others derived by using similar methods, have dominated research on group differences in mental abilities. Until recently, there has been little interest in demonstrating possible underlying biological differences, but we are currently seeing a revival of interest in this question.[39]

Current theories of innate sex differences in mental abilities, unlike their nineteenth-century counterparts, usually do not claim that women are less intelligent than men. Instead, they focus on sex differences that are observed in tests of different types of mental abilities. This shift in interest is a result of peculiarities in IQ tests. Although they do show class and race differences when total scores are compared, IQ tests do not yield overall sex differences. Instead, women do better on certain kinds of questions asked on these tests (those identified as more verbal) while men do better on more mathematical ones and on questions involving visual-spatial ability.[40] Historically, on early tests women scored higher than men on tests that combined both types of questions, which prompted the test designers to include appropriate numbers of each type of question so as to eliminate the sex difference in the overall score.

Recently, the sex differences in verbal and spatial-visual abilities have become the focus of attention, and a number of theories have tried to tie these differences in performance to innate differences in the brains of women and men.

One widely publicized group of theories suggests that the human brain is programmed before birth by sex hormones. These prenatal influences are said to result later in sex differences in the kinds of mental abilities mentioned above as well as in aggressiveness and reproductive behavior.[41] This claim is based on studies in which differences can be seen in the patterns of nerve fiber connections in certain regions of the brains of female and male *mice*—regions that are specifically associated with reproductive behavior. Studies of correlations between hormone levels and mating behavior in laboratory animals are also cited as evidence. Other kinds of studies involve people with various genetic and hormonal abnormalities. The many problems with the data used and their interpretation have been thoroughly discussed by a number of people.[42] The most important problem with the human behavioral studies is that they have failed to control for any effect of environmental differences encountered by people with and without observed abnormalities.[43]

Another attempt to explain the origin of sex differences in mental abilities is provided by theories suggesting that innate sex differences exist in the way the two halves or hemispheres of the brain are used. In humans, although the interactions of the two sides of the brain are complex, the right and left halves are to some extent specialized for different functions and are used to a somewhat different extent depending on the tasks that are being performed. (This is called lateralization.)[44] The left hemisphere is usually more active during verbal tasks and during tests that involve analytical and linear thought patterns, whereas the right hemisphere is more active when a person is visualizing or manipulating objects in space, or thinking intuitively or associatively. Anatomical differences have been reported between the left and right sides of the human brain, but it has not been established that these are related to such lateralization of function.[45] It is clear that even though there must be some relationship between anatomy and function, lateralization of functions cannot be rigidly connected with brain anatomy. For one thing, after lesions to one side of the brain that result in losses of function such as speech, recovery sometimes occurs by the other side taking over the function.[46] Children usually experience complete recovery, which makes it appear likely that the two halves of the brain are initially equipotential, or nearly so, and that they specialize only after the first few years of life. Studies of the electrical activity of the brain also provide evidence that brain lateralization can be affected by environment and behavior, and that this is by no means a fixed or "wired-in" aspect of brain functioning.[47] What these studies also show is that the type and degree of asymmetry appears to be affected by mood, motivation and training. For example, after biofeedback training, people are able to produce more or less asymmetry at will on tasks for which they preferentially used one or the other side of the brain before the training. Numerous investigators have claimed that they can demonstrate a connection between sex differences in brain lateralization and sex differences in verbal and spatial abilities,[48] but such claims are largely fact-free speculation, since there is no evidence for simple, clear-cut sex differences in brain lateralization. There are some studies that indicate sex differences on various tasks designed to probe lateralization,[49] but the relationship of performance on these behavioral tasks to actual differences in brain lateralization is speculative at best and in large part is based on the very assumptions that the investigators are trying to prove. Thus, observations that the verbal skills of women are less disrupted by damage to the left hemisphere than are those of men have been taken as an indication of sex differences in lateralization.[50] However, such a straightforward deduction is unwarranted. Studies of electrical activity of the brain show some kinds of sex differences in hemispheric assymmetry during the performance of particular tasks, but these differences are complex and seem to be affected by personality and behavioral factors, such as motivation or strategies chosen for performing the task, or by biofeedback training.[51]

Given the ambiguous evidence, it is likely that investigators who propose theories connecting sex differences in mental abilities with brain lateralization simply start by assuming that there *is* a sex difference in the degree of lateralization. They then couple the fact that men perform spatial tasks better than women while women do better on verbal tasks than do men with the observation that the brain is lateralized for these functions. They end up where they began, by concluding that sex differences in the performance of verbal and spatial tasks must be connected with sex differences in brain lateralization. In fact, the two most widely accepted theories come up with opposite assumptions as to what the sex differences in brain lateralization might be.[52] Both theories can persist simultaneously only because there is no direct evidence to support either one.

There is a general difficulty with trying to make a straightforward connection between sex differences in brain lateralization and sex differences in behavior. The way skills are assumed to be divided between the two halves of the brain is not the same way that skills are divided up between women and men. The left half of the brain is thought to be more specialized for verbal skills, at which women are better, and for analytical and logical skills, at which men are widely held to be superior. The right half of the brain, on the other hand, appears to be more involved in spatial-visual ability, at which men are better, and in intuitive thinking, a trait usually assigned to women. Therefore, explanations that try to fit brain lateralization to sex stereotypes necessarily end up with major contortions.

Work on brain lateralization illustrates some of the difficulties of trying to find causal relationships between anatomy, physiology, and function of the brain. Even if sex differences in brain lateralization were to be demonstrated more clearly than is the case at present, and if some correlation were to be established with sex differences in verbal and spatial skills, we still would have no information on the origins of the differences. Given the evidence that experience and training affect brain lateralization, it is perfectly possible that differential socialization and learning can lead to differences in brain lateralization instead of, or in addition to, differences arising from innate factors. Some work has suggested, for example, that since girls seem to acquire verbal skills earlier than boys, they may be predisposed to develop verbal strategies. This early acquisition of verbal ability could come from some innate biological difference, perhaps related to lateralization, or it could possibly come from the fact that parents, teachers, and others talk differently to little girls and boys, which might (or might not) lead to differences in lateralization; or it could come from a combination of the two. At present, however, there is no way of separating the effects of these two factors or of knowing their relationship to brain lateralization—or even of knowing whether they are the right ones to consider, for that matter.

Any theories that postulate that innate differences in the brain lead to behavioral differences involve the same kinds of difficulties as do the brain lateralization theories or the sex hormone theories. The relationships among anatomy, physiology, and behavioral functions are not at all clear, and the extent to which all of them are influenced by the environment is largely unknown. We do, however, have a number of research findings indicating that environmental effects on the human brain cannot be neglected.

Like other parts of our body, the human brain is responsive to the environment—it is *plastic*, and its structure, physiology, and performance may be modified in response to changing external circumstances. It is this plasticity of the brain that makes learning and memory possible. At the level of the neurons, the brain's modified experience may be expressed by modifications of biochemistry, of connections between cells, and of their individual and collective electrical responses.[53] Obviously, there is also *specificity* in the brain in that the human brain is different from the brains of other species and works in ways that are specific to human beings. It has different capabilities and responds to stimuli in different ways. But we do not know the extent to which environment can influence brain structure and function.

Over the past few decades, the plasticity of the brain in response to a variety of environmental stimuli has become widely recognized. The impact of changes in the environment on both the developing and, to a lesser extent, the mature brain has been demonstrated anatomically, biochemically, and physiologically. There is a growing body of evidence that lack of stimulation and experience can affect brain function as profoundly as can physical damage, and also that enriched and stimulating environments can augment brain development.[54]

Perhaps the most dramatic finding is the extent of plasticity of animal brains.[55] Rats raised in an environment in which they have a variety of "toys" and a high degree of social contact show marked changes in the brain when compared with rats raised in an isolated environment with little stimulation.[56] In general, in these circumstances environmental stimulation in animals leads to an increase in brain weight and size—an increase most marked in the cerebral cortex. Within the cortex, changes are found in cell bodies, dendrites, synapses, protein, RNA, and certain brain enzymes. As little as one hour a day of exposure to an environment with some sources of stimulation is enough to produce measurable differences in the brain. (It is important to note that laboratory environments are extremely unstimulating and bear little resemblance to the ways normal rats live. That such extreme changes in environment show correlated brain changes need not mean that usual life differences do the same.)

Other studies have also demonstrated that brain structure can be modified by sensory or social deprivation. Thus, when animals are reared in the dark a number of structural and biochemical differences in the visual cortex can be observed between them and siblings reared in the light.[57] Such effects are also produced by less drastic environmental differences. Fish that are reared without being able to see or touch other members of their own species, but that are otherwise exposed to a variety of visual stimuli, show similar kinds of changes.[58]

There are indications that environmental effects on brain structure or function may also occur in humans. There is a report, for example, that a group of Indians raised in tepees has a different sensitivity to vertical, horizontal, or slanted lines than do Europeans raised in rectangular rooms, and it was postulated that the environmental difference had led to differences in the visual cortex, since an effect of this kind is well established in animals.[59] I wish to emphasize that, in all of these cases, the fact that one finds changes in the brain does not necessarily imply that there are corresponding differences in behavior, even for animals. Visually deprived kittens learn to behave like normal kittens after they begin to see, even when the "deprivation" has led to permanent anatomical changes in the visual cortex.

More evidence about plasticity comes from examining the effects of destruction of brain tissue. It has been found that both loss and recovery of function in animals with brain lesions are affected by experiences both before and after the lesion occurs. Rats raised in "enriched" environments have different impairments than "deprived" rats when brain lesions are made in the same area. There is often a certain amount of recovery of function after brain lesions, though the exact mechanism for this is uncertain. Physiological changes occur in animals and these differ depending on post-lesion experiences.[60] Studies of recovery of function after lesions in non-human primates seem to indicate that function may be restored by nearby regions somehow taking over the functions previously performed by the damaged region. This is most likely to happen if the brain is stilll developing. Also, experience does seem to play some role in recovery.[61]

When working with people, it is obviously not possible to experiment with controlled environments or experimental brain lesions followed by examination of the brain. While experiments with animals can be suggestive, they obviously are not conclusive. Observations of people who have had strokes or other kinds of brain damage provide some indications of environmental effects. The recovery of function after brain damage indicates that just as in non-human primates, different parts of the brain, at least to some extent, can take over functions of lost or damaged parts. As we saw earlier, if brain damage that results in loss of ability to speak, for example, occurs during childhood, there is usually complete recovery of language ability. Furthermore, various forms of therapy can affect recovery from brain injury, indicating that environment can play a part.

None of this work on the brain is definitive but it illustrates the problems with *any* attempts to attribute human behavioral differences to innate differences in brain structure or function. There is no way to eliminate the possibility that brain differences may have resulted from behavioral or environmental differences rather than the other way around.

Sex Hormones

Work on sex hormones is another politically loaded area, since interpretations of results often end up reinforcing sex-role stereotypes. The article by Birke in this collection discusses some of the research on sex hormones in detail, so that we will look here only at some of the general problems in interpretation.

It has been widely assumed that average differences in sex hormones must contribute to average differences in behavior. In an effort to prove this, scientists have tried to show that variations in hormone levels within each sex are tied to variations in behavior. One of the primary difficulties with these attempts is that the hormone levels themselves are affected by behavior and by the environment, so that no causal relationships can be established. For example, it is claimed that levels of the "male" hormone, testosterone, determine aggression and social ranking among male non-human primates and among people.[62] The case of testosterone and social ranking in non-human primates provides a classic example of how wrong it is to make assumptions about causes when all that one observes are correlations. Although hormone levels do indeed *correlate* with rank in dominance hierarchies among male rhesus monkeys, for example, studies indicate that hormone levels do not *determine* position in the hierarchies.[63] When the social rank of individuals changes, the hormonal activity of those individuals also changes. It is not clear whether the experience of dominance increases testosterone production, or whether being in a subordinate position decreases it, or perhaps both. In general, it is found that psychological, physical, or social stress can lower both social rank and testosterone levels.

In humans, no correlations have been found between levels of testosterone and social rank or aggression. However, social environment, exercise, psychological states, and physical stress all affect testosterone levels.[64]

Another difficulty arises with trying to connect hormones and behavior. There is evidence that the relationship between hormonal states and behavioral and emotional states varies in different situations. For the same person, the same hormonal state can be associated with a number of different behaviors, depending on the social setting. For example, one experiment indicated that if the adrenal system is aroused through injections of adrenal

Figure XII. Rugby.
(Photo: Ree Giorgi)

hormones in a situation where anger is an appropriate response, then the response comes out as anger. If euphoria is a more appropriate response to a given situation, then people become euphoric.[65]

Hormonal contributions to behavior, then, depend in part on the levels of hormones at a particular moment, which are themselves determined by a person's past interactions with the social environment, and in part on the details of the current social environment. It is not possible to abstract behavior from its social context.

What Is the Point of All of This?

Biological determinism is a particular way of viewing causes of social structures. It offers a specific, scientifically based model for the existence of social hierarchy and social inequality by postulating that differences in innate biology lead to differences in behavior, which in turn lead to differences in social position.

In the previous sections we have seen some of the flaws in the scientific basis of the biological determinist model. First, biology cannot be taken as something fixed and immutable; and second, it is in general not possible to show that a given biological state *causes* a specific behavior. Biology does affect human behavior, but the examples I have discussed make it clear that there is no way to separate the contributions biology and culture make to differences in behavior.

It is the second step in the biological determinist model that is of political importance, since this step gives the model its specific political character. It is the desire to explain observed social differences that leads biological determinists to look at differences between the sexes, races, or classes. The implications for social change are what impel others to listen to what biological determinists have to say. However, this model is no better at establishing the connection between behavior and social position than it is at showing the connection between a' genetically programmed biology and behavior.

The argument that it is primarily ability or talent that determines social position in our society is contradicted by a number of studies of social opportunity and mobility in the United States.[66] Furthermore, if we look specifically at sex differences, we find immediate evidence to refute the argument. Women and men are in general much more alike than different, and the magnitude of observed average differences in behavior or abilities (which may in fact be overestimated) is much smaller than are the differences between women's and men's social roles. For example, differences in mathematical ability are often cited as the reason for the small number of women scientists and engineers. But the differences in participation rates of women and men in scientific fields in our society are so much larger than the reported sex differences in mathematical ability that differences in ability could not begin to account for the different numbers of women and men scientists.[67] In general, observed sex and race differences in behavior would explain only a small part of our social hierarchy, even if merit and ability were the most important factors.

It is apparent that the importance attached to biological explanations of sex, race, or class differences in ability or behavior comes from the political content of the theories rather than from their scientific merit. If race, sex, and class were not politically and economically significant categories it is likely that no one would care very much about biological differences between members of these groups. To pay attention to the study of sex differences would be rather peculiar in a society where their political importance was small. Biological determinist theories, which pretend to explain why the world is the way it is, are built on the myth that social structures are determined by the biological nature of human beings. As we have seen, the political messages are usually quite explicit. The theories would have us believe that the only way women and men, or Blacks and whites, could have equal social positions would be if opportunities were rigged so as to compensate for intrinsic deficiencies.

The proponents of these theories often accuse their critics of reading too much into their ideas. They emphasize that difference does not necessarily imply inequality; but what they ignore, of course, is the reality of our society, where social inequality has been *constructed* along biological lines and indeed even contributes to biological differences.

Biodeterminist theories have reappeared at this particular time because we are in a period of major social change, the eventual outcome of which is uncertain. The movement of women into the wage labor force during the last decades has resulted in important changes in everyone's lives. Meanwhile, the appearance of the Women's Movement has brought out explicit discussion of issues raised by these changes and has helped to raise women's expectations for equal rights and opportunities. However, these increased expectations pose a threat to the maintenance of many existing social structures. Therefore, it is not surprising that we now see women's demands for equal opportunity being met by arguments about biological limitations. The current theories about the biological basis of social structures—the theories that say, for example, that women are "naturally" disadvantaged—are of use to those who want to preserve and strengthen the dominant political and economic interests. One result is that a great deal of media attention is given to biological theories that offer naturalistic explanations for the distribution of wealth and power in this society.

Since these theories are so persuasive and so politically loaded, we cannot ignore them or allow them to go unchallenged. But we must be clear about why feminists should examine the question of the possible existence of biologically based sex differences. We should do so only *in response to* the claims of biological determinists who say that these differences have social significance and that a knowledge of them provides a guide to social policy and to the limits of possible social change. We should not try to make a knowledge of the origins of sex differences the basis of our own vision of the future and, indeed, we could not even if we wanted to. Contrary to the claims of biological determinists, studies of the contributions biological factors make to human behavior can *at most* give only very limited information about the origins of present differences in human behavior (and probably no information about the origins of present social structures). Such studies offer *no* insight into the effects of social change on behavior. Whatever sex differences in behavior exist now and whatever their origins, we have no reason to assume that they would be barriers to any egalitarian societies we may want to build.

Notes

1. *See* for example E. O. Wilson, *Sociobiology, The New Synthesis* (Cambridge, MA: Harvard University Press, 1975); A. S. Rossi, "The Biosocial Side of Parenthood," *Human Nature* 1, no. 6 (1978), pp. 72–79.
2. *See* M. Lowe and R. Hubbard, "Sociobiology and Biosociology," *Genes and Gender II: Pitfalls in Research on Sex and Gender*, R. Hubbard and M. Lowe, eds., (New York: Gordian Press, 1979) pp. 91–111 and references therein.
3. J. W. Berry, "Tenine and Eskimo Perceptual Skills," *International Journal of Psychology* 1 (1966), pp. 207–229; A. G. Goldstein and J. E. Chance, "Effects of Practice on Sex-related Differences in Performance on Embedded Figures," *Psychomonic Science* 3 (1965), pp. 361–362; J. Conner, L. Servin and M. Schackman, "Sex Differences in Children's Response to Training on a Visual-Spatial Test," *Developmental Psychology* 13 (1977), pp. 293–294; B. Whiting and C. Pope Edwards, "A Cross-Cultural Analysis of Sex Differences in the Behavior of Children Age 3–11," *The Journal of Social Psychology* 91 (1973), pp. 171–188; K. P. Bradway and C. W. Thompson, "Intelligence at Adulthood: A 25 Year Followup," *Journal of Educational Psychology* 53 (1962), pp. 1–14.
4. This is one of the factors in the resistance to women's participation in athletics. It becomes harder to maintain our cultural stereotypes of

female fragility and passivity when women play contact sports, run fifty miles or lead Himalayan expeditions.

5. The acceptance of the inevitable superiority of male strength on the part of both women and men also contributes to men's violence against women and may in part explain women's hesitation to resist.

6. E. Maccoby and C. Jacklin, *The Psychology of Sex Differences* (Stanford, CA: Stanford University Press, 1974), pp. 307–311.

7. *Statistical Abstracts of the United States* (U.S. Department of Commerce, 1978), p. 121.

8. R. Malina, "Quantification of Fat, Muscle and Bone in Man," *Clinical Orthopaedics* 65 (1969), pp. 9–38; D. Mathews and E. Fox, *The Physiological Basis of Physical Education and Athletics*, 2nd ed. (Philadelphia: W. B. Saunders Co., 1976).

9. B. Lofstedt, *Human Heat Tolerance* (Lund: Department of Hygiene, University of Lund, Sweden, 1966).

10. D. Mathews and E. Fox, *Physical Education*, pp. 452–460.

11. In measures of performance, effects of expectations by both experimenters and students may not be negligible. The actual strength of women, particularly in the United States, may have been underestimated for this reason. For women, the social stigma attached to being strong may well affect our actual performance on strength tests. Women have been conditioned throughout our lives not to appear strong and we may not ignore these social expectations, particularly when the test is administered by a male, as is usually the case.

12. D. Mathews and E. Fox, *Physical Education*, pp. 460–466.

13. D. Mathews and E. Fox, *Physical Education*, pp. 447–450; J. Douglas and J. Miller, "Record Breaking Women," *Science News* 112 (1977), pp. 172–174; E. Gerber, J. Felshin, P. Berlin, W. Wyrick, *The American Woman in Sport* (Reading, MA: Addison-Wesley, 1974), pp. 403–418; J. Ullyot, *Women's Running* (Mountain View, CA: World Publications, 1976), p. 91.

It should not surprise us that most of the money-making professional sports emphasize areas in which men have the advantage. Football depends on physical size and strength. Among sports, it is the epitome of male brute strength. Basketball depends on height, baseball on upper-arm strength and speed, and hockey on willingness to use physical violence. For a further discussion of the role of sports in maintaining cultural stereotypes, see M. Albert and R. Hahnel, Unorthodox Marxism (Boston, MA: South End Press, 1978), pp. 218–220.

14. J. Ullyot, *Women's Running*, p. 96.

15. J. Douglas and J. Miller, "Record Breaking Women," p. 172.

16. J. H. Wilmore, "Alterations in Strength, Body Composition and Anthropometric Measurements Consequent to a 10-week Weight Training Program," *Medicine and Science in Sports* 6 (1974), pp. 133–138;

J. H. Wilmore, "Inferiority of Female Athletes, Myth or Reality," *Journal of Sports Medicine* 3 (1975), pp. 1–6.

17. J. Peterson and D. Kowal, *Project 60: A Comparison of Two Types of Physical Training Programs on the Performance of 16–18 Year Old Women* (West Point: Office of Physical Education, 1977).

18. M. B. Wardle and D. Gloss, "Women's Capacity to Perform Strenuous Work," *Women and Health* 5 (1980), pp. 5–15.

19. E. Jokl, *Physiology of Exercise* (Springfield, IL: Charles C. Thomas, 1964), p. 62.

20. J. H. Wilmore, "Body Composition and Strength Development," *Journal of Physical Education and Recreation* 46, no. 1 (1975), pp. 38–40; C. Brown and J. H. Wilmore, "The Effects of Maximal Resistance" *Medicine and Science in Sports* 6 (1974), pp. 174–177.

21. E. Gerber *et al.*, *American Woman in Sport*, pp. 455–485.

22. D. Harris, "Survival of the Sweatiest," *Women's Sports*, Nov. 1977; B. L. Drinkwater and S. M. Horvath, "Heat Tolerance and Aging," *Medicine and Science in Sports* 11 (1979), pp. 49–55.

23. J. Kaplan, *Women and Sport* (New York: Viking, 1979), p. 40; E. Gerber *et al.*, *American Woman in Sport*, p. 440.

24. E. Gerber *et al.*, *American Woman in Sport*, p. 421.

25. G. Gorer, *Bali and Angkor* (London: Michael Joseph, 1936).

26. J. M. Tanner, "Growing Up," *Scientific American* 229 (September 1973), pp. 34–43.

27. A. Oakley, *Sex, Gender and Society* (New York: Harper Colophon, 1972), p. 28.

28. S. Suzuki, S. Oshima, E. Tsuji, K. Tsuji and F. Ohta, "Interrelationships Between Nutrition, Physical Activity and Physical Fitness," *Nutrition, Physical Fitness and Health*, J. Parizkova and V. Rogozkin, eds., (Baltimore, MD: University Park Press, 1978).

29. R. Frisch, "Population, Food Intake and Fertility," *Science* 199 (1978), pp. 22–30.

30. T. Landauer and J. Whiting, "Infantile Stimulation and Adult Stature of Human Males," *The Development of Behavior*, V. Denenberg, ed., (Stamford: Sinauer Associates, 1972).

31. K. Kuoppasalmi, H. Naveri, S. Rehunen, M. Harkomen and H. Adler-creutz, "Effect of Strenuous Anaerobic Running Exercise on Plasma Growth Hormone, Cortisol, Luteinizing Hormone, Testosterone, Androstenedione, Estrone and Estradiol," *Journal of Steroid Bio-chemistry* 7 (1976), pp. 823–829; J. Jurkowski, N. Jones, W. Walker, E. Younglai and J. Sutton, "Ovarian Hormonal Responses to Exercise," *Journal of Applied Physiology* 44 (1978), pp. 109–114; L. Levi, "Sympatho-adrenomedullary and Related Biochemical Reactions Durring Experimentally Induced Emotional Stress," *Endocrinology and Human Behavior*, R. Michael, ed., (New York: Oxford University Press, 1968); S. Reichlin, "Hypothalamic Control of Growth Hormone Secre-

tion and the Response to Stress," *Endocrinology and Human Behavior*, R. Michael, ed., (New York: Oxford University Press, 1968), pp. 256–283.

32. L. Gardner, "Deprivation Dwarfism," *Scientific American* 227 (July 1972), pp. 76–82.

33. E. Powell, A. Frantz, M. Rabkin and R. Field, "Growth Hormone in Relation to Diabetic Retinopathy," *New England Journal of Medicine* 275 (1966), pp. 922–925. It is important to be cautious in interpreting this kind of work. Children in "emotionally deprived" homes also may have been nutritionally deprived, either because they did not get sufficient food or because they were too upset to eat. Thus, even though hormonal abnormalities were noted, one cannot easily assign causes.

34. G. Rareck, "Exercise and Growth," *Science and Medicine of Exercise and Sport*, W. R. Johnson and E. Buskirk, eds. (New York: Harper and Row, 1973). This study needs to be interpreted with care as do all studies comparing athletes to the general population. It is possible that there is a self-selection process, by which girls with certain kinds of physical characteristics are more likely to be found training as athletes. Such caution is especially indicated when looking at changes in age or onset of menstruation. It may be that late onset means, for girls, a postponement of the social pressures associated with puberty and thus more freedom to continue with athletics in a serious way.

35. Even in measuring physical differences experimenter bias can distort observed characteristics.

36. E. Fee, "Nineteenth Century Craniology: The Study of the Female Skull," *Bulletin of the History of Medicine* 53 (1979), pp. 415–433.

37. S. J. Gould, "Morton's Ranking of Races by Cranial Capacity," *Science* 200 (1978), pp. 503–509.

38. R. McCall, *Intelligence and Heredity* (Homewood, IL: Learning Systems Co., 1975).

39. A. Barfield, "Biological Influences on Sex Differences in Behavior," *Sex Differences: Social and Biological Perspectives*, M. Teitelbaum, ed. (Garden City, N.Y.: Anchor Books, 1976).

40. E. Maccoby and C. Jacklin, *Psychology of Sex Differences*, p. 349–360.

41. J. Money and A. Ehrhardt, *Man and Woman, Boy and Girl* (Baltimore, MD: Johns Hopkins University Press, 1972); A. Rossi, "A Biological Perspective on Parenting," *Daedalus* 106 (1977), pp. 1–31; D. Broverman, E. Klaiber and W. Vogel, "Gonadal Hormones and Cognitive Functioning," *The Psychology of Sex Differences and Sex Roles*, J. Parsons, ed. (New York: McGraw-Hill, 1980).

42. E. Ramey, "Sex Hormones and Executive Ability," *Annals of the New York Academy of Sciences* 208 (1973), pp. 237–245; *see* also the articles by F. Salzman, R. Bleier and M. Lowe and R. Hubbard in R. Hubbard and M. Lowe, *Genes and Gender II: Pitfalls in Research on Sex and Gender* (New York: Gordian Press, 1979).

43. See the article by B. Fried in this volume.
44. J. Levi-Agresti and R. Sperry, "Differential Perceptual Capacities in Major and Minor Hemispheres," *Proceedings of the National Academy of Sciences of the USA* 61 (1968), pp. 1151–1162; R. Sperry, "Lateral Specialization in the Surgically Separated Hemispheres," *The Neurosciences: Third Study Program*, F. Schmitt and R. Wardon, eds. (Cambridge, MA: M.I.T. Press, 1974).
45. N. Geshwind, "The Anatomical Basis of Hemispheric Differentiation," *Hemisphere Function in the Human Brain*, S. J. Dimond and J. G. Beaumont, eds. (New York: Wiley, 1974).
46. H. Hecaen, "Acquired Aphasia in Children," *Brain and Language* 3 (1976), pp. 114–134.
47. R. J. Davidson and G. E. Schwartz, "Patterns of Cerebral Lateralization during Cardiac Biofeedback versus the Self-regulation of Emotion: Sex Differences," *Psychophysiology* 13 (1976), pp. 62–68; Y. Matsumiya, "The Psychological Significance of Stimuli and Cerebral Evoked Response Symmetry," *Behavior Control and Modification of Physiological Activity*, D. Mostofsky, ed. (Englewood Cliffs, NJ: Prentice-Hall, 1976).
48. S. L. Star, "Sex Differences and the Dichotomization of the Brain," in R. Hubbard and M. Lowe, *Genes and Gender II*, and references therein.
49. D. Waber, "Sex Differences in Mental Abilities, Hemispheric Lateralization and Rate of Physical Growth at Adolescence," *Developmental Psychology* 13 (1977), pp. 29–38; S. Witelson, "Sex and the Single Hemisphere: Specialization of the Right Hemisphere for Spatial Processing," *Science* 193 (1976), pp. 425–427.
50. J. Bogan, "The Other Side of the Brain II: An Appositional Mind," *Bulletin of the Los Angeles Neurological Societies* 34 (1969), pp. 135–162; H. Lansdell, "A Sex Difference in Effect of Temporal-Lobe Neurosurgery on Design Preference," *Nature* 194 (June 1962), pp. 852–854.
51. Y. Matsumiya, "Psychological Significance of Stimuli."
52. W. Buffery and J. Gray, "Sex Differences in the Development of Spatial and Linguistic Skill," *Gender Differences: Their Ontogeny and Significance*, C. Ounsted and D. Taylor, eds. (Edinburgh: Churchill Livingstone, 1972); J. Levi, "Lateral Specialization of the Human Brain: Behavioral Manifestations and Possible Evolutionary Basis," *The Biology of Behavior*, J. A. Kiger, ed. (Corvallis: Oregon State University Press, 1972).
53. S. Rose, *The Conscious Brain* (New York: Random House, 1976).
54. R. Walsh and W. Greenough, *Environments as Therapy for Brain Dysfunction* (New York: Plenum Press, 1976).
55. A great deal of the work on the relationship between behavior, hormones and the environment, as well as a great deal of work on the brain, has been done by looking at animals. Such work has extremely limited

usefulness in understanding anything about human behavior. The physiology of all mammals has a fair amount in common, as we acknowledge when we use animals to test drugs. However, since there are major species differences, extrapolations from one species to another is a risky business. Nevertheless, given the fact that plasticity is a particularly strong characteristic of the human brain, when we find evidence of plasticity in various aspects of brain function in other animals, it should alert us that something similar (though probably quite different in its actual expression) may be going on in humans. Behavior is another matter entirely. Species differences in behavior are immense and since extreme behavioral plasticity is one of the distinguishing features of the human species, extrapolations of conclusions or theories about specific behaviors from other animals to humans seems unjustified under any circumstances.

56. M. R. Rosenzweig, "Effects of Environment on Development of Brain and of Behavior," *The Biopsychology of Development*, E. Tobach, L. Aronson and E. Shaw, eds. (New York: Academic Press, 1971).

57. W. Greenough, F. Volkmar and T. Fleischmann, "Environmental Effects on Brain Connectivity and Behavior," *Behavioral Control and Modification of Physiological Activity* (Englewood Cliffs, NJ: Prentice-Hall, 1976).

58. R. Coss and A. Globus, "Spine Stems on Tectal Interneurons in Jewel Fish are Shortened by Social Stimulation," *Science* 200 (1978), pp. 787–790.

59. N. Calder. *The Human Conspiracy* (New York: Viking, 1976), p. 94.

60. W. Greenough, B. Fass and T. Devoogd, "The Influence of Experience on Recovery Following Brain Damage in Rodents: Hypotheses Based on Development Research," *Environments as Therapy for Brain Dysfunction*, R. Walsh and W. Greenough, eds. (New York: Plenum Press, 1976); R. Walsh and R.Cummins, "Neural Responses to Therapeutic Environments," in R. Walsh and W. Greenough.

61. P. S. Goldman, "Developmental Determinants of Cortical Plasticity," *Acta Neurobiologica* 32 (1972), pp. 495–511.

62. E. Maccoby and C. Jacklin, *Psychology of Sex Differences*, p. 243, p. 368; S. Goldberg, *The Inevitability of Patriarchy* (New York: Morrow, 1973).

63. A. Kling, "Testosterone and Aggressive Behavior in Man and Non-human Primates," *Hormonal Correlates of Behavior*, B. Eleftheriou and R. Sprott, eds. (New York: Plenum Press, 1975).

64. C. Hale, "Physiological Maturity of Little League Baseball Players," *R   ·ch Quarterly* 27 (1956), pp. 276–284; L. E. Kreuz, R. A. Rose   R. Jennings, "Suppression of Plasma Testosterone Levels and   ɔgical Stress," *Archives of General Psychiatry* 26 (1972),   -482; K. Kuoppasalmi, H. Naveri, S. Rehunen, M. Harkonen   dlercreutz, "Effects of Strenuous Anaerobic Running Exercise

on Plasma Growth Hormone, Cortisol, Luteinizing Hormone, Testosterone, Androstenedione, Estrone and Estradiol," *Journal of Steroid Biochemistry* 7 (1976), pp. 823–829.
65. S. Schachter and J. Singer, "Cognitive, Social and Psychological Determinants of Emotional State," *Psychological Review* 69 (1969), pp. 379–399.
66. C. Jencks, *Who Gets Ahead: The Determinants of Economic Success in America* (New York: Basic Books, 1979); P. Blau and O. Duncan, *American Occupational Structure* (Riverside, N.J.: Macmillan, 1978); R. de Leone, *Small Futures: Children, Inequality and the Limits of Liberal Reform* (New York: Harcourt Brace Jovanovich, 1979).
67. M. Kimball, "A Critique of Biological Theories of Sex Differences," *International Journal of Women's Studies*, in press.

[Marian Lowe *is Associate Professor of Chemistry and Director of the Women's Studies Program at Boston University. She has long been interested in the social implications of scientific research, with a particular focus on the role science plays in shaping ideas about women and women's role in society. She has also worked on environmental issues, primarily issues of energy policy. She is coeditor, with Ruth Hubbard, of* Genes and Gender II: Pitfalls in Research on Sex and Gender *(New York: Gordian Press, 1979). As an identical twin, she considers herself an expert on twin studies.*]

Jeanne M. Stellman and
Mary Sue Henifin

No Fertile Women Need Apply: Employment Discrimination and Reproductive Hazards in the Workplace

In late 1977, five women underwent sterilization rather than lose their jobs in the lead pigment department of the American Cyanamid plant in Willow Island, West Virginia. The women made this difficult decision after the company announced that it would exclude fertile women from jobs with lead exposure due to the chance of damaging a fetus if the women proved to be pregnant. None of these women was in fact pregnant and none was planning a pregnancy. One of the women was married to a man who had had a vasectomy. Another woman was supporting her family after being widowed. Company representatives, nevertheless, informed the women that they had the option of leaving jobs which paid $225 per week plus substantial overtime and transferring to janitorial jobs which paid $175 per week with no extras. Instead, these women chose sterilization. Barbara Cantwell, one of the sterilized employees said, "I wish I had never done it, but I was scared."

In 1980, Lillian Reese, a trained laboratory technologist, applied for a job with the Midway Coal Mining Company, a subsidiary of the Gulf Oil Corporation, to work in a laboratory doing research on coal tars (known to be human carcinogens). Ms. Reese states in her complaint to the Washington State Human Rights Commission that she was told, "It is against company policy to employ women of childbearing capacity in their laboratories. Mr. Schuller [the Personnel Officer] stated that the government did not permit the company to hire women of childbearing capacity in the laboratory. He then offered to find me a clerical job. I declined the offer, since I am a fully trained, experienced laboratory technician. . . . The next day . . . I spoke again with Mr. Schuller and asked if he would consider my application if I had a tubal ligation and therefore could not become pregnant. He said that I could not have surgery to get the job. . . . The company engages in discrimination against women by failing and refusing to hire women in the plant, by assigning women into relatively lower paying clerical positions, and by denying employment to women of childbearing capacity in plant, laboratory, and perhaps other positions."[1]

Figure XIII. Women workers exposed to lead, late 1800s. (Connecticut Historical Society)

Table 1

Instances in which Either Civil Suits, Arbitration, or Statutory Violations Were Filed to Protest Exclusion of Fertile Women from Certain Jobs.

- Five women were laid off from the specialty chemicals division of Allied Chemical Corporation in Danville, Illinois, on the grounds that fluorocarbon-22 might interfere with future pregnancies. Three of these women elected sterilization, the other two women were laid off permanently.

- The B. F. Goodrich plant in Avon Lake, Ohio, excluded women of childbearing capacity from jobs involving exposure to vinyl chloride, a known carcinogen and mutagen. Only women with medical sterilizations were eligible for hiring.

- American Cyanamid Lederle labs at Pearl River, New York, excluded six fertile women from jobs involving exposure to diomex and methotrexate, both cancer-therapy drugs with extreme systemic toxicity. Women unable to bid on other jobs were laid off.

- Avtex Fibers, Inc., of West Virginia, excluded fertile women from jobs involving high risk of exposure to carbon disulfide which has been associated with a two- to five-fold increase in coronary heart disease among viscose rayon workers. It is also known to affect the brain, the eyes, the liver, and the kidneys, as well as to induce neurological disorders and psychosis.

- At Environmental Protection and Aeration Systems of Memphis, Tennessee, a woman hired by the company and sent to a work site in Florida was refused entry on the site due to possible lead exposure. She returned to Tennessee and was fired.

- Since 1972, no fertile women have been employed in the battery plants of General Motors' Delco Remy division.

- The EPA (Environmental Protection Agency) has proposed that women of childbearing age be prohibited from contact with two pesticides—lindane and ferriamicide—which have other adverse health effects.

These two examples and those given in Table 1 demonstrate a trend in industry toward excluding fertile women from jobs in which they are exposed to certain chemicals, because of the chance that the chemicals may harm their fetuses if they should prove to be pregnant.[2] This is being done even when women are not pregnant and even when the agents, such as lead and the other chemicals listed in Table 2, are known to harm the male reproductive system as well, and/or threaten the health of male and female workers themselves. Furthermore, the exclusions are only selectively applied to women. For instance, women are not being barred from *traditional* "women's jobs" such as pottery painting, in which they are exposed to lead. In addition, when it was discovered—and highly publicized—in 1977 that men producing the pesticide dibromochloropropane (DBCP) in a chemical factory were becoming sterile, these men were not excluded from the workplace. Instead, the production of DBCP was banned in the United States.[3] There are other clear instances of discrimination. Companies have responded to research showing that cadmium and vinyl chloride are toxic to fetuses by excluding women from jobs with exposure to these chemicals. In contrast, when these substances were found to harm men's reproductive capabilities,[4] industry moved to solve the problem by lowering exposure levels. The growing momentum to exclude fertile women from certain better-paying industrial jobs is another example of using women's biology, in this case our childbearing *potential*, as the basis for discrimination in employment. Although the rationale is relatively new, the consequences are not. Women historically have been denied full employment rights because they were supposedly rendered sick by their raging hormone cycles or because they were deemed not sufficiently intelligent due to their evolutionary adaptation for childbearing. (Those traditional arguments, like the more recent one that occupational exposures may harm a fetus should women be pregnant, are all based on the selective use of biased research findings.) But, as other articles in this book show,[5] much research on women's biology is based on faulty hypotheses and methodologies. Denial of employment rights because of reproductive hazards in the workplace is another example of the use of so-called scientific facts specifically to keep women out of higher-paying jobs.

Industry representatives argue that this new employment discrimination against women of childbearing capacity is based on the desire to protect "potential fetuses." But what are "potential fetuses" other than an invention that converts women's childbearing capacity into an imaginary permanent state of pregnancy—and one which enables industry to treat all women of childbearing age as though they were always pregnant? As will be shown later in this chapter, very few blue-collar women, in fact, become pregnant after the age of thirty. Furthermore, many of the hazards to women's reproductive health adversely affect men's as well, yet few policies to protect "fertile men," "potential fathers," or "potential sirers" have been formulated. Only women, pregnant or not, are completely identified with their childbearing function.

The issue needs to be redefined: the problem is not how to protect "potential fetuses," but how to ensure that *all* workers have a safe and healthy workplace. Every worker is entitled to a job that not only protects but in fact promotes all aspects of their health,[6] including the right to bear healthy children. The goal is to eradicate the hazard, not to be diverted by arguments over exclusion from particular jobs.

A good model toward a solution is in the policy guidelines adopted by the Canadian Advisory Council on the Status of Women.[7] The Council urges that single standards for each hazard be determined which would ensure maximum protection for the most susceptible worker of any age and either sex. The guidelines also urge that workers of either sex, when confronted with a reproductive hazard, be granted the right "to refuse work and leave the hazardous work area immediately without loss of income or job security."

Industry Fears

Many industry representatives have argued that, since chemicals such as those listed in Table 2 may harm a fetus, *all* fertile women must be excluded from contact with them.[8] For this purpose, "fertile women" are defined as women of ages fifteen to fifty-four, or between the onset of menstruation and menopause, or women who have not been sterilized. They explain that a woman might be pregnant unintentionally and that her embryo therefore might be exposed before she knows that she is pregnant. It is true that embryos are particularly susceptible to damaging environmental influences at the time

when their organs begin to develop, which for humans occurs from about the eighteenth to the sixtieth day after conception, and that they are most sensitive between the twentieth and thirtieth days. At that point, many women do not yet know that they are pregnant, especially if they tend to have irregular menstrual cycles.

Rather than adopting exclusionary hiring practices, some industries have instituted pregnancy-testing programs to make sure that women employees are not pregnant or have required women to report a missed menstrual period. When eleven of thirty-one women employees at the Kellogg, Idaho, Bunker Hill Company lead smelter refused to take pregnancy tests in 1976, they were fired. After the United Steelworkers of America filed a union grievance on their behalf, the women were reinstated but the company then instituted a policy that no fertile women could work in the lead smelter.

Table 2. *

Women have been excluded from jobs with exposure to:

Acrylamide	Hexafluoroacetone (HFA)
Cadmium	Hydrazine Hydrate
Carbon Disulfide	Hydrazine Sulfate
Coal Tar Distillates	Lead
Dimethylacetamide	Mercury
Dimethyl Formamide	Methotrexate
Estrogen	Tetramethyl Thiourea
Ethylene Thiourea	Thiotepa
Fluorocarbon-22	"Unspecified Solvents"
Formamide	Vinyl Chloride

Some of these agents are commonly used in industrial processes, exposing many workers. Others are used only in the manufacturing of specific products, exposing relatively few workers. For example, methotrexate is a very specialized drug with severe health effects, used in cancer therapy. Some chemical companies bar women from manufacturing this drug, yet nurses who administer it to patients are exposed. Fluorocarbons are used widely in plastics manufacturing. Although fertile women are being excluded from jobs with exposure to these agents, most of them also harm the male reproductive system as well as hurting the general health of both men and women workers.

*Based on Information from the Women's Legal Defense Fund and the Coalition for the Reproductive Rights of Workers.

Olga Madar, past president of the Coalition of Labor Union Women (CLUW), remarked in testimony on the U.S. Occupational Safety and Health Administration's (OSHA) proposed lead standard: "Pregnancy testing is included in the list of medical screening tests. . . . This is an outrageous invasion of privacy to women workers and their families. . . . Such testing has great potential for abuse by supervisors or fellow workers. Requiring pregnancy testing also has an impact on female workers not felt by male workers, making it highly suspect under Title VII of the Civil Rights Act's ban on sex discrimination in employment."[9]

Since most reproductive hazards are also hazards to the health of male and female workers, why is there such a heavy emphasis on the health of the fetus? Why not simply make sure that workers are not exposed to these hazards at levels that can harm a fetus—a policy that would protect the health of workers as well.

Some of management's reasons are economic. Excluding fertile women is much less expensive than lowering levels of exposure. The Lead Industries Association (LIA), for example, argued against a sex-neutral lead standard and for one that excluded women from certain lead-exposure jobs. LIA's points were summarized in an appeals court decision on the legality of the standard:

> Because fertile women require such low blood-lead levels, no feasible [meaning cost-effective] lead standard could possibly protect them, and that any standard that did so protect them would keep virtually all workers out of the workplace. . . .
>
> LIA had essentially argued that only the reproductive functions of women were threatened at low blood-lead levels, and that the only feasible means of protecting fertile women was to exclude them from the workplace or to counsel them out on a case-by-case basis.[10]

Many in industry fear expensive liability lawsuits.[11] Under Worker's Compensation laws, workers who are harmed by workplace hazards may not bring civil suit against their employers unless gross negligence can be demonstrated—an extremely difficult legal task. But a deformed child who could prove that the birth defects resulted from a parent's exposure to a reproductive hazard on the job is not covered by compensation laws and could sue the company for an unlimited amount. Because so little is known about the effects of toxic exposures of fathers on fetuses, it has been assumed rather too hastily that most such lawsuits would argue maternal exposure.[12]

From Homunculus to the New Myth of "Immaculate" Conception

Ironically, the assumption of some legal theorists and industrialists that danger to the fetus is mostly due to the environment of the pregnant female reverses earlier ideologies. In the 1600s, many believed that women played a minor role in reproduction. The animalculists thought that each male sperm contained a complete miniature human, the homunculus. Therefore, the child's characteristics were believed to come mainly from the father. Of course, we now know that each parent contributes half the genetic material. However, sophistication about genetics does not extend to some environmental toxicologists,[13] who often act as though only women are responsible for damage to the fetus's genes or chromosomes (that is, genetic damage) or for actual damage to the fetal organs (which is called *teratogenesis*).[14] From looking at the woman's function as that of passive vessel for the father's seed, we have gone to the opposite extreme, so that now the woman is made entirely responsible for the health status of the fetus and newborn. Thus, while women are denied employment opportunities in order to prevent potentially adverse effects on the fetus, men's reproductive role is not considered.

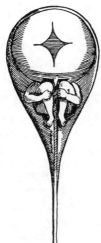

Figure XIV. Homunculus in human sperm, drawn by the animalculist Hartsoeker, originally published in Essai de Diatropique (1694).

The Science of Eggs, Sperm, and Fetuses

Reproductive hazards are caused by agents[15] that can interfere with normal female or male reproductive functioning, for example, by producing abnormalities in eggs or sperm or by damaging the fetus directly.[16] These agents can be *mutagens* or *teratogens* (for explanation of these terms see footnote 14). Both can result in structural or functional abnormalities in offspring and there is overlap between these kinds of reproductive hazards. For example, a substance like estrogen may interfere with normal female or male reproductive functioning and also act directly on the fetus. Or, mutagenic substances such as DBCP may cause direct damage to the chromosomes as well as destroy the male germinal epithelium (the cells in the testes that are responsible for producing sperm).

There is a growing concern that mutagens (see Table 3 for examples) may significantly increase the numbers and kinds of human diseases by increasing the frequency of genetic damage.[17] A substance capable of reacting with genes can affect reproduction by inducing mutations in the sperm or eggs, or in the sperm- or egg-producing cells of men and women, or else by acting directly on the developing fetus. While the potential number of genetically harmful agents is large, it is too soon to know exactly the extent to which they produce birth defects and other adverse reproductive outcomes. Indeed, many mutagens have not yet been tested to see whether they are also teratogens or exert their toxic effect on the reproductive system.[18]

Table 3

Selected Occupationally Significant Agents Observed to Cause Chromosomal Aberrations and/or Induce Mutations

alkylating agents	ethylene oxide
arsenic	hair dyes
b-naphthylamine	halogenated anesthetic gases
benzene	ionizing radiation
benzidine	lead
bis-chloromethyl ether	nitrosamines
chlorinated hydrocarbons	polycyclic aromatic hydrocarbons
chloroprene	tris (flame retardants)
dioxin	vinyl chloride

Some scientists argue that eggs are more susceptible to mutagens because the cells in the ovaries (called *oocytes*) from which eggs develop are formed before birth and may therefore be repeatedly and cumulatively exposed to mutagens during the woman's entire life. However, the reproductive organs of both women and men develop from the same embryonic tissue and by similar developmental paths. There is no evidence to suggest differences in their sensitivity to mutagens. New sperm cells are generated continuously from the age of puberty onward, whereas eggs are produced from precursor cells that are laid down in the embryo. However, the cells in men's testes from which sperm cells mature also are present from before birth and therefore can be cumulatively exposed to mutagens. Like women's eggs, men's sperm also mature cyclically, only men's cycles, unlike women's, start at close intervals and run concurrently (sperm take seventy to eighty days to develop). Some even argue that sperm may be *more* susceptible to genetic damage than are eggs because rapidly proliferating cells generally are more sensitive. Furthermore, the egg precursor cells are embedded in layers of protective, connective tissue within the ovaries inside the woman's body, whereas sperm precursor cells in the testicles may be more exposed.

Several kinds of agents have been shown to affect male reproduction. They include ionizing and microwave radiations, anesthetic gases used in operating rooms, and vinyl chloride used in plastics manufacturing. And indeed, the unexposed wives of *men* exposed to anesthetic gases and vinyl chloride have experienced excessive numbers of miscarriages and stillbirths.[19]

To date, quantitative relationships between defective sperm and adverse reproductive outcomes have not been established. However, current evidence certainly does not justify the widespread assumption that defective births do not arise from damage to testicles or sperm (see Table 4). On the contrary, a common laboratory test, the "dominant lethal assay," demonstrates that chemically treated male animals will sire unviable pups when mated with untreated females.[20] (Similar tests are not usually done for birth defects.) Tris(dichloropropyl)phospate, which is commonly used as a flame retardant in such products as Dacron and polyurethane foam, has been found in seminal fluid. It is known that tris can be absorbed into the body in a very short time from fabrics treated with flame retardant. Tris is highly mutagenic and several types of tris have been found to cause sterility in male experimental animals.[21] Another example is male exposure to dinitrotoluene (DNT) and toluene diamine (TDA). Men exposed to these substances at the Olin Chemical Company plant in Brandenburg, Kentucky, were found to have sperm abnormalities; their wives were found to have high miscarriage rates; and two of the men fathered sons with growth abnormalities and mental retardation.[22]

Table 4

Selected Agents Toxic to the Male Reproductive System which May Be of Occupational Significance. *

Hazards	Species Effect Observed (h = humans, a = animals)	Examples of Occupations where Hazards May Occur
alcohol	h	Social hazard
alkylating agents	h,a	Chemical and drug manufacturing
anesthetic gases; nitrous oxide	a,h	Medical, dental and veterinary workers
cadmium	h,a	Storage batteries; smelter workers
carbon disulfide	h,a	Viscose rayon manufacture; soil treaters
carbon tetrachloride	a	Chemical laboratories; dry cleaners
diethylstilbesterol	a,h	DES manufacturers
chloroprene	h,a	Rubber Workers
ethylene oxide	a	Health care workers (disinfectants); users of epoxy resins
hair dyes	a	Cosmetic manufacturers; hairdressers and barbers
lead	h,a	Storage batteries; policemen; smelter workers
manganese	h	Welders, ore smelters and roasters
nickel	a	Smelters, welders
organic mercury comp'ds.	a	Pesticide workers
tris (flame retardants)	a	Clothing and textile work
Pesticides (e.g.): dibromochloropropane kepone DDT DBCP carbaryl DDVP malathion	a	Farm workers; pesticide manufacture; spray applicators; exterminators
vinyl chloride	h	PVC manufacture and processing
elevated carbon dioxide	a	Brewery workers; chemical manufacture
elevated temperatures	h,a	Bakers; glassblowers; foundry and oven workers
microwaves	h,a	Radar operators; air crewmen; transmitter operators
x-irradiation	h,a	Health workers; radiation workers

*Adapted from Manson.[4]

Some researchers have suggested that sperm with genetic or structural defects are less likely to fertilize eggs, so that damage to the sperm may not lead to fetal damage. However, little is known about this supposed "selection of the fittest sperm." And this kind of "selection of the fittest" (if it occurs) could also mean that genetically damaged egg precursor cells do not mature into viable eggs. This could easily happen and not be noticed, since girls are born with 300,000 to 400,000 egg-producing follicles, most of which cease functioning without producing mature eggs. By age thirty, for example, a woman has only about 25,000 egg precursor cells left, and only 400 follicles release mature eggs during all of a woman's reproductive years.[23] Another unanswered question is how genetic repair mechanisms act on genetically damaged cells from which eggs and sperm mature. Little information is available to answer these questions, but there is no reason to *assume* that the cells from which sperm mature are more able to repair genetic damage or that "selection of the fittest" occurs for sperm more than for eggs.

In addition to mutagens, teratogenic substances acting on the mother can cause structural or functional alterations in the normal development of the fetus by interfering with developmental biochemistry (see Table 5).[24] Examples of teratogens include cadmium, organic mercury, and ionizing radiation. The possibility of such adverse effects in susceptible embryos is the reason most often given for excluding fertile women from certain jobs. But the relationship between maternal exposure to a teratogen and the amount that reaches the fetus is not straightforward. For example, animal experiments show that the percentage of cadmium detected in a fetus is a function of the stage of pregnancy rather than of the dose.[25]

Other reproductive hazards include substances that can cause a loss of libido. This has been seen in male welders exposed to manganese or male workers exposed to estrogen in drug manufacturing.[26] Disturbances in the menstrual cycle have been noted among several kinds of female employees, such as women involved in the manufacture of diethylstilbesterol (DES) and airline flight personnel.[27] Exposure of men can also affect reproductive outcome when the testicular fluid is contaminated with agents that penetrate the physiological barrier than ordinarily restricts the entry of substances from the blood stream. There may also be a possibility that toxic agents present in seminal fluid can contaminate the uterine environment. For example, male rabbits exposed to the well-known teratogen, thalidomide, have been found to have the drug in their semen and have sired litters with congenital defects, low birth weights, and poor survival rates. Similarly, the unusual chemical polonium-210 has been found in semen, which is indicative of exposure to lead (including the unusual isotope, lead-210).[28]

Table 5

Substances of Occupational Significance Observed to Induce Adverse Reproductive Outcomes with Exposure During Pregnancy*

	Some Occupations where Exposure May Occur
alkylating agents	Drug workers
anesthetic gases	Operating room personnel (incl. dental and veterinary)
arsenic	Agricultural workers
benzene	Chemical workers; laboratory technicians
carbon monoxide	Outside workers; offices with smokers
chlorinated hydrocarbons	Laboratory workers; craft workers
diethylstilbesterol	Drug workers
dimethyl sulfoxide	Laboratory workers
dioxin	Agricultural workers

Infectious agents:
  rubella virus
  cytomegalovirus
  herpes virus hominis
  toxoplasma
  syphilis
	Health care workers; social workers; teachers; animal handlers; meat cutters and inspectors; laundry workers

Ionizing radiation	X-ray technicians & technologists; atomic workers; drug workers
organic mercury compounds	

organophosphate pesticides
  DFP
  parathion
  captan
  carbaryl
  theram
	Agricultural workers

polychlorinated biphenyls	Electrical workers; microscopists

*adapted from Wilson[24]

The Myth of Unique Female Susceptibility:
The Case of Inorganic Lead

The differential treatment given to men and women exposed to reproductive hazards, which we have described above, is well illustrated by the example of lead. Lead is widely used in many industrial processes, with more than 800,000 men and women working in over 120 different occupations exposed to it regularly. They include those working with ceramics, enamel, insecticides, and other chemical products. It has been known for almost a century that absorption of excessive lead can cause damage to the kidneys and the central nervous system. It is also known that lead can cause blood dysfunction, gastrointestinal disturbances, weight loss, lassitude, and other symptoms.

Both human and animal evidence exists showing that the male and female reproductive systems are affected by lead, but there are little data about the ability of lead to act as a teratogen. Yet lead has been the most controversial substance in this area of social and scientific debate.

In 1976, the Occupational Health and Safety Administration (OSHA) asked for comments on whether the proposed new lead standard should consider fertile women as particularly susceptible to lead poisoning because of the effect lead might have on the fetus. The Coalition of Labor Union Women (CLUW) presented testimony disputing the concept that all women of childbearing age represent a uniquely "susceptible" subgroup of the population. The purpose of the testimony was not to show that there are no adverse effects of lead on women or that there are no adverse effects of lead on reproduction. Rather, it was to show that there is much evidence against treating women's occupational exposure to lead differently than men's, instead of setting the standard so as to avoid adverse effects in any worker, irrespective of sex.

Susceptibility is defined as the state of being readily affected or acted upon.[29] Some studies that argue for particular female susceptibility are based on judging female biological parameters by male norms. For example, when discussing the relationship between blood lead levels and serum iron levels, a group of researchers refers several times to "the relative iron deficiency which may exist in females."[30] The absurdity of such an argument becomes apparent if we reverse it. Does it make sense to assume that all males suffer from a relative *excess* of iron in their blood. It makes sense to assume that normal females and males have enough iron for their respective healthy functioning. It naturally follows that if a given level of lead interferes with blood functions in women it is not because we have *too little* iron; it is because that lead level is *too high.*

Fallacious Historical References to Female Susceptibility

The use of fallacious data or a biased perspective for evaluating the incidence and effect of occupational diseases and exposures in women is not new. Perhaps the most notable instance is the work of Dr. Thomas Oliver, which doggedly persists in making its way into the literature. In 1902, Oliver wrote that "females contract lead poisoning more readily, the symptoms are usually more acute, they suffer more severely and succumb to it more quickly than males."[31] In 1946, Dr. Anna Baettjer laid Oliver's data to rest. She analyzed the data and arguments of Oliver and others and concluded that "on the basis of the data there appears to be no very convincing evidence that women are more susceptible to lead poisoning than men." She also noted that "for the most part, the theory that women are more susceptible than men to occupational disease has arisen by the repeated quoting in the literature of statements to the effect made by one or two industrial health authorities. In many cases the statements represented only a personal opinion."[32]

Baettjer pointed out that Oliver himself qualified his statements on the particular susceptibility of females. Oliver wrote, "no doubt much of the greater prevalence of lead poisoning, hitherto observed in women who have worked in white lead factories, is to be explained by the fact that they have until recently worked in larger numbers than the men in the dangerous processes. After males replaced females the number of cases of lead poisoning in men increased, whereas those in women decreased." Yet this part of Oliver's analysis has been omitted when later researchers have quoted his work.

These same unfounded generalizations asserting women's greater susceptibility, as well as other poor studies, are still being cited in the scientific literature. As recently as 1975, they were referred to uncritically in a review of the literature on the reproductive hazards of lead:

> Lead has been known to affect women during pregnancy for more than a century. Teratogenesis has been reported and documented by animal experimentation. In addition to lead's direct adverse effect on the course of pregnancy, it has an indirect effect by its toxic action on the male germ cell. There is evidence that women, particularly during specific periods, i.e., adolescence, pregnancy, etc., are more susceptible to toxic effects of lead. Oliver has noted that women are more susceptible to lead poisoning between ages of 18 and 23, and after shorter exposure. He noted that women poisoned by lead have a higher incidence of encephalopathy, and a lower incidence of paralysis and colic, than males.[33]

The paper citing Oliver has, in turn, already been cited by at least one source.[34] Once a fallacious statement or idea makes its way into the literature it is difficult to remove it, especially if it serves other useful ends that are in the interest of powerful groups in society.

Susceptibility as a Function of Reproductive Ability

We have pointed out why it is wrong to assume a so-called "special susceptibility" because of women's biological role in reproduction. With regard to lead, there is good evidence that male reproductive abilities are also affected. Furthermore, the extreme cases of stillborns, miscarriages, spontaneous abortions and other reproductive mishaps due to maternal lead exposure are associated with very high levels of lead that are known to have many other toxic effects, and hence are high enough to make virtually the entire workforce "susceptible." Thus, women of childbearing age do not represent a uniquely susceptible subgroup of the population.

Thank God! The women and children are safe from lead poisoning

Lancranjan *et al.* studied the semen quality of 150 male workers exposed to lead in a storage battery facility.[35] In lead-poisoned workmen (mean blood lead = 5.28 μg/100 ml) an "obvious and highly significant" decrease in semen quality was observed. Results showed significant sperm changes including decreased motility, and increased malformations. Even in workers with moderate lead absorption (mean blood lead = 4 μg/100 ml), significant decreases in sperm numbers and motility were observed. The most significant and frequent alterations revealed by the semen analysis were sperm malformations. *Children* of *male* lead workers have also been shown to have excessive lead levels in their blood most likely due to the contamination of the workers' clothing that is brought home.[36]

In summary, then, there is no scientific justification for considering all women of childbearing age a susceptible subgroup of the working population. There are sufficient data to show that a significant proportion of the population is at risk from the effects of exposure to lead, and hence must be deemed susceptible. Further, if the intent of the OSHA standard is to protect workers from hazards to reproduction there is not justification for treating women differently from men. In all probability, blood levels that would be sufficiently low to prevent adverse effects on sperm formation are comparable to the level necessary to prevent injuries to fetal development.

The testimony presented by CLUW and other feminist and labor groups convinced OSHA to present a sex-neutral policy in its final rule-making on the standard. This standard was vigorously challenged by industry groups but equal safeguards for both sexes were upheld by the appeals court in 1980 and the Supreme Court in 1981.

Radiation: Enter at Your Own Risk

In contrast to industrial jobs involving lead, where many women have been denied employment in traditional "men's jobs" with lead exposure, there has been little attempt to exclude fertile women from "women's jobs" that involve exposure to ionizing radiation. Yet ionizing radiation is a known teratogen (as well as causing sperm abnormalities and other adverse human health effects). Perhaps the lack of attention to radiation exposure is due to the fact that it has occurred most commonly in a traditionally low-paid sector of the health-care industry. However, increasing numbers of women are now being employed in the nuclear power industry, and at present 10 to 15 percent of the workforce in nuclear power plants, fuel processing, and by-product manufacturing is female. It was not until August 1980 that the Environmental Protection Agency tentatively proposed regulations for the exposure of women workers to ionizing radiation. Prior to the proposed EPA regulations, the only government policy concerning radiation and women were guidelines suggested by the Nuclear Regulatory Commission. These guidelines are stated in terms of vague generalities: it is the responsibility of the employer to take all practical steps to reduce a pregnant woman's radiation exposure, and it is the responsibility of the pregnant woman to decide whether the exposure she receives on the job is low enough to be safe for the unborn child.[37]

The EPA is currently proposing four alternatives, one of which is that women of childbearing age shall be excluded from jobs for which the whole-body dose rate is more than 0.2 rem per month. Another alternative suggests the above dose limit for fertile women, but does not make it mandatory. The other two alternatives are sex neutral.

The Coalition for the Reproductive Rights of Workers (CRROW) in its comments on the proposed standards, points out the sex discrimination in EPA's approach: "EPA's justification for suggesting an exclusionary policy for women workers exposed to radiation is to protect the offspring. However, if the offspring is the concern, men would be the justified target of exclusion. According to EPA's own background report, the risk to future generations from mutational effects is five times greater for paternal exposure than maternal exposure." The United Nations Scientific Committee on the Effects of Atomic Radiation discusses the greater sensitivity of sperm compared to eggs in their 1977 report, and estimates: "from 2 to 10 congenitally malformed liveborn children per million conceptuses, per rad of paternal irradiation, with about five times this number of recognizable abortions and about 10 times the number of losses at the early embryonic stage. The corresponding risk from maternal radiation is likely to be small" (per rad of radiation exposure). Further, because effects on the fetus from maternal irradiation have only been seen at very high levels of irradiation, the U.N. Committee states, "no satisfactory data are yet available for deriving reliable quantitative estimates of the risk from pre-natal irradiation at comparable developmental stages, particularly at the low doses and dose rates."[38] In spite of this, the EPA has proposed protecting a fetus by excluding women of childbearing capacity when it has made no parallel suggestion to exclude fertile males.

Why Exclude Women?

The exclusion of fertile women has come about in recent years as more women have moved into traditionally "male" job areas and as the more primitive rationales, such as the "raging hormone" theories, have been generally discredited as a basis for excluding women. For example, only 1 percent of American Cyanamid's workforce is women.[39] Many of the jobs from which women are now being excluded are the higher-paying industrial jobs that have traditionally been male territory. There has not been an attempt to exclude fertile women from lower-paying, *traditional* "women's" jobs where they are exposed to known reproductive hazards. For example, nurses, x-ray technologists, and nurses' aides continue to be exposed to radiation; and ceramic and jewelry workers to lead.

A historical perspective on female "protective" legislation can help us understand why women are being excluded today. In the early 1900s, many groups supported protective legislation for women that limited the number of hours in the workday first to ten and then to eight hours per day. Early protective legislation was socially progressive in that such protections were often later extended to men. However, there was a sexist basis to most protective legislation.

For example, a U.S. Supreme Court decision in 1908 (Muller v. Oregon) provided the impetus to establish the constitutionality of limiting the number of hours in a workday. The basis for this decision, however, was the supposed biological vulnerability of women. The court decision stated:

> That woman's physical structure and the performance of maternal functions place her at a disadvantage in the struggle of existence is obvious. This is especially true when the burdens of motherhood are upon her. Even when they are not, by abundant testimony of the medicine fraternity continuance for a long period of time on her feet at work, repeating this from day to day, tends to injurious effects upon the body, and as healthy mothers are essential to vigorous offspring, the physical well-being of woman becomes an object of public interest and care in order to preserve the strength and vigor of the race. . . . Differentiated by these matters from the other sex, she is properly placed in a class by herself, and legislation designed for her protection may be sustained, even when like legislation is not necessary for men, and could not be sustained. It is impossible to close one's eyes to the fact that she still looks to her brother and depends on him. Even though all restrictions on political, personal, and contractual rights were taken away, and she stood, so far as statutes are concerned, upon an absolutely equal plane with him, it would still be true that she is so constituted that she will rest upon and look to him for protection; that her physical structure and a proper discharge of her maternal func- tion—having in view not merely her own health, but the well-being of the race—justify legislation to protect her from the greed as well as the passion of man. The limitations which this statute places upon her contractual powers, upon her right to agree with her

employer as to the time she shall labor, are not imposed solely for her benefit, but also largely for the benefit of all.[40]

The driving wedge in forwarding the constitutionality of laws requiring only a ten-hour day were based on this decision. Within seven years the ten-hour day was found constitutional for all workers. But the gain was made on the basis of sexist generalities about the special biological weaknesses and roles of women and about our subservience to the interests of "the race."

Later, protective legislation forbade women jobs with night hours or certain lifting requirements *except* when those jobs were traditional women's

Females — the weaker sex

work, such as waitressing, nursing, and housework. For example, in California, women were forbidden night work except during harvesting seasons when canneries operated twenty-four hours per day, and low-paid women were indispensable at the assembly lines. Another example is that of low-paid office maids. Since their services were needed at night, so that their cleaning would not "interfere" with daytime office work, they were routinely exempted from state restrictions on women's night work. Now most protective legislation has been struck down as discriminatory by federal statute. But this same type of "protection" of women workers (and their "potential fetuses") is being resurrected now in the area of reproductive hazards.

The Myth of Permanent Pregnancy

Exclusionary policies against fertile women also lend emphasis to the myth of "permanent pregnancy." Such policies treat *all* women as if they were always pregnant, whereas data in fact show that most women plan their pregnancies and complete their families by the age of thirty.

Data from interviews on randomly selected samples collected by the National Center for Health Statistics demonstrates that the vast majority (80 percent) of all children born to married women living in families with three or fewer children are planned. Families with three or fewer children in turn constitute the vast majority of working-class families with employed mothers. Further, the Bureau of the Census reports that only approximately 1 percent of all blue-collar working women aged thirty or above *plan* to become pregnant.[41]

Using these data on birth expectations and on degree of family planning and child "wantedness," and estimating a very liberal rate of 50 percent "unexpected" childbearing for the very small percentage of women employed outside the home who report "planning" to have a child after age thirty, one can still conclude that fewer than 2 percent of all blue-collar women who are employed outside the home will bear children after the age of thirty. That is, even when we take into account an "unexpected" childbearing rate two and one-half times greater than that reported by the National Center for Health Statistics, we still come up with an extremely low rate of birthing for blue-collar women in their thirties. Clearly, it is not reasonable to treat all fertile women as if they were permanently pregnant.

What are the Solutions?

It is difficult to know the number of women that have been affected by exclusionary policies since many industries have not made public their discriminatory employment practices and many women may not know when their civil rights are being violated. Joan Bertin, staff attorney on the American Civil Liberties Union's Women's Rights Project, who is litigating the Willow Island American Cyanamid case and the Lillian Reese case, has remarked:

> . . . there are many situations where women have been transferred within companies or never hired in the first place. These situations have often not been challenged. . . . If women don't pursue the case or if they are unaware of their civil rights that's the end of it. . . . I know of at least one instance in a chemical company where a woman was demoted. Then the company stopped hiring women altogether; no challenges have been brought against the company.[42]

Fertile women are currently being denied employment opportunities because of potential reproductive hazards in the workplace and there is no legislative or social solution to the problem in sight. The U.S. Equal Employment Opportunity Commission (EEOC) in February of 1980 proposed guidelines under Title VII of the Civil Rights Act, for employment policies based on reproductive hazards.[43] These proposed guidelines were problematic because they stipulated that fertile women could be excluded from certain jobs if there were known reproductive effects on women, and no research had yet been done on reproductive effects on men.

Feminist and labor groups protested the proposed EEOC guidelines because they allowed for discriminatory exclusion of fertile women under certain circumstances. Industry representatives also vehemently protested the guidelines because they called for careful evaluation of reproductive effects on men and mandated that the industry conduct research on male reproductive effects if there were known effects on women and no research had yet been conducted on men. Both groups breathed a sigh of relief when the proposed guidelines were withdrawn prior to the beginning of the Reagan administration. It must be noted that guidelines are hardly ever withdrawn, because they expire automatically after a certain length of time if they are not ruled on. It is now up to the courts to set legal precedents in this area—a very slow process. A better model for a solution is in the previously mentioned policy guidelines adopted by the Canadian Advisory Council on the Status of Women that urge for each hazard a single standard that would ensure maximum protection for the most susceptible worker of any age and either sex.

Neither women nor men should have to work under conditions that endanger their health or the health of future generations. The greatest emphasis should be on cleaning up the workplace. The vast majority of reproductive hazards also have adverse effects on the workers' own health. Where there is special concern about reproductive effects, a humane solution is to grant women and men planning families job transfers or "parental leave" with full seniority and salary security.

Only if these provisions are followed will both workers and their future children be protected—without discrimination against potential fathers who must continue to work while being exposed to reproductive hazards and fertile women who do not plan to become pregnant. To invoke myths about women's biology and special reproductive liability imposes hardships on women, men, and their children.

Notes

1. Washington State Human Rights Commission Complaint by Lillian Reese, Case Number TEA 0111-80-1 (October 17, 1980).
2. These cases are excerpted from descriptions in an article that presents an overview of scientific and legal issues: Jeanne M. Stellman and Gary Epler "Scientific and Policy Aspects of Reproductive Health in the Workplace," *Toxic Substances Journal* 1, no. 1.
3. For a description of the male reproductive effects associated with DBCP *see* D. Wharton *et al., Lancet* 2 (1977), p. 1259; and *Journal of Occupational Medicine* 21 (1971), pp. 161-166. A front-page article in the *New York Times* on September 11, 1977, reports on U.S. policy regarding DBCP production.
4. *See* J. Manson and N. Simons, "Influence of Environmental Agents on Male Reproductive Failure," *Work and the Health of Women*, Vilma Hunt, ed. (Boca Raton, FA: CRC Press, 1979).
5. See chapters by Birke and Best, R. Hubbard, and M. Walsh in this collection.
6. The International Labor Organization and the World Health Organization begin their joint definition of Occupational Safety and Health: ". . .The promotion and maintenance of the highest degree of physical, mental, and social well-being of workers in all occupations; the prevention among workers of departures from health caused by their working conditions; the protection of workers in their employment from risks resulting from factors adverse to health. . . ."
7. Canadian Advisory Council on the Status of Women, Position Paper: "Reproductive Health Hazards in the Workplace, 1980," available from the Council, Box 1541, Station B, Ottawa, K1P, Ontario, Canada.
8. Revealing interviews with industry representatives are included in Gail Robinson's article, "The New Discrimination," *Environmental Action* (March 1979), pp. 4-9. Further industry views are found in *Women,*

Work, and Health: Challenges to Corporate Policy, Diana Chapman
Walsh and Richard H. Egdahl, eds. (New York: Springer-Verlag, 1980);
L. Krause, "Pregnant Women in the Workplace: A Company Program
to Control Toxic Exposures," *National Safety News* (February 1979),
pp. 52–55; and K. Rao, "Protecting the Unborn: The Dow Experience,"
Occupational Health and Safety (March 1981), pp. 53–61.

9. From Testimony on Behalf of the Coalition of Labor Union Women
(CLUW) by CLUW President Olga M. Madar on the Occupational Safety
and Health Administration's Proposed Lead Standard (March 11,
1977).

10. U.S. Court of Appeals for the District of Columbia Circuit, No. 79–1048,
United Steelworkers of America, AFL–CIO–CLC, Petitioner v. F. Ray
Marshall, Secretary of Labor, U.S. Department of Labor, and Dr. Eula
Bingham, Assistant Secretary for Occupational Safety and Health,
Respondents, pp. 125, 127.

11. N. Stillman, industry lawyer, writes about the possibility of such lawsuits in "A Legal Perspective on Workplace Reproductive Hazards,"
Walsh and Egdahl; eds., *Women, Work, and Health.*

12. N. Stillman makes this assumption in Walsh and Egdahl, eds., *Women,
Work, and Health*, pp. 177–178. However, a lawsuit *has* been filed by
fathers to recover damages for their sons' birth defects. The suit, filed
in March 1981, charges that the fathers, who were employees of the
Long Island Railroad, were exposed to weedkillers containing dioxin, a
highly toxic contaminant. The suit charges that their exposure to
dioxin caused hip defects in their children that required corrective surgery. *See Coalition for the Reproductive Rights of Workers Newsletter*
1, no. 1 (Spring 1981), 1917 Eye St., N.W., Washington, D.C. 20006.

13. Dr. L. Krause, Director of Environmental Hygiene and Toxicology, Olin
Corporation, emphasizes women's responsibility for damage to the
fetus (*see* his article referenced in footnote 8). Review articles and
bibliographies summarizing scientific research on reproductive hazards
also illustrate the lack of research on male reproductive hazards. *See*
K. Kurzel and C. Cetrulo, "The Effect of Environmental Pollutants
on Human Reproduction, Including Birth Defects, *Environmental
Science and Technology* 15 (1981), pp. 626–654; and Wendy Chafkin
with Laurie Welch, *Occupational Hazards to Reproduction: An Annotated Bibliography* (published by the Program in Occupational Health,
Montefiore Hospital and Medical Center, 111 E. 210th St., Bronx,
N.Y. 10467, 1980). A welcome improvement is the recent National
Institute of Occupational Safety and Health *Proceedings of a Workshop on Methodology for Assessing Reproductive Hazards in the
Workplace* (Cincinnati, Ohio: NIOSH, 1981) which proposes male and
female reproductive toxicology testing.

14. The terms *mutagen* and *teratogen* and the concepts of mutagenesis and
teratogenesis can be confusing. Mutagens are agents that can produce changes (mutations) in the genetic material (DNA) contained in

the chromosomes of cells. The action of mutagens is known as muta-genesis. When mutations occur in sperm, eggs, or their precursor cells, they may interfere with the transmission of genetic information from parents to children. Fetal death or birth defects may result. Agents that disrupt or transform the normal development of the fetus are known as teratogens. The process is called teratogenesis and the study of these processes is called teratology. Some substances act as mutagens, others are teratogens, yet others are both. For example, mutagens may act directly on the fetus, causing death or altering normal development.

There is growing concern over the increasing number of chemical mutagens in our environment. This concern is, in general, due to the number of mutagens that are known to cause cancer in humans and/or laboratory animals, and the fact that mutagens can produce defects in offspring of exposed parents. However, the exact relationship between mutagenesis, carcinogenesis, and teratogenesis is not well understood.

15. The word *agent* is used to cover the variety of forms in which occupational and environmental reproductive hazards occur. They may be chemical substances such as dibromochloropropane (DBCP), biological organisms such as the viruses causing German measles and hepatitis, or physical agents such as microwave radiation and excessive heat.

16. For a more complete description of the science of male and female reproductive hazards see the chapter by Jeanne Stellman, "The Occupational Environment and Reproductive Health," W. Rom and J. Lee, eds., *Occupational Medicine* (Boston: Little, Brown and Co., in press).

17. Tables 3–5 are from an article by Jeanne Stellman, Effects of Toxic Agents on Reproduction," *Occupational Safety and Health* (April 1979), pp. 36–46.

18. Approximately 7 percent of all live-born infants have recorded defects (Raymond Harbison, "Teratogens," *Toxicology*, J. Doull *et al.*, eds. [New York: Macmillan Publishing Co., 1980].) Approximately 60–70 percent of these birth defects are of unknown origin. Robert Dixon estimates that approximately 33 percent of early human embryos die, before the women even are aware that they are pregnant, and approximately 15 percent of recognized pregnancies abort spontaneously ("Toxic Responses of the Reproductive System," *Toxicology*, Doull *et al.*, eds., p. 332).

It is difficult to quantify the effect of environmental/occupational exposures on these reproductive outcomes for a variety of reasons. Mothers and fathers are both exposed to a wide variety of substances previous to, and during pregnancy. It is difficult to associate any one of these exposures with outcomes. However, workers are often exposed to much higher quantities of particular substances in the workplace than is the general population. Thus workers often become the guinea pigs on which the toxic effects of particular substances are first noted.

19. Descriptions of the outcomes in offspring of males exposed to reproductive hazards is found in G. R. Strobino, J. Klein, and Z. Stein, "Chemical and Physical Exposure of Parents: Effects on Human Reproduction of Offspring," *Early Human Development* 1 (1978) p. 371; and in Mason and Simons, "Influence of Environmental Agents."

20. The dominant lethal assay is a laboratory test that exposes male rats or mice to substances to test for genetic defects. Although there are different methodologies for this test, a common procedure is to expose male mice to the substance for a short period of time. These males are then mated with unexposed virgin female mice at 7 days, 14 days, 21 days, 28 days, etc. The female mice are then sacrificed during their pregnancies and the number of non-viable fetuses are counted. It is assumed that the non-viability of the fetus is the result of lethal genetic damage to the sperm or to the cells that produce sperm. By tabulating how many days after exposure of the males the greatest number of non-viable fetuses occur in unexposed mated females, it is possible to tell whether the substance affects sperm cells themselves or the spermatogonia that generate them. For example, if the greatest number of non-viable fetuses occur shortly after the males are exposed, then mature sperm are affected. However, if the greatest number of non-viable fetuses occurs thirty-five days after exposure of the males (in mice) this means that the spermatogonia are affected.

21. T. Hudec *et al.*, *Science* 211 (1981), p. 951.

22. *Occupational Safety and Health Reporter* 10 (1980), p. 616.

23. Dixon, "Toxic Responses," p. 335–336.

24. Table 5 is adapted from James Wilson, *Pathophysiology of Gestation*, N. Assali, ed. (New York: Academic Press, 1972).

25. R. B. Sonawane *et al.*, "Placental Transfer of Cadmium in rats: Influence of Dose and Gestational Age," *Environmental Health Perspectives* 17 (1975), p. 139.

26. P. Schuler *et al.*, "Manganese Poisoning," *Industrial Medicine and Surgery* 26 (1957), p. 167. J. Harrington *et al.*, "The Occupational Exposures of Formulating Oral Contraceptives," *Archives of Environmental Health* 33 (1978), p. 12.

27. F. Preston *et al.*, "Effects of Flying and of Time Changes on Menstrual Cycle Length and on Performance in Airline Stewardesses," *Aerospace Medicine* 44 (1973), pp. 438–443. J. Harrington *et al.*, "Occupational Exposure to Synthetic Estrogens," *American Industrial Hygiene Association Journal* 39 (1978), p. 139.

28. For references for all these chemical exposures see Stellman, "Occupational Environment and Reproductive Health."

29. The issue of whether or not industry should screen out susceptible workers is the subject of growing debate. A series of articles in the *New York Times* by Richard Severo (February 3–6, 1980, available as a reprint from the Women's Occupational Health Resource Center, 60

Haven Avenue, B-1, New York, NY 10032) outlines the basics of genetic screening. Workers have been excluded from certain jobs because they have sickle cell trait, G-6-P-D deficiency, or thalassemia. A recent article by T. R. Leach *et al.* ("Occupational Health Survey: An Evaluation of Potential Health Hazards in the Workplace," *American Industrial Hygiene Journal* 42 (1981), pp. 160–164) of the BASF Wyandotte Chemical Corporation, suggests that certain workers, "because of medical history, accidents, sex, or aging, may be at a greater risk or more adversely affected by exposure to specific chemical or physical agents." Jobs are then to be rated "unrestricted, controlled, and restricted." The growing discomfort with such screening is based on concern over its use to cover up racial and sexual employment discrimination and because it takes emphasis away from protecting workers by cleaning up the workplace and instead places emphasis on protecting industry's liability for job-related ill health.

30. R. Zielhuis and A. Wibow, "Susceptibility of Adult Females to Lead," *Proceedings of the 2nd International Workshop on Occupational Exposure to Lead* (Amsterdam 1976); A. Wibow *et al.* "Interaction Between Lead and Iron Metabolism," *Ibid*.

31. T. Oliver, *Diseases of Occupation* (New York: Dutton, 1916).

32. Anna Baettjer, *Women in Industry* (New York: W. B. Saunders Co., 1946).

33. Sidney Lerner, "Blood Lead Analysis—Precision and Stability," *Journal of Occupational Medicine*, 17 (1975), pp. 153–154.

34. Samual Epstein (OSHA Oversight Hearings, 1976).

35. I. Lancranjan *et al.*, "Reproductive ability of Workmen Occupationally Exposed to Lead," *Archives of Environmental Health* 30 (1975), pp. 396–401.

36. E. Baker *et al.*, "Lead Poisoning in Children of Lead Workers," *New England Journal of Medicine* 296 (1977), pp. 260–261; and Center for Disease Control, "Increased Lead Absorption in Children of Lead Workers—Vermont," *Morbidity and Mortality Weekly Report* 26, no. 8 (1977).

37. The actual wording of the NRC's guidelines are as follows:

"a. If you are now pregnant or expect to be soon, you could decide not to accept or continue assignments in these areas.

b. You could reduce your exposure, where possible, by decreasing the amount of time you spend in the radiation area. . . .

c. If you do become pregnant, you could ask your employer to reassign you to areas involving less exposure to radiation. If this is not possible, you might consider leaving your job. If you decide to take such steps, do so without delay. The unborn child is most sensitive to radiation during the first three months of pregnancy.

d. You could delay having children until you are no longer working in an area where the radiation dose to your unborn baby could exceed 0.5 rem.

You may also, of course, choose to:

e. Continue working in the higher radiation areas, but with full awareness that you are doing so at some small increased risk for your unborn child."

38. United Nations Scientific Committee on the Effects of Ionizing Radiation, *Sources and Effects of Atomic Radiation* (New York: United Nations, 1977), pp. 8–9.
39. "Chemical Firms Move to Protect Women from Substances that May Harm Fetuses," *Wall Street Journal* (November 7, 1977), p. 7.
40. Muller v. Oregon 208 U.S. 412 (1908).
41. R. H. Weller, "Wanted and Unwanted Childbearing in the U.S.," *Vital and Health Statistics*, Series 21:32, (Department of Health Education and Welfare publication, Public Health Service, 78:1978); and U.S. Bureau of the Census, *Current Population Reports*, Series P-20, no. 325, "Fertility of American Women: June, 1977," (U.S. Government Printing Office, Washington, D.C., 1978).
42. Interview with Joan Bertin by Mary Sue Henifin (January, 1981).
43. U.S. Equal Employment Opportunity Commission, Proposed Interpretive Guidelines on Employment Discrimination and Reproductive Hazards *45 Federal Register* 7514, (February 1, 1980).

[Mary Sue Henifin *is past Coordinator of the Women's Occupational Health Resource Center at Columbia University School of Public Health where she also received her Master's of Public Health Degree in Environmental Science. She speaks and writes frequently about issues affecting women's health and has coproduced a monthly radio program, "The Women's Occupational Health Radio Hour." She is completing a book on the health hazards of office work with Jeanne Stellman.*]

[Jeanne M. Stellman *is Associate Professor in the Faculty of Medicine, Division of Environmental Science at Columbia University School of Public Health, where she is also Executive Director of the Women's Occupational Health Resource Center. Her research includes studies of the health effects (including reproductive hazards) of microwaves, dioxin contamination of Agent Orange (a herbicide used in Vietnam), ethylene oxide (a hospital sterilant), and health hazards to office workers. She has written two landmark books:* Work is Dangerous to Your Health *with Susan Daum (New York: Pantheon Books, 1973); and* Women's Work, Women's Health: Myths and Realities *(New York: Pantheon Books, 1977). She has written extensively on women's occupational health as well as other occupational health issues and is a noted lecturer in the United States and Canada.*]

Figure XV. Sterilization Abuses Women
(Photo: Helen Rodríguez-Trías, M.D.)

Helen Rodrigues-Triaz, M.D.

Sterilization Abuse*

Recent events have shown that sterilization is a procedure freely chosen by some people in family planning but demanded of others against their will. Consider, for example, what happened in India. Indira Gandhi's government was defeated—an event many attribute to the mass forced sterilization program it sponsored—but before the final ousting, at least 300 Indians died in riots protesting the assault of forced sterilizations on both men and women. The Indian sterilization experience showed the world that some population control programs mean ugly coercion.

A second situation exists right here at home: large numbers of Native American women have been sterilized by the Indian Health Services, a United States government agency. A report from the General Accounting Office, produced at the request of Senator James Abourezk from South Dakota, reveals 3,406 sterilizations of American Indian women between the ages of fifteen and forty-four. These were performed in Aberdeen, Albuquerque, Oklahoma City, and Phoenix between 1973 and 1976.[1] Evidence that the basic elements of informed consent were not communicated to the patients lends credence to Dr. Connie Redbird Uri's many public statements charging the United States government with having a deliberate genocide policy against her people, who number under one million.[2] Although there has been a moratorium since April 1974 on government financing of sterilization of women under twenty-one years of age, there were thirteen violations by the Indian Health Service in two years.[3] To date, there has been no action against the violators.

Sterilization has long been an acceptable procedure in the United States for eugenic purposes. The first law empowering a state to sterilize unwilling and unwitting people was passed in 1907 by the Indiana Legislature. The Act was intended to prevent procreation of "confirmed criminals, idiots, rapists, and imbeciles" who were confined to state institutions. The law was clear in its premise that heredity plays an important part in the transmission of crime, idiocy, and imbecility.[4] After World War I, a model federal law was proposed by Dr. Harry Hamilton Laughlin, Superintendent of the Eugenics Record Office, and copies were widely distributed in large quantities to governors,

*First presented November 10, 1976, as one of two Reid Lectures at the Barnard College Women's Center.

legislators, newspaper and magazine editors, clergymen, and teachers. According to the model law, the following ten groups were labeled "socially inadequate" and were therefore subject to sterilization: (1) feeble-minded; (2) insane (including the psychopathic); (3) criminalistic (including the delinquent and wayward); (4) epileptic; (5) inebriate (including drug habitues); (6) diseased (including the tuberculous, the syphilitic, the leprous, and others with chronic, infectious, and legally segregable diseases); (7) blind (including those with seriously impaired vision); (8) deaf (including those with seriously impaired hearing); (9) deformed (including the crippled); and (10) dependent (including orphans, "ne'er-do-wells," the homeless, tramps, and paupers).[5] Laws such as this, known as the eugenics laws, were passed in 30 states and as of 1972 were still on the books in 16.[6]

It is shocking to many to learn that between 1907 and 1964, more than 63,000 people were sterilized under these eugenics laws in the United States and one of its colonies, Puerto Rico.[7] Practices sanctioned inside institutions often become commonly accepted practices in the larger community. It is therefore important to keep in mind this long history of legally sanctioned, forced sterilization as a framework for understanding current hospital practices.

The labeling of people as mentally retarded, insane, criminal, or indigent is an act we must examine closely. This sort of labeling is a peril in itself, but when it is used as grounds for sterilization, it is doubly dangerous. The groups considered undesirable may change, but they always include people who work for wages or are unemployed—inevitably the most exploited and therefore the poorest. In the United States, the labeling process has additional racial overtones, because most Third World people are in the least remunerated strata of the working class and are definitely poor. It is a cruel irony that people with preventable diseases, due almost solely to poverty, are included in groups deemed fit for sterilization.

Under the eugenics laws, many Black women had been sterilized without challenge. The challenge came only when in 1924 Carrie Buck, a poor, white, eighteen-year-old woman institutionalized for mental retardation, was threatened. Although judged retarded, Buck had completed six grades of school in five years. But she had defied the norms by bearing an illegitimate child and was about to be sterilized, when members of a religious group in Virginia challenged the law all the way to the Supreme Court. Justice Oliver Wendell Holmes handed down his well-known Buck v. Bell opinion in favor of her sterilization in which he stated: "the principle that sustains compulsory vaccination is broad enough to cover cutting the fallopian tubes." He concluded that "three generations of imbeciles are enough."[8]

Implicit in Justice Holmes's opinion was the belief that Carrie Buck's alleged mental retardation was hereditary. Today, mental retardation is often determined on such questionable evidence as inability to cope with the school system, the discredited IQ tests, or even evidence of cultural differences.

Perhaps an even more serious consequence of these infamous laws was the social legislation they inspired. At least ten states have proposed compulsory sterilization of people on welfare.[9] No state has passed such legislation, but the very existence of such proposals should make us question the prevailing social climate. In a country plagued by chronic unemployment, such proposals reveal virulent feelings toward women who cannot earn a living because they must care for children, the elderly, or others.

Physicians play an important role in implementing the view that poor people have no right to decide on the number of their children. A survey of obstetricians showed that although only 6 percent favored sterilization for their private patients, 14 percent favored it for their welfare patients. For welfare mothers who had borne illegitimate children, 97 percent of the physicians favored sterilization.[10] Similarly, a number of polls of the public-at-large show that the idea of sterilization of welfare recipients is very much accepted. In a 1965 Gallup poll, about 20 percent of the people surveyed favored compulsory sterilization for women on welfare.[11]

We are witnessing a resurgence of the Malthusian ideas that proclaimed the poor unfit to receive the knowledge and hygienic measures that might decrease their death rate.[12] The more sophisticated modern version calls for a decrease in the social, medical, educational, and other resources allotted to poor people and for an offer of sterilization instead. In lieu of social changes to provide a decent living for every American, the population planners choose to curtail population. In the words of Dr. Curtis Wood, past president of the Association for Voluntary Sterilization:

> People pollute, and too many people crowded too close together
> may cause many of our social and economic problems. As physi-
> cians, we have obligations to the society of which we are a part.
> The welfare mess, as it has been called, cries out for solutions: one
> of these is fertility control.[13]

The use of the phrase "fertility control" is itself deceptive. In reality, it often means only one thing: permanent control—that is, sterilization. Therefore it should not surprise us that a 1973 survey revealed that 43 percent of women sterilized in federally financed family programs were Black.[14]

Hysterectomy, now the most frequent major operation, with rates in the United States four times greater than in Sweden, is an indication of still another way of sterilizing women without their consent.[15] Black women on welfare suffer the most abuse. According to the *New York Times*:

> In New York and other major cities, a hysterectomy which renders a patient sterile costs up to $800, while a tubal ligation (the tying off of the fallopian tubes), which does the same thing, pays only $250 to the surgeon, increasing the motivation to do the more expensive operation. Medicare, Medicaid, and other health plans for both the poor and the affluent will reimburse a surgeon up to 90 percent for the costs of any sterilization procedure, and sometimes will allow nothing for abortion. As a consequence, "hysterilizations"—so common among some groups of indigent blacks that they are referred to as "Mississippi appendectomies"—are increasingly popular among surgeons despite the risks.[16]

Several lawsuits since 1973 around the country provide evidence of the widespread nature of abuse as well as of the rising resistance on the part of people.

Most notorious is the case of two sisters, Mary Alice, then fourteen, and Minnie Lee Relf, who was twelve at the time of their sterilizations in Montgomery, Alabama, in June 1973. As described in court by their mother, two representatives of the federally financed Montgomery Community Action Agency called on her requesting consent to give the children some birth control shots. Believing that the Agency had the best interests of her daughters' health in mind, she consented by putting an X on paper.[17]

Judge Gerhard Gesell, who heard the case, declared:

> Although Congress has been insistent that all family planning programs function purely on a voluntary basis there is uncontroverted evidence in the record that minors and other incompetents have been sterilized with federal funds and that an indefinite number of poor people have been improperly coerced into accepting a sterilization operation under the threat that various federally supported welfare benefits would be withdrawn unless they submitted to irreversible sterilization.[18]

In another case, a number of women from Aiken, South Carolina, sued Dr. Clovis Pierce, a white former army physician, for his coercive tactics in obtaining consent including threats to refuse to deliver their babies. In 1973, Black women were subjects of sixteen of the eighteen sterilizations paid for by Medicaid and performed by that physician.[19]

Norma Jean Serena, a Native American mother of three children, has been the first to raise sterilization abuse as a civil rights issue. She charges that in

1970 health and welfare officials in Armstrong County, Pennsylvania, conspired to have her sterilized when her youngest child was delivered.[20]

Ten Mexican-American women are currently suing the Los Angeles County Hospital for obtaining consent in English when they spoke only Spanish. Some were in labor at the time, others even under anesthesia. A few reported being told such things as "Sign here if you don't want to feel these pains anymore" while a piece of paper was waved before their eyes.[21]

Largely as a result of the pressure mounted when the Relf case came to light, the Department of Health, Education and Welfare (HEW) decided to write guidelines on sterilization procedures during 1974. In effect, these established a moratorium on sterilizations of people under twenty-one years of age, and on those who for other reasons could not legally consent. In addition, the guidelines stipulated that there must be a seventy-two-hour waiting period between granting consent and carrying out the sterilization. They also required an informed consent process, including a written statement, to the effect that people would not lose welfare benefits if they refused to be sterilized, and they included the right to refuse sterilization later, even after granting initial consent.

Although HEW promulgated the guidelines early in 1974, a study conducted in 1975 by the Health Research Group, a renowned Washington-based organization,[22] and later corroborated by Elissa Krauss of the American Civil Liberties Union, showed that only about 6 percent of teaching hospitals provided only the vaguest of consent forms without proper explanations.[23] A still more recent study by the Center for Disease Control, an HEW agency, revealed that widespread non-compliance continued to be the rule. The study attributed the fact to ignorance of the guidelines.[24]

Early in 1975, those of us who were concerned about the issue of abuse formed a committee which we called the Committee to End Sterilization Abuse (CESA). We were faced with some hard realities: First, HEW can only regulate for federally funded procedures, and although it is true that the primary targets of sterilization abuse have been women on welfare, there are still many other vulnerable groups who are not welfare recipients, including the recently unemployed, undocumented workers, and workers whose earnings are just barely above the poverty line. Second, it seemed obvious that without a national monitoring system, it is impossible to determine what is happening to whom or whether guidelines are being followed. Third, those who control information often manipulate people's behavior. For example, the inclusion of hysterectomy as one form of sterilization in an HEW information pamphlet[25] tends to grant legitimacy to that mutilating operation in the eyes of the reader. Finally, the need for strong enforcement mechanisms became clear. There is no way that well-established actions and practices can be uprooted without the use of some measure of enforcement, particularly when the practices are profitable and socially sanctioned.

These facts, coupled with the increase in the number of sterilizations observed in the New York City hospital system, particularly in those hospitals serving black and Puerto Rican communities, prompted a number of concerned people from the Health and Hospitals Corporation, the New York agency responsible for the municipal hospitals, and from citizens' groups, to form an *ad hoc* Advisory Committee on Sterilization Guidelines early in 1975. The Committee to End Sterilization Abuse, Healthright, Health Policy Advisory Center, the Center for Constitutional Rights, the community boards of the hospitals, and many other organizations and individuals were represented on this new Committee. Most of the members were women involved in patient advocacy and who at the same time represented New York's various ethnic communities.

Our goal was to write new guidelines for the municipal hospitals. We met initially to ascertain the facts and to analyze the processes by which abuse takes place. Then we compiled the information in a report.[26] We identified existing weaknesses in the HEW regulations by using women's experiences as the touchstone for the drafting of stronger guidelines.

We found that many consents are obtained around the time of abortion or childbirth. The philosophy behind this practice is exemplified in the words of one doctor who said, "Unless we get those tubes tied before they go home, some of them will change their minds by the time they come back to the clinic."[27] When consent had been obtained at a time of great stress, the waiting period of only seventy-two hours allowed no opportunity for the woman to discuss the matter with friends, family, or neighbors to assure herself that she really wanted a sterilization. The time of abortion was particularly hazardous, because many teaching hospitals offer abortions as a "package deal" together with sterilization.[28] What kind of information was given to women was also key since both the vocabulary and the amount of information can clarify or confound. This reflects another weakness inherent in both the structure of the health-care system and the doctor-patient relationship—the coercive nature of medical advice when given in a patriarchal setting.

Our coalition of concerned groups drafted guidelines to remedy these weaknesses. We called for a thirty-day waiting period; an interdiction of consent at time of delivery, abortion, or of hospitalization for any major illness or procedure; the requirement that full counseling on birth control be available so that alternatives are offered; the stipulation that the suggestion of sterilization should not originate with the doctor; and the provision that any information or educational materials must be in the language best understood by the woman.

The guidelines also stated that if she wished, the woman could bring a patient advocate of her choosing to participate at any stage of the process. She could also have a witness of her choice sign the consent form. Perhaps the most important point we made was that a woman should express in her own words, in writing on the consent form, her understanding of what the sterilization entailed, particularly its permanence.

We were unprepared for the ferocity of the opposition to our guidelines. Our files, replete with angry letters from obstetricians, organizations involved in family planning and population control, and other groups, attest to the length and difficulty of our struggle. The chiefs of the obstetrical services in the municipal hospitals marshalled many objections, especially to the extended waiting period and the prohibition of consent around the time of abortion or childbirth. Some based their arguments on dramatic stories about the "habitual aborter" and the "grand multipara." The "habitual aborter" was described as a young woman who is using repeated abortions rather than contraceptive methods; the "grand multipara" as a woman who could only consent to sterilization while in hospital for childbirth, this being the only time she sought services. Our response was to continue to bring testimony of how abuse takes place and to negotiate on the provisions of the guidelines until they were acceptable to our Committee, its constituency, and to the obstetricians and other staff. It was a massive outreach effort that gained the support of community groups, boards of hospitals, health organizations, and legal groups. And it was this broad-based support, backed by several thousand letters and petitions, three or four demonstrations, hundreds of speeches, and dozens of meetings that finally overcame the still strong opposition in medical circles.

New York City's Health and Hospitals Corporation had once more been responsive to public wishes, illustrating that even imperfect institutions can respond, provided consumers find the channels through which they can fight for change.

The guidelines became effective November 1, 1975. Barely three months later, six professors of obstetrics and gynecology representing six major medical schools filed suit opposing the guidelines issued by HEW, New York State, and the New York City Municipal Hospital System. They claimed that the guidelines interfered with the rights of two women specifically: one a mentally retarded nineteen-year-old, the other a woman about to have a third cesarean section, both of whom requested sterilization. In their own behalf, the doctors claimed infringement of their right to free speech, since they were mandated to discuss sterilization only in the context of other methods of birth control.[29] Since their objections had been overruled by a vigilant public, the obstetricians carried their protest to the court.

During the same period, a nineteen-year-old Black woman, detained on a criminal charge at Rikers Island Prison, arrived for an abortion at King's County Hospital, largest of the municipals, and the one that had publicly refused to follow the regulations. She stated that she had been asked whether she desired pregnancies in the immediate future. When she said no, she was offered an operation for contraceptive purposes, which, she was told, could be reversed "when she became a normal citizen." She consented. Her uterus was removed during the operative procedure. This young woman is currently suing for gross malpractice. It was painful to see the need for enforcement of the guidelines through her suffering. The observance of just *one* single stipulation would have prevented that tragedy: that consent cannot be obtained during admission for abortion.

We began to see that there were some critical problems to be solved. First, how could we implement and enforce the new regulations? Second, how could we establish a monitoring system to know about sterilizations on an ongoing basis? Third, how could we apply the guidelines to private hospitals, not just the municipal hospitals? At present, the medical schools contract with the municipal system to deliver medical services. Doctors work at both public and private hospitals and can carry out their programs at either place. Often, doctors prefer the private setting as long as the fees for the services are forthcoming. We realized that they could circumvent the guidelines simply by admitting Medicaid recipients or other insured patients to their private hospitals.

A fourth concern was the definition of "elective" in any surgical procedure. By specifying that the guidelines applied only to "elective" procedures, large loopholes were left for what was "medically indicated." Doctors define "medical indications" on the basis of their experience or preference. Entrenched in their positions of power, doctors often resent any questioning. Attempts by patients to raise questions in the sacrosanct areas of "medical indications" are invariably vigorously repelled.

A fifth and extremely important problem, mentioned previously, was the control of information. It is possible to sell a procedure by giving distorted accounts of its benefits and by downplaying the risks. We strongly maintained that the mental, physical, and social hazards of sterilization should be discussed in the informational materials that were distributed. We felt that these materials should be different from those of HEW, whose off-hand information can easily mislead.

Finally, the concern over the abuse of hysterectomy was still paramount. Since excessive hysterectomies are performed not only for reasons of money, custom, poor medical practice, and hostility to women, but also as a covert method of sterilizing women without their informed consent, we recognized the need for guidelines on hysterectomies also.

These problems are as yet unresolved. Our approaches have been to continue to organize coalitions of people from within and without the hospitals in order to monitor what is happening and to continue to press in whatever ways we can to have some impact upon these practices. For although one group of people managed to write and pass the guidelines, many more people are needed to see that they are honored.

The lessons from these battles have been invaluable. We have learned that we *can* organize coalitions of community groups and health workers, and that these coalitions can be effective in sharing information as well as in applying pressure.

More important to me have been the experiences shared with women and community groups. We managed to identify some of the ways in which racist ideology keeps us from acknowledging our common oppression as women. Within the women's movement we sometimes found a denial of the experiences of others in statements such as "I had a hard time getting a sterilization five years ago. I can't see the need for a waiting period." And certainly it is true that in the not-too-distant past many middle-class women were denied sterilizations by physicians. The issue became clear only when women understood that the same people who would deny a white, middle-class woman her request often were the ones who were sterilizing working-class Whites, Blacks, Puerto Ricans, Native Americans, and Mexicans without even bothering to obtain consent.

We examined social class attitudes of superiority which can lead people to accept coercion of "others" such as welfare recipients, and we dealt with them in open discussion. We learned to identify our friends from the ranks of women's groups, Third World people, health workers, and church groups. We likewise identified our opponents from among gynecologists, board members of the organizations dedicated to population control—which promulgate the "people pollute" ideology—and those who favor a coercive society that oppresses people.

We also learned that there are many organizations that mask their ideology of population control by providing needed services in the areas of health, education, and family planning. These organizations are often linked to the large corporations and to a small number of private foundations in the United States.

In the process of study, we analyzed the case of Puerto Rico. There during the last thirty years the government, with United States funding, has sterilized over one-third of the women of childbearing age.[30] This was achieved by providing sterilizations free at a time when women were joining the workforce in large numbers. The lack of family support services, of legal and safe abortions, of alternative methods of contraception, and of full information about the permanency of sterilization have all combined to produce those startling numbers.

An analysis of the complex situation of Puerto Rican women showed us that there are many coercive factors in society that easily lead to sterilization abuse. Freedom of choice requires that there be real alternatives. We have deepened our understanding of the connections between the current denial of abortion rights to poor women, the dearth of child-care facilities, the cuts in welfare, and the sterilization abuse.

We are now confident that we will halt sterilization abuse in New York City and that our example will serve as a model to groups of like-minded people who are springing up across the country to combat the same problem in their communities.

Epilogue

Since November 10, 1976, the date of this lecture, many important developments have taken place. My note of optimism about the effectiveness of coalitions to win protective legislation was justified. As consciousness developed, a movement grew to support further legislation in New York City. Public Law #37 was passed by the City Council with a vote of thirty-eight to zero in April of 1977. The law embodies the principles of the guidelines on sterilization of the New York City Health and Hospitals Corporation, applying them to all New York City health facilities, *both public and private.* The law also regulates sterilization of *both women and men.* [31]

The success of this coalition effort forced the Department of Health, Education and Welfare to promulgate guidelines embodying the same provisions as the New York law, which were officially adopted in November 1978. Just as in the case of New York City, the wide support of groups such as the Committee for Human Rights, Community Boards of Methodist and St. Luke's Hospitals, Committee of the United Neighborhood Houses of New York, New York City Coalition for Community Health, Committee to End Sterilization Abuse, New York Civil Liberties Union, Physicians Forum, Women United for Action, National Black Feminist Organization, and New York National Organization for Women was developed on the national level. Hundreds of organizations and individuals testified in ten regional hearings. Over 80 percent of the testimony was overwhelmingly in favor of guidelines, and many testified to the need for stringent enforcement to prevent abuse. Provisions of the New York City law and HEW guidelines are:

1. Informed consent in the language spoken or read by the person.
2. Extensive counseling to include information as to alternatives.
3. A prohibition on consent at times of delivery or at any other time of stress and of overt or veiled pressures on welfare patients.

4. The right to choose a patient advocate throughout the counseling or any other aspect of the process.
5. A thirty-day waiting period between consent and procedure.
6. No sterilization of people under twenty-one years of age.

On the negative side, however, there are some pervasive problems. Passed each year since June 1977, the Hyde Amendment and corresponding budgetary limitations in most cases effectively cut off Medicaid funding for abortion in all cases except where pregnancy would endanger the life of the mother or is due to rape or incest. As a result, Medicaid-funded abortions have been reduced by 99 percent. These measures have imposed incalculable hardship on women on welfare, making them vulnerable to the risk of illegal abortion, and opening the door to the sale of the "package deal" of abortion/sterilization since the federal government continues to pay 90 percent of the cost of a sterilization procedure. For institutions eager for these funds, it becomes an additional incentive to coerce women desperate for an abortion into consenting to sterilization.

In response to this setback, numerous organizations joined the issues of abortion rights and sterilization abuse into effective campaigns.

In June 1980, the U.S. Supreme Court ruled that the Hyde Amendment was constitutional. In its decision the Court ignored all the facts showing the devastating impact that loss of funding for abortion services has on the lives and health of poor women and their families. Although Medicaid covers most medically necessary care, the Court ruled that abortion funding could be excluded because the procedure involves potential life.

The ruling constitutes an imposition of government's values over those of women to the point of risking women's health and lives.

Additional restrictions may in the future dictate funding for sterilization only or for hysterectomy only, thus further imperiling women.

Another development causes us great concern. In 1979, some women employed by the American Cyanamid Corporation in West Virginia charged that the company had threatened to fire them from the lead pigments division of the plant unless they underwent sterilizations. Lead exposure is harmful to unborn children and the company feared potential pregnancies.

They complied in order to keep their jobs, although lead exposure is also harmful to male reproduction and the general health of workers.[32]

As of October 1980, their suit in Federal Court on the grounds of sex discrimination is still pending, but knowledge of the case has reached many people who have become indignant.

This clear attack on workers' rights to a safe workplace and on their reproductive freedom has spurred trade unions to unite with women's organizations. A new exciting coalition is emerging.

The increasing numbers of sterilizations of Native American women has been documented by Women of All Red Nations (WARN). "Lee Brightman, United Native Americans president, estimates that of the Native population of 800,000, as many as 42 percent of the women of childbearing age and 10 percent of the men have already been sterilized."[33] This, together with the effort to deprive Native Americans of their right to their land, particularly that which contains uranium, has placed the Native American people at the very head of a broad movement for ecology and a safer environment for us all.

This is the challenge for today: a unification of the forces for a safe workplace, and environment, for women's rights, for Native Americans' rights to their lands, for the rights of Third World people. May we meet the challenge with strength.

Notes

1. U.S. General Accounting Office Report to Hon. James G. Abourezk, B 164031 (5) November 1976, p. 3.
2. "Uri Charges I.H.S. with Genocide Policy." *Hospital Tribune* 11, no. 13 (August 1977).
3. U.S. General Accounting Office Report B 164031, p. 4.
4. Allan Chase, *The Legacy of Malthus: The Social Costs of the New Scientific Racism* (New York: Knopf, 1977), pp. 15–16.
5. H. H. Laughlin, *Eugenics Sterilization* (1922), pp. 446–447, quoted in *ibid.*, p. 134.
6. Gena Corea, *The Hidden Malpractice, How American Medicine Treats Women as Patients and Professionals* (New York: Morrow, 1977), p. 128.
7. Chase, *Legacy of Malthus*, p. 16.
8. *Ibid.*, p. 315.
9. James E. Allen, "An Appearance of Genocide: A Review of Governmental Family Planning Program Policies." *Perspectives in Biology and Medicine* (Winter 1977).
10. "Physicians' Attitudes: MDs Assume Poor Women Can't Remember to Take the Pill," *Family Planning Digest* (January 1972), p. 3.
11. Chase, *Legacy of Malthus*, p. 2.
12. *Ibid.*, p. 6.
13. H. Curtis Wood, Jr., "The Changing Trends in Voluntary Sterilization," *Contemporary Obstetrics and Gynecology* 1, no. 4, (1973), pp. 31–39.

14. Denton Vaugham and Gerald Sparer, "Ethnic Group and Welfare Status of Women Sterilized in Federally Funded Family Planning Programs," *Family Planning Perspectives* 6 (Fall 1974), p. 224.
15. Joann Rodgers, "Rush to Surgery." *The New York Times Magazine* (September 21, 1975), p. 34.
16. *Ibid.*, p. 40.
17. Jack Slater, "Sterilization: Newest Threat to the Poor," *Ebony* (October 1973), p. 150.
18. Relf v. Weinberger, 372 Federal Supplement 1196, 1199 (D.D.C. 1974).
19. Slater, "Sterilization," p. 152.
20. Joan Kelly, "Sterilization and Civil Rights," *Rights* (publication of the National Emergency Civil Liberties Committee), (September/October 1977).
21. Claudia Dreifus, "Sterilizing the Poor," *The Progressive* (December 1975), p. 13.
22. Robert E. McGarraugh, Jr., "Sterilization Without Consent: Teaching Hospital Violations of HEW Regulations: A Report by Public Citizens' Health Research Group," January 1975 (available from Public Citizens' Health Research Group, 2000 P Street, Washington, D.C.).
23. Elissa Krauss, "Hospital Survey on Sterilization Policies: Reproductive Freedom Project," *ACLU Reports* (March 1975).
24. Carl W. Tyler, Jr., "An Assessment of Policy Compliance with the Federal Control of Sterilization," June 1975 (available from the Center for Disease Control, Atlanta, GA).
25. U.S. Department of Health, Education and Welfare, "Your Sterilization Operation, Hysterectomy" (Washington, D.C.: U.S. Government Printing Office, 1976).
26. New York City Health and Hospitals Corporation, "Why Sterilization Guidelines Are Needed," (1975) (available from Office of Quality Assurance, 125 Worth Street, New York City 10013).
27. Bernard Rosenfeld, Sidney Wolfe, and Robert McGarren, "A Health Research Group Study on Surgical Sterilization: Present Abuses and Proposed Regulations," (Washington, D.C.: Health Research Group, October 1973), p. 22.
28. *Ibid.*
29. Gordon W. Douglass, M.D. *et al.* and John L. S. Holloman, Jr. *et al.* Civil Action File no. 76, CW 6, U.S. District Court (January 5, 1976).
30. Jose Vázquez-Calzada, "La Esterilización Feminina en Puerto Rico," *Revista de Ciencias Sociales*, (San Juan, Puerto Rico) 17, no. 3 (September 1973), pp. 281–308.
31. Carter Burden, Testimony upon Introduction of Bill #1105, April 18, 1977, now Public Law #37 (available from New York City Council).
32. See the article by Stellman and Henifin in this collection.
33. *Akwesasne Notes* (late winter 1979), p. 29.

General References

1. Clara Eugenia Aranda *et al.*, *La Mujer: Explotación, Lucha, Liberación.* (Mexico 20 D. F.: Editorial Nuestro Tiempo S.A., 1976) (available from Avenida Copilco 300, Locales 6 y 7, Mexico 20 D. F.).
2. Barbara Caress, "Sterilization," *Health Pac Bulletin* 62 (January/February, 1975).
3. _____. "Sterilization Guidelines," *Health Pac Bulletin* 65 (July/August 1975).
4. Linda Gordon, *Woman's Body: Woman's Right* (New York: Grossman/Viking, 1976).
5. Terry L. McCoy *et al.*, *The Dynamics of Population Policy in Latin America* (Cambridge, MA: Ballinger, 1974).
6. Bonnie Mass, *Population Target* (Brampton, Ontario, Canada: Charters, 1976).
7. Barbara Seaman and Gideon Seaman, *Women and the Crisis in Sex Hormones* (New York: Rawson Associates, 1977).

[Dr. Helen Rodriguez-Trias *lived in Puerto Rico until she was ten; she spent the remainder of her formative years in New York City. She returned to her homeland for college, marriage, the beginning of her family, and then, after an interruption of several years, premedical and medical study. She graduated from the University of Puerto Rico Medical School with the highest honors, first in her class.*

Dr. Rodriguez-Trias' career has been one of activism, teaching, and medical care. She has been director of the Department of Pediatrics at Lincoln Hospital in the South Bronx and a member of the Faculty of the Biomedical Program at City College. In 1978, Dr. Rodriguez-Trias became Associate Director of Pediatrics for Primary Care at the St. Luke's-Roosevelt Hospital Center. She is also an Associate Clinical Professor of Medicine at Columbia University College of Physicians and Surgeons.]

Lynda I. A. Birke with Sandy Best*

Changing Minds: Women, Biology, and the Menstrual Cycle

This chapter is concerned with scientific statements about women's menstruation and menstrual cycles. There is a vast medical literature dealing with various problems associated with menstruation, such as period pain, cessation of periods, irregularity, and excessive bleeding, as well as premenstrual changes in mood. However, to review this literature from a feminist perspective would be an immense task; so we have restricted ourselves to only a limited area. We begin by discussing the ways in which menstruation is seen as an illness and look briefly at some of the consequences of this view. We then consider the various changes in mood and behavior that have been grouped together as the "premenstrual syndrome," and question whether these are as dependent upon the fluctuations in hormones as is often suggested. Finally, we consider the ways in which women are portrayed in the advertising of products related to menstruation. One type of advertising portrays women as victims of an ever-changing biology; another portrays the product as in some way able to "liberate" women from this biology. Yet women are increasingly finding that menstrual products are often not quite as wonderful as they first appeared; even tampons, which seemed such a relief from napkins, have now been associated with disease.

Although this chapter is primarily a critique of the medical literature, which continues to view menstruation as some kind of pathology, it also grows out of our experience as feminists involved in the Women's Movement—particularly with issues surrounding women's health care. In that sense, although we remain responsible for what is written, many of the ideas are derived from the experiences of all the women with whom we have worked or discussed issues of women's health.

*The chapter was written by L.B. It is, however, based on work and endless discussions with S.B., as well as being partly based on jointly written work that has already appeared.

Advertisements for tampons, pads, and nostrums from the collection of Emily E. Culpepper.

Figure XVI. Women students in the laboratory of the International Institute for girls in Spain (1920). The laboratory was founded by Mary Louise Foster (1865–1960), Associate Professor of chemistry at Smith College (1908–1933). (M.I.T. Historical Collections)

Menstruation: The Monthly Illness

Women's monthly bleeding is, in most societies, feared and distrusted. The manner in which this is expressed varies from culture to culture, but in some form it is almost universal.[1] Most frequently, the fear is manifested in certain taboos that restrict women's activities during menstruation. Women, for example, may be expected to enter menstrual huts, or be proscribed from touching food throughout their menstrual flow. One consequence of this has been that menstruation is labeled as pathological by many cultures. The Koran, for instance, refers to menstruation as an "illness."[2] In the Hebrew Bible, it is a form of pollution. In many Western medical writings, particularly during the nineteenth century, menstruation was regarded as "pathological—proof of the inactivity and threatened atrophy of the uterus."[3]

In a curious sense, to menstruate month after month and year after year, *is* "pathological." For most of human history, women have had a short life expectancy and probably had rather few menstrual periods owing to repeated pregnancies followed by often prolonged breast-feeding, as well as very different diets and activity patterns from those that most women experience now.[4] But this cannot explain the extremely negative way in which menstrual periods are viewed in most societies. In the mythology of many cultures, menstruation is associated with powerful, demonic forces which, if loosed, will wreak havoc on the world.[5]

In Western society we have (arguably) lost touch with such mythology since mythology and religion—as ways of organizing our understanding of the world—have in general been superseded by scientific explanations. The negative evaluation, however, remains and can be discerned in much medical writing, especially that concerned with what has been called "premenstrual tension" (PMT), or premenstrual syndrome. We will address this medical literature later in the chapter. To begin, however, we consider the consequences of labeling menstruation as pathological.

One not unexpected consequence of labeling menstruation as a disease is that it might be viewed as a source of contagion. In fact, there is plentiful evidence that many cultures view menstrual blood as contagious and, in particular, as a source of danger for men.[6] In 1875, for example, one medical paper suggested that menstrual blood was the source of male gonorrhea,[7] while during the early twentieth century, some medical papers suggested that menstrual blood was the source of a toxic substance referred to as "menotoxin."[8] Underlying such suggestions is the idea that menstrual blood might harm male genitals, an idea that still persists, for example, in the common avoidance of intercourse during menstruation. While some may rationalize this as related to the "uncleanness" of menstrual blood, others point to possible untoward effects on men. Adrienne Rich describes one man who rationalized his behavior by saying that "intercourse with a menstruating woman did not appal *him*—but that it resulted in irritation of 'the' penis."[9]

A second consequence of a disease label is that the process becomes medicalized and therefore a legitimate concern of the medical profession. Physicians, in their turn, contribute to the further development of the "illness" by a proliferation of learned and not-so-learned papers on the topic. This was particularly prevalent, for example, in the mid-nineteenth century, when innumerable medical papers appeared expounding the sickliness of woman, especially during her menstrual period.[10] And, of course, these arguments had certain social consequences: the sickliness of women was proposed as a reason for keeping women at home and thus out of the professions and other prestigious occupations (at least if they were middle

My husband won't live with me five days each month

class; the capacity of working-class and poor women to carry on as normal during menstruation was never questioned).

A third consequence of the disease label is that the concept of disease can be carried over to associated bodily states. For example, the labeling of menstruation as sick has contributed to the labeling of the menstrual *cycle* as sick, and therefore as subject to medical intervention. Prior to the 1930s, the premenstrual syndrome did not exist as a *medically* defined phenomenon.[11] That is not to say that before that women did not sometimes feel discomfort premenstrually; simply that the premenstrual period had not until then been seen as problematic for large numbers of women. Since then, the literature describing this "illness" and appropriate medical treatments for it has mushroomed, as we shall see.

While menstruation itself has been regarded as a sickness, so, often, has its absence. Unless menstruation has ceased for some obvious reason, such as pregnancy, its cessation, or diminution, is often feared by women and seen by them and their physicians as a sign of sickness. A widespread belief among women of several different cultures is that to retain health it is necessary to remove "bad blood" through menstruation.[12] Thus, while menstrual blood itself represents sickness, the process of menstruation also is seen as necessary to *remove* sickness. A corollary of this belief is that anything that is thought to reduce menstrual flow (whether it actually does or not) such as oral contraceptives, getting cold, or a hot bath, is perceived as dangerous since, if the blood does not flow freely, it must "go somewhere else." And if it does this, it is liable to cause *other* forms of illness. Thus, if the "bad blood" goes to the brain, for example, it is feared that it might cause insanity or a stroke.[13] Such fear of "migrating sickness" may be the basis of the exhortations that young girls are often still given not to wash their hair, or take a bath, during a menstrual period.

It is interesting that the few studies that have been done of such folklore have focused on women's beliefs—on stories women tell each other. It would be interesting to compare the details of women's and men's beliefs concerning menstruation as illness in different cultures, although to our knowledge this has not been done. Our guess is that men would tend to view menstruation *per se* negatively, focusing on menstrual blood as the source of sickness. Women, on the other hand, are perhaps more ambivalent toward menstruation, and this is reflected in the sets of beliefs they pass on to each other. While patriarchal dislike of menstruation inevitably has become part of women's beliefs, women tend also to view menstruation in positive ways. Thus, for many heterosexual women menstruation is positive and important in that it confirms that they are not pregnant. An ambivalence may be seen in cultural beliefs among women that menstruation helps to "get rid of" something unwanted and harmful.

This is, of course, somewhat speculative. Our point, however, is that women may not evaluate their own menstruation as negatively as is often assumed. It is perhaps an error that much feminist writing on menstruation (including, we might add, our own)[14] has tended to exhort women to think positively about their menstrual cycles and implied that women evaluate menstruation wholly negatively. It would appear that women have been more ambivalent, and it might be better to build in the future on the good feelings that already exist than to stress the presence of negative ones that ought to be relinquished.

Menstrual Cycles: The Domination of Raging Hormones

There are two significant ways in which menstruation is viewed culturally. First, it is seen as something shameful, to be kept hidden at all costs. Second, as we have indicated, it is viewed as some sort of illness. That is not to say, of course, that women are imagining the discomforts that we feel when we menstruate, but rather that our society defines menstruation in such a way that all menstruation carries pathological connotations. This is reflected in the terms used to describe it, such as "the curse," "feeling poorly," "being off." The important point here is not that women sometimes feel discomfort while menstruating, but that the tendency to view menstruation negatively and with shame inevitably affects women's experiences of our own menstruation.

For example, we often attribute feelings of ill health to menstruation, even if they have nothing to do with it. This is apparent in the tendency to ascribe, retrospectively, any feelings of being below par that *happen* to occur premenstrually, to the subsequent menstruation.

There is recent evidence, for example, that women attribute unpleasant feelings to menstruation if they *believe* themselves to be premenstrual. In one study by a woman psychologist,[15] women were told that it was now possib e scientifically to predict when they were to start menstruating. Those who were told that they were premenstrual consistently reported feeling worse than those who were told that they were intermenstrual. The author commented: "It appears that learned associations or beliefs might lead a woman to overstate what she is actually experiencing or to perceive an exaggeration of naturally fluctuating bodily states . . . when she believes she is premenstrual."

What this implies is that the negative evaluation of menstruation can have a considerable effect upon the ways in which we interpret the way we feel. That is not the same as saying that we "imagine" our symptoms, as so many doctors have told women complaining of menstrual problems. Rather, what it means is that we may interpret particular bodily sensations according to what we believe about menstruation. The problem of interpretation of bodily states is not one unique to menstruation. In studies of stimulant drugs, for example, researchers have usually found that people who have been given the drug claim to feel different things according to the mood of the people around them.[16] Thus, if they are surrounded by people who are laughing and joking, they are likely to report feeling happy, while if surrounded by unhappy faces, they are likely to report feeling sad. The difference is not so much in the way the drug works, but in how the effects are interpreted by the person who takes it.

But interpretation can only be part of the story for menstruation. Many women, for example, experience acute pain associated with their periods (dysmenorrhea), which cannot be entirely explained in terms of psychology. While beliefs may exacerbate the discomfort or pain one feels, there is recent evidence that at least some menstrual pain is associated with specific changes in uterine biochemistry. Thus, women who experience acute pain, it seems, often have exceptionally high levels of certain hormones (called *prostaglandins*) in the uterus. Prostaglandins are known to increase uterine contractions (which is why they are used to induce mid-trimester abortions), and so are likely to increase pain. As a result, drugs that inhibit the synthesis of prostaglandins have been found to be helpful to many women with severe menstrual pain.[17]

Just as many women's experiences of menstrual pain have been dismissed by the medical profession as "just" part of being a woman, so have women's experiences of premenstrual problems. It has, perhaps, become less fashionable to take this attitude toward female patients in recent years, as few doctors can now be unaware of the vast literature on what has come to be known as premenstrual tension. The impression given by much of this, and certainly by drug advertisements related to it, is that PMT is a definable entity that is, in most cases, amenable to some form of drug therapy. Whether this is an improvement in the treatment that women receive is debatable, as we shall see; but first we need to consider what is meant by the term.

What is Premenstrual Tension?

So far, we have taken as a starting point the social definition of menstruation—and, by inference, of the premenstrual period—as illness. In doing so, we have referred to premenstrual tension as though it *is* a unitary phenomenon. But whereas many women have some idea of what it means for them individually, comparisons between women reveal enormous variability. Headaches, dizziness, inability to concentrate, fluid retention, and many other sensations or body states are considered symptoms of PMT. Different problems distress different women, but the most common problems include anxiety, depression, irritability, and bloating. Irritability and depression are the problems most frequently referred to, perhaps because they are the ones most likely to affect the people in close contact with the woman, such as husband, lover, or children.

There have been several attempts to explain PMT, and particularly the often violent premenstrual mood swings of which some women complain. But, unlike period pain, which seems in many cases to be associated with high uterine levels of prostaglandins, PMT has been much harder to pin down to physiology, at least in part because it is less easy to define. Indeed, we already noted that the concept of a premenstrual syndrome did not exist as a medically defined entity until the 1930s, when Frank wrote a paper entitled "The Hormonal Causes of Premenstrual Tension."[18] During the nineteenth century, there had been little understanding of the physiology of the menstrual cycle, and only hazy notions of the relationship between menstruation and ovulation. The significance of the ovarian hormones in the physiology of menstruation finally began to emerge during the early twentieth century.[19] The discovery of these hormones allowed medical writers to pin the premenstrual period to specific endocrine events—namely, the declining levels of estrogens and progesterone that occur premenstrually.[20] The premenstrual period had already been regarded as a period of impending illness; all that was needed was some physiological correlate to which to nail it. Hormones provided an answer, and in the years since the publication of Frank's paper there has been a plethora of papers attempting to find some hormonal abnormality with which to associate premenstrual tension, as well as a vast array of hormonal therapies.

Of course, many changes in bodily function occur during the menstrual cycle in addition to changes in the output of hormones from the ovaries. Epileptic seizures are often associated with menstruation because of changes in brain function.[21] There are changes in such body biochemistry as carbohydrate metabolism, thyroid function, mineral and water balance, resting temperature,[22] and sensitivity to smells.[23] Along with all of these go the better-known hormonal swings. Many of the subtler changes may go unnoticed by the individual woman; mood and behavior change, on the other hand, are easier to observe. This may be why most of the published theories of the origin of premenstrual tension focus on behavioral changes.

These theories commonly involve the assumption that women with PMT suffer from some hormonal imbalance. There are, for instance, theories that PMT results from too little progesterone;[24] or that the adrenal hormones may be involved;[25] or that prolactin, the hormone primarily responsible for milk production during breast-feeding, might be a factor.[26] However, the fact that, for many women, changes of mood and behavior during the cycle are a distressing reality and often a source of immense misery, does not necessarily mean that they are the *direct* result of some biological change such as the fluctuations in hormone levels. Yet much of the literature dealing with PMT seems to imply that hormonal fluctuations *cause* the behavioral changes. And for this reason, any intervention aimed at smoothing those fluctuations is thought of as a potential cure.

You just told the whole party it's about that time of the month.

The attempt to pinpoint a unique, preferably molecular, "cause" ignores the immense complexity of both the menstrual cycle and the various factors that influence it. We have already noted the extent of the menstrual taboo, even in our own society. Its very existence, coupled with possible anticipation of discomfort, may be sufficient to create symptoms, or to make them worse than they would otherwise be.[27] The fact that women from other cultures report rather different symptoms during the premenstrual and menstrual phases[28] would tend to support the notion that biology is not the only, or perhaps even primary, "cause" of PMT.

It is quite customary in present-day scientific practice to look for molecular "causes" of behavioral "problems" (however these are defined). Unfortunately, this ignores many other factors that might be involved, as well as the immensely complex ways in which these factors might interact. For example, not only is social context likely to be important in influencing women's subjective *experiences* of menstruation, as we have implied; it also may well affect measurable hormone levels. It is extremely hard, and in fact may be impossible, to unravel *all* the factors that could influence a woman's experience. Faced with that potentially insoluble puzzle, scientists not surprisingly are attracted to hormonal theories which, because they over-simplify, seem to provide straightforward answers. Also, and perhaps most importantly, hormonal theories of PMT suggest medical solutions.

Finally, our negative evaluation of our physiological cycles and mood swings in part may stem from the Western ideal of remaining constant, ever productive, and "not changing our minds."[29] We are all familiar with the long-standing stereotype of woman as fickle, unreliable, and irritable, whereas men are stereotyped as constant (even though neither stereotype corresponds to reality). It is scarcely surprising that the deviations from the myth of constancy that we tend most to notice in ourselves are precisely those that agree with the negative stereotype. Thus, we report when we become dis-tractible and tearful premenstrually; but we are less likely to mention that we feel energetic and creative at mid-cycle.

In short, then, the changes that women tend to say they experience are the ones society deems bad. Furthermore, we come to accept the myth of con-stancy and judge our monthly cycles against it. But this is absurd, a denial that, as women, we *do* have cycles and that all people—women and men—experience changes in mood and behavior. But we are in even worse trouble because some of the behaviors said to appear premenstrually are generally considered inappropriate for the "feminine" role: the truly feminine woman is never irritable, depressed, or aggressive; she is the epitome of selfless calm, devoted to husband and children. In this connection, it is interesting that in one study precisely the women who conformed most closely to our society's stereotype of femininity were found to suffer more from menstrual and premenstrual problems.[30] But like all such studies, the results should prob-ably be viewed with some skepticism.

Many women say that in the second half of the cycle they feel more with-drawn, more in tune with their subjective selves: then, as menstruation approaches, women are expected (and perhaps therefore allow themselves) to become more testy and difficult to live with. One British book has suggested that this is when the "truth flares into consciousness." The authors suggest that, throughout most of the cycle, a woman represses dissatisfactions—whether of friends, lovers, or society in general—but that this repression breaks down before a period, and the anger and hurt flare up: "It is the 'moment of truth,' which in a society which refuses woman her true place may become the moment of despair."[31]

Whatever the merits of this idea, it serves to remind us that our experiences of menstruation are likely to be a product of the society in which we live; and at this time, our society gives us some leeway to be moody premenstrually. In a society where women are profoundly oppressed and devalued, we have every reason to be both angry and depressed. Perhaps rather than ask why a woman "suddenly" becomes aggressive and irritable before her period, the theorists should ask how she so successfully suppresses her anger for the rest of the month.

Attempts to look at positive aspects of the menstrual cycle are largely lacking in medical papers which, in general, mirror the cultural emphasis on the undesirable changes. Of course, the more popular magazines produce such headlines as "Cycles of Misery;"[32] while books appear with titles such as "PMT—the Unrecognized Illness."[33] They usually draw their evidence of the

terrible changes that can happen to women premenstrually from one of two sources. First is the suggestion, largely from the work of Dr. Katherine Dalton in London, that bad accidents, severe illness, and suicide attempts are more prevalent premenstrually than at other times.[34] Much of this has been criticized on statistical grounds.[35] It is entirely possible that in some of these examples, the traumatic event itself might be sufficiently stressful to *bring on* a period, in which case the woman would have only appeared to be "premenstrual" when the accident took place.

A second source of information is often questionnaire-based research, such as that carried out by Moos and his colleagues.[36] These questionnaires, and particularly the Moos Menstrual Distress Questionnaire, attempt to analyze the different moods and behavioral changes that women report, and to understand their relationships to the cycle. The questionnaires include questions about distractibility, anxiety, insomnia, confusion, depression, decreased efficiency, excitement, and feelings of suffocation and nausea, as well as about several physical changes such as fluid retention.

Most questionnaire-based studies of the menstrual cycle have, however, been criticized on methodological grounds.[37] Generally, the way the studies are done leaves much to be desired.[38] For example, there is nothing to guard against a woman answering the way she thinks she ought to answer, rather than how she really feels. To explore this hidden bias in the data, Mary Brown Parlee asked women to complete such a questionnaire, and also asked men to fill it out by reporting what they thought women felt. The two sets of answers were nearly identical, which led Parlee to suggest that the "premenstraul syndrome" reflects what women (and men) felt was *likely* to occur. In other words, there appears to be a social consensus about what happens to women before their periods. But if that is so, then women are, of course, much more likely in fact to experience those feelings as a self-fulfilling prophecy.

Women's supposed changeability has often been used as justification to keep us out of jobs that require responsiblity and mental effort, and that carry prestige and good salaries. Women are supposedly the victims of their raging hormones, which render them periodically unstable and incapable of sensible decisionmaking.[39] The perpetrators of such tales do not, of course, consider the large numbers of men in similar jobs who fail to make the right decisions. Instead, they simply *assume* that women's fluctuating hormones jeopardize their intellectual capabilities.[40]

Medical reports have not challenged this dominant ideology. While medical writing does not often make the quantum leap from talking about changes within women to saying that we must therefore be excluded from particular jobs,[41] it tends to underwrite the notion that premenstrual women are less capable of meeting intellectual demands. "Confusion," "inability to concentrate," "lowered judgments," and "forgetfulness" are some of the terms used to describe the premenstrual period. And not surprisingly, this can become the way women describe themselves.[42]

Fortunately, not all studies of possible relationships between the menstrual cycle and behavior have used questionnaires. In more "objective" studies, there often is little or no correlation between behavior and time in the cycle. For instance, in studies of reaction time, in which a subject is required to perform a particular task as quickly as possible following a signal, no systematic variation of reaction time with the cycle has been found.[43] The consistent failure to find any variation in a test which clearly requires concentration suggests that the problem of "inability to concentrate" is nowhere near as universal among women as is often suggested.

There is also evidence that intellectual performance does not vary with the cycle in a predictable way.[44] In studies of American college women, for instance, intellectual performance has been found not to change with menstruation, nor do these women believe that their menstruation disrupts their intellectual life. Indeed, if anything, there is evidence that some women try *harder* premenstrually, possibly in the belief that their performance may be impaired at this time. Certainly, college women appear to compensate for what they perceive as potential "off-days."[45] Another study found that women who believed themselves to be premenstrual actually scored better on psychological tests than women who did not.[46]

There is, therefore, not much evidence to support the assertion that women's intellects are at a low ebb premenstrually—or indeed at any other time of our cycles. Pseudoscientific research that portrays menstruation as a disease provides one example of the way in which "biological" arguments can operate to perpetuate the notion of women's inferiority. The scientists who write these papers may believe themselves to be objective and simply studying natural phenomena. But that belief would be naive. Hypotheses generated, interpretations given, even the type of data collected, are all products of the scientists' lived experience. Part of that experience may have involved a scientific training and understanding of the natural world; but a large part consists of her or his experiences within the human social world. And in our society, that world contains a powerful set of negative assumptions and attitudes toward menstruation and female cycles.

We have implied above, in discussing women's attitudes toward their menstruation, that there might be a strong "psychological" component in states such as PMT. Saying this does not mean that hormones are never involved; simply that they probably do not directly *determine* the behavioral changes. Many women have highly irregular cycles, and yet perceive some mood changes a day or two before their period. This might be taken to support a hormonal theory, rather than one implying anticipation, but it need not. It is quite possible that women become sensitive to the slight changes in our bodies that precede menstruation. We might, for example, become aware of minor changes in fluid balance that lead to feelings of bloatedness. If we do, then anticipation may be a factor, despite irregularity.

Hormones certainly influence a number of bodily states, as we have indicated. But that is a far cry from saying that they "determine" behavior. Reducing the menstrual cycle to hormonal swings ignores the effects psychological states and environmental factors can have on hormone output. It is well known, for example, that women's menstruation can cease during periods of stress. Similarly, environmental factors may influence cyclicity. Some suggestions have been made that moonlight can synchronize menstrual cycles;[47] and many women have had the experience that living with other women can produce menstrual synchrony.[48]

A further problem with simplistic notions of hormonal determinants is that, while some writers may pay lip service to environmental factors, they tend to focus on the biological changes as primary and causal. Any environmental effects can then be seen as merely added onto the basic biology. This kind of view has a long history in Western science, but it does little to explain the constant, mutual interactions between the individual's biology and her or his environment. If we focused upon these interactions, the question of whether, and how much, this or that bit of biology is primary becomes meaningless. Yet rather than seeing PMT as a product of many interacting factors, medical personnel are encouraged to view it as a hormonal imbalance for which, of course, the treatment is hormonal—a situation that is continually encouraged by advertisements for hormonal drugs. The danger of this is that hormones that *may*, if taken over long periods of time, prove hazardous[49] are being given to large numbers of perfectly healthy women.

Some recent research has suggested that PMT may be related to an increased need for pyridoxine (Vitamin B_6),[50] and this is now used in PMT treatment. While still a simplistic explanation, at least it has the advantage that, since pyridoxine is an easily excreted vitamin, it is unlikely to present health risks when taken in moderate doses. It is also readily obtainable without a prescription, and therefore a woman herself can decide whether she wants to try it without needing a medical consultation.

Advertising Images: the "Liberated" Woman

That the portrayal of women in advertising is sexist is well known; images of women as "feminine" or unable to cope abound. There are, however, other ways in which women's oppression is used in media images to sell goods. One particular image seems to be that of the "liberated woman." When advertisements are aimed specifically at women, they often adopt a tone of "buy such-and-such, and you will be freed from drudgery." This is particularly apparent in advertising of household products, which are said to "liberate" women from the toil of housework. The flaws in this concept of "liberation" with regard to housework have been pointed out by others.[51]

However, notions of liberation are used no less as marketing devices when it comes to the technological changes that are occurring in relation to women's biology. The Pill, for example, was hailed as a great advance for women's emancipation.[52] In a similar vein, hormonal therapy for menopausal problems and for PMT is commonly advertised and popularized as emancipating women from the constraints of our biology, and as keeping us eternally

Why wear more than you have to?

feminine.[53] These "advances" have often proved to be less than wonderful as the wonder-drugs are shown to be hazardous to health. Women are much more cautious now of the contraceptive pill, for example, than they were ten years ago since we have come to know more about its risks to health.[54] Such wariness of hormonal preparations has been further enhanced as the story has emerged that girl babies exposed to hormones *in utero* are at risk of developing a rare form of vaginal cancer during adolescence.[55]

Tampons are a further example of advances that may seem less progressive now than when they were first introduced. At that time, they were generally considered to be liberating. Advertisements commonly point out that, by using a tampon, a woman can be free to lead a normal life, rather than being encumbered by pins and pads. Advertisements also stress that tampons make it easier to conceal that one is menstruating, and in this way perpetuate the idea that periods are something shameful that is best not talked about.

Now, it is undeniable that most women find tampons far preferable to the available alternatives. They are less messy, and certainly less of a hindrance. But that is not to say that they are unequivocally beneficial. Recently there has been evidence that tampons are implicated in the occurrence of Toxic Shock Syndrome (TSS). This is a serious illness, involving collapse, high temperature, diarrhea, and vomiting or even death, and apparently results from a toxin produced by a microorganism that grows readily on blood-soaked tampons.[56]

Tampons have been available for some years, so why has TSS only recently been discovered? Part of the answer again must lie in the problem of naming. Until the disease has been named and described, doctors confronted with a case of it may simply call it, and treat it, as something else. But that is not the whole story. It seems that the recent rise in the number of cases of TSS has coincided with the addition of a number of chemicals to tampons, particularly the "super-absorbent" ones, which may encourage the growth of these microorganisms. Tampons have been shown to cause tiny ulcerations in the vaginal lining, and these apparently allow the toxin produced by the multiplying microorganism to get into the bloodstream and produce TSS.

The Women's Movement has directed campaigns against manufacturers of tampons, particularly those who market deodorized tampons, which not only can cause unnecessary vaginal irritation but also serve to maintain the view that menstruation is rather nasty, and must be hidden away with perfumes. Perhaps the point about tampons that raises the greatest feminist anger is the large profits reaped at our expense by the companies manufacturing tampons and pads. As one possible solution, women's health groups have advocated the use of natural sponges, which can be reused and therefore are also less wasteful. The problem with this response, of course, is that it is very much a personal solution. The magnitude of the menstrual taboo is such that most women would feel totally unable to cope with, say, changing a sponge in a public lavatory. And there is a certain irony in feminist acceptance of natural sponges as a "safe" alternative. At the time of writing, there has just been a report that sea sponges also can be hazardous, since nowadays they contain toxins and pollutants that they have absorbed from the sea.[57]

Solutions to Our Problems?

We have argued in this chapter that the medical emphasis on biological determinants for various menstrual problems is misplaced since it ignores the social context in which menstruation is considered as pathology and is met with shame and denial. In such a context, the experience of menstruation readily becomes laden with problems, for which women naturally seek help. The common response of most doctors is to advocate some form of drug—usually hormonal—the nature of which may well depend on which particular theory she or he espouses. Thus, acute pain (dysmenorrhea) is often treated with estrogens, while the natural hormone, progesterone, as well as its synthetic derivatives (progestins) have been tried for some cases of PMT.[58] For women with depression, who are likely to feel worse during the premenstrual period than at other times of the cycle,[59] antidepressants and major tranquilizers are sometimes prescribed. As we have indicated before, such drugs present risks, some of which may be as yet unknown. Certainly, they reinforce the notion that premenstrual problems are of biological origin *and* represent some kind of pathology.[60]

The Women's Movement has generated a number of self-help remedies for a variety of health problems, including menstrual problems. They include herbal treatments (for example, raspberry leaf tea for cramps), menstrual extraction and consciousness raising. These remedies are to be welcomed as they allow women to make personal decisions about the best way to deal with menstrual problems, even if such solutions do still tend to be rooted in the idea of menstruation as illness and as *likely* to cause problems. We suspect that this attitude will continue and that women will have to experiment much more with alternatives to professional medicine before we can begin to move toward a conceptualization of menstruation as health rather than illness.

We can begin to work toward this by, among other things, trying to adopt a more positive view of our own menstruation and cyclicity, as feminists have often pointed out.[61] But, that said, we must also warn that exhortations to view our bodily processes more positively carry certain dangers. First, it is very much a personal response, and hence does not directly challenge the dominant ideology. Simply viewing *ourselves* more positively does little to

When you're wearing a tampon you don't worry about odor. But should you?

make the rest of the world change its mind. Second, there is a tendency in some women's writing to discuss menstruation with reference to some past Golden Age, in which female deities were worshipped, and to see the origins of the present-day menstrual taboo in hypothetical prehistorical rites in which menstrual blood was sacred. It is always possible, of course, that the taboo did indeed have such origins. But, as Hilary Standing has pointed out:

> Some of the recent literature [on menstrual cycles] has been point-
> ing to anthropological literature to illustrate the claim that "primi-
> tive' [that is, pre-patriarchal] woman was treated as 'sacred', as being
> 'in touch with nature', and therefore powerful in relation to men. But
> there is no evidence at all for these assertions. The sacred and the pollu-
> ting are two sides of a coin. They signify social control as well as power.
> The exhortation to be positive about our bodily cycles is in danger of
> producing new mystifications based on myths about women's psychic
> communion with the moon or with nature.[62]

We need to be concerned, not with appeals to some past Golden Age, but with our own experiences of menstruation, shaped as they are by a society that consistently devalues and exploits our activities and our bodies. We need to recognize that we experience our bodies and interpret those experiences in ways that are shaped by that society rather than by our "raging hormones." In the nineteenth century, it was thought that women's moods and behavior needed to be made more controllable by removing the ovaries or uterus—that is, through surgery. In the late twentieth century, we are still being controlled and diminished by attributions of alarming moods that must be eradicated rather than understood. The treatment may now be hormonal, but the effort at control is the same. Scientific inquiry has done little to dispel the prejudices and mystifications surrounding our monthly bleedings; rather, it has itself contributed to those prejudices by focusing on physiology to the exclusion of nearly everything else. Society condemns women's anger and hostility, whether it occurs premenstrually or not, and has looked to science to provide explanations and rationalizations. But, as we have argued, explanations that focus on our hormonal cycles only serve to limit women's anger. And there remains much for us to be angry about.

Notes*

1. *See* for example Paula Weideger, *Menstruation and Menopause* (New York: Knopf, 1977).
2. E.G.M. Seklani, "Fecundity in Arab Nations: Attitudes," *Muslim Attitudes toward Family Planning*," O. Schieffelin, ed. (New York: The Population Council, 1969).
3. V. L. Bullough and M. Voght, "Women, Menstruation and Nineteenth Century Medicine," *Bulletin of the History of Medicine* 47 (1973), p. 67.
4. Roger Short has suggested that, in this sense, we may be ill-adapted to the swings of the ovarian hormones that accompany each cycle. *See* R. V. Short, "The Evolution of Human Reproduction," *Proceedings of the Royal Society of London* 195 (1976), pp. 3–24.
5. *See* Weideger, *Menstruation and Menopause*, especially pp. 95–123.
6. *Ibid*.

*Many of the books to which we have referred have been published only in the United Kingdom and not in the United States, or we have had to refer to a British edition of a book which will be familiar to American readers. We have stuck to the British references, if only because page numbers may differ between editions.

7. A. F. A. King, "A New Basis for Uterine Pathology," *American Journal of Obstetrics* 8 (1875), pp. 242–243.

8. C. S. Ford, *A Comparative Study of Human Reproduction* (New Haven, CT: Yale University Press, 1945).

9. A. Rich, *Of Woman Born: Motherhood as Experience and Institution* (London: Virago, 1977), p. 106.

10. *See* Bullough and Voght, "Nineteenth Century Medicine;" and B. Ehrenreich and D. English, *For Her Own Good: 150 Years of the Experts' Advice to Women* (New York: Anchor Books, 1978). *See* also Mary Roth Walsh's article in this collection.

11. *See* R. T. Frank, "The Hormonal Causes of Premenstrual Tension," *Archives of Neurological Psychiatry* 26 (1931), pp. 1053–1057.

12. *See* V. Skultans, "The Symbolic Significance of Menstruation and the Menopause," *Man* 5 (1970), pp. 639–651; and L. F. Snow and S. M. Johnson, "Modern Day Menstrual Folklore," *Journal of the American Medical Association* 237 (1977), pp. 2736–2739.

13. *Ibid.*, p. 2738.

14. L. Birke and S. Best, "The Tyrannical Womb: Menstruation and Menopause," *Alice Through the Microscope: The Power of Science Over Women's Lives* (London: Virago, 1980).

15. D. N. Ruble, "Premenstrual Symptoms: A Reinterpretation, *Science* 197 (1977), pp. 291–292.

16. *See*, for example S. Schacter and J. E. Singer, "Cognitive, Social and Psychological Determinants of Emotional State," *Psychological Review* 69 (1962), pp. 379–399.

17. For example, D. R. Halbert and L. M. Darnell, "Dysmenorrhea and Prostaglandins," *Obstetric and Gynecological Survey* 31, no. 1 (1976), pp. 77–81.

18. Frank, "Hormonal Causes."

19. *See* for example M. Borrell, "Organotherapy, British Physiology, and the Discovery of the Internal Secretions," *Journal of the History of Biology* 9 (1976), pp. 235–268.

20. *See* for example K. Dalton, *The Menstrual Cycle* (Harmondsworth: Penquin, 1969).

21. D. B. Dusser and F. A. Gibbs, "Variations in the Electroencephalogram During the Menstrual Cycle," *American Journal of Obstetrics and Gynecology* 44 (1942), pp. 687–690.

22. A. L. Southam and F. P. Gonzaga, "Systemic Changes During the Menstrual Cycle," *American Journal of Obstetrics and Gynecology* 91 (1965), pp. 142 *et seq.*

23. J. S. Vierling and J. Rock, "Variations in Olfactory Sensitivity to Exaltolide During the Menstrual Cycle," *Journal of Applied Physiology* 22, pp. 311–315.

24. Dalton, *Menstrual Cycle.*

25. For example, D. Janowsky, S. C. Berens, and J. M. Davis, "Correlations

Between Mood, Weight, and Electrolytes During the Menstrual Cycle," *Psychosomatic Medicine* 35 (1973), pp. 143–154.

26. For example, D. F. Horrobin, *Prolactin: Physiology and Clinical Significance* (Lancaster, England: Medical and Technical Publishing, 1973).

27. We use the term "symptoms" here because that is the way in which most medical papers, as well as popular women's magazines, tend to describe changes occurring premenstrually that present problems. However, we are unhappily aware that such a term also maintains the notion of sickness which we ourselves have criticized in this chapter.

28. O. Janiger, R. Riffenberg, and R. Karsh, "Cross-cultural Study of Premenstrual Symptoms," *Psychosomatics* 13 (1972), pp. 226–235.

29. Against this ideal of constancy, of course, the ever-changing hormones of women can be seen in contrast. Men's hormones, on the other hand, are popularly—and erroneously—supposed to be unchanging.

30. P. Slade and F. A. Jenner, "Attitudes to Female Roles, Aspects of Menstruation and Complaining of Menstrual Symptoms," *British Journal of Social and Clinical Psychology* 19 (1980), pp. 109–113.

31. P. Shuttle and P. Redgrove, *The Wise Wound: Menstruation and Everywoman* (London: Victor Gollancz, 1978), pp. 59–60.

32. W. Cooper, "Cycles of Misery," *World Medicine* (November 2, 1977), pp. 47–50.

33. J. Lever, M. Brush and B. Haynes, *P.M.T. – The Unrecognised Illness* (London: Melbourne House, 1979).

34. *See* K. Dalton, "Menstruation and Accidents," *British Medical Journal* 2 (1960), pp. 1425–1426; and Dalton, *Menstrual Cycle*.

35. M. B. Parlee, "The Premenstrual Syndrome," *Psychological Bulletin* 80 (1973), pp. 454–465.

36. For example, R. K. Moos *et al.*, "Fluctuations in Symptoms and Moods During the Menstrual Cycle," *Journal of Psychosomatic Research* 13 (1969), pp. 37–44; and P. H. Moos, "Typology of Menstrual Cycle Symptoms," *American Journal of Obstetrics and Gynecology* 103 (1969), pp. 391–402.

37. *See* for example J. M. Abplanalp, A. F. Donnelly, and R. M. Rose, "Psychoendocrinology of the Menstrual Cycle I: Enjoyment of Daily Activities and Moods," *Psychosomatic Medicine* 41 (1979), pp. 587–604. But also see R. A. Markum for a defense of the Moos Questionnaire in "Assessment of the Reliability of and the Effect of Neutral Instructions on the Symptoms Ratings on the Moos Menstrual Distress Questionnaire," *Psychosomatic Medicine* 38 (1976), pp. 163–172.

38. M. B. Parlee, "Stereotypic Beliefs about Menstruation: A Methodological Note on the Moos Questionnaire and Some New Data," *Psychosomatic Medicine* 36 (1974), pp. 229–240; and M. B. Parlee, "Social Factors in the Psychology of Menstruation, Birth and Menopause," *Primary Care* 3, no. 3 (1976), pp. 477–496.

39. *See* for example the quotations given by Karen Paige, "Women Learn to

Sing the Menstrual Blues," *Psychology Today* (September 1973), pp. 41–46.

40. A notable instance of this is provided by the hypothesis put forward by D. M. Broverman, E. L. Klaiber, Y. Kobayashi, and W. Vogel, "Roles of Activation and Inhibition in Sex Differences in Cognitive Abilities," *Psychological Review* 75 (1968), pp. 23–50.

41. At least, this is not done in contemporary medical writing—it was clearly done in nineteenth-century writing, as we have indicated. Further, it may well be implied in some of the current medical articles, even if it is rarely made explicit.

42. Several feminist scholars have documented the ways in which women's use of language is limited and shaped by the society in which they live. If women, who are devalued, choose words of negative connotation to describe their experience of bodily functions that are also devalued, should we really be surprised? *See* for example Dale Spender, *Man Made Language* (London: Routledge & Kegan Paul, 1980).

43. W. R. Pierson and A. Lockhart, "Effect of Menstruation on Simple Reaction and Movement Time," *British Medical Journal* 5 (1963), pp. 796–797; and B. S. Kopell, D. T. Lunde, R. B. Clayton, and R. M. Moos, "Variations in Some Measures of Arousal During the Menstrual Cycle," *Journal of Nervous and Mental Disorders* 148 (1969), pp. 180–187.

44. For example, S. Golub, "The Effect of Premenstrual Anxiety and Depression on Cognitive Function," *Journal of Personal and Social Psychology* 34 (1976), pp. 99–105; and R. Hommes, "Menstrual Cycle Changes and Intellectual Performance," *Psychosomatic Medicine* 34 (1972), pp. 263–269.

45. B. E. Bernstein, "Effects of Menstruation on Academic Performance Among College Women," *Archives of Sexual Behavior* 6, no. 4 (1977), pp. 289–296.

46. J. Rodin, "Menstruation, Reattribution and Competence," *Journal of Personal and Social Psychology* 33 (1976), pp. 345–353.

47. *See* Shuttle and Redgrove, *Wise Wound*, pp. 155–162. It is also a suggestion that we have heard from several women's health groups in Britain, based on the experiences of women's health collectives.

48. M. McClintock, "Menstrual Synchrony and Suppression," *Nature* 229 (London: 1971), p. 244; and C. A. Graham and W. C. McGrew, "Menstrual Synchrony in Female Undergraduates Living on a Coeducational Campus," *Psychoneuroendocrinology* 5 (1980), pp. 245–252.

49. It is worth remembering the story of diethylstilbesterol (DES), which has been associated with vaginal cancer in the daughters of women who were given it during pregnancy. *See* K. Weiss, "Vaginal Cancer: An Iatrogenic Disease?" *International Journal of Health Services* 5, no. 2 (1975), pp. 235–251; and B. Seaman and G. Seaman, *Women and The Crisis of Sex Hormones* (Hassocks, England: Harvester Press, 1978).

50. *See* A. W. Clare, "The Treatment of Premenstrual Symptoms," *British*

Journal of Psychiatry 135 (1979), pp. 576–579; and L. Birke and K. Gardner, *Why Suffer: Periods and Their Problems* (London: Virago, 1979).

51. *See* for example Ehrenreich and English, *For Her Own Good*; and C. Bose, "Technology and Changes in the Division of Labor in the American Home," *Women's Studies International Quarterly* 2 (1979), pp. 295–304.

52. *See* V. Walsh, "Contraception: The Growth of a Technology," *Alice Through the Microscope: The Power of Science Over Women's Lives*, Brighton Women and Science Group, eds. (London: Virago, 1980).

53. For example W. Cooper, *No Change* (London: Hutchinson, 1975).

54. *See* Walsh, "Contraception."

55. *See* Weiss, "Vaginal Cancer"; and Seaman and Seaman, *Crisis in Sex Hormones*.

56. For example K. M. Shands *et al.*, "Toxic Shock Syndrome in Menstruating Women," *New England Journal of Medicine* 303 (1980), pp. 1436–1442; and J. P. Davis *et al.*, "Toxic Shock Syndrome. Epidemiological Features, Recurrence, Risk Factors and Prevention," *New England Journal of Medicine* 303 (1980), pp. 1429–1435. *See* also report in *Morbidity and Mortality Weekly Report*, 30 (January 30, 1981), pp. 25–36.

57. Report in Women's Information, Referral and Enquiry Service, W.I.R.E.S. 106 (March 1981), p. 13 (Newsletter of the Women's Liberation Movement, printed in Nottingham, England).

58. See Dalton, *Menstrual Cycle*; and G. A. Sampson, "Premenstrual Syndrome: A Double-Blind Controlled Trial of Progesterone and Placebo,". *British Journal of Psychiatry* 135 (1979), pp. 209–215.

59. T. Kashiwagi, J. N. McClure and R. D. Wetzel, "Premenstrual Affective Syndrome and Psychiatric Disorder," *Disorders of the Nervous System* 37 (1976), pp. 116–119.

60. Cooper, "Cycles of Misery," for example, concluded an article which is remarkable for its assumptions of biological "causes" as well as for its strange concept of evolution by natural selection shown in the statement: ". . . PMT may simply be an adverse response to hormonal fluctuations. In view of the fact that it may take evolution a few more million years *to get the faulty mechanism* ironed out, it is just as well that research into this common and neglected problem is beginning to provide answers to the root cause" (our emphasis).

61. *See* for example, Emily Culpepper, "Exploring Menstrual Attitudes," *Women Look at Biology Looking at Women*, R. Hubbard, M. S. Henifin, and B. Fried, eds. (Cambridge, MA: Schenkman Publishing Co., 1979).

62. H. Standing, "Sickness is a Woman's Business?: Reflections on the Attribution of Illness," *Alice Through the Microscope: The Power of Science Over Women's Lives*, Brighton Women and Science Group, eds. (London: Virago, 1980), p. 138.

[Sandy Best *was a member of the Brighton Women and Science Group that edited* Alice Through the Looking Glass: The Power of Science over Women's Lives *(London: Virago, 1980). She was a graphic artist, but has more recently been involved with community work, as well as catering (specializing in vegetarian cookery). She has lately been on a Women's Studies Tour to China, and has been active in both feminist and lesbian groups.*]

[Lynda Birke *is a research Fellow in the Biology Department at the Open University in Milton Keynes near London. She is involved with the Open University's Women's Studies course and teaches and does research on animal behavior. Her current research interests include the effects of hormones on behavioral development. With Sandy Best, she was a member of the Brighton Women and Science Group, which collectively edited* Alice Through the Microscope: The Power of Science over Women's Lives *(London: Virago, 1980). She has been actively involved in feminist and lesbian groups.*]

[Emily E. Culpepper *is a feminist scholar and activist interested in developing radical feminist philosophy. She is currently on the boundary of Harvard University as a doctoral candidate at the Divinity School. Her long-standing fascination with menstruation has led her to collect a variety of women's stories, myths, anecdotes, art work, health information and customs, and to produce* Period Piece, *a short film exploring attitudes toward menstruation.*]

Figure XVII. Pumping Iron.
(Photo: D. Wald)

Marlyn Grossman and
Pauline B. Bart

Taking the Men Out of Menopause

They call me Grace.
Yesterday I went
to the grocery store.
I had filled up
the cart
and was half way through
the check stand
before I realized
I had shopped for the whole family.
The last child left
two years ago.
I don't know what
got into
me.
I was too embarrassed
to take things back
so I spent the week cooking
casseroles.
I feel like one of those
eternal motion machines
designed for an
obsolete task
that just keeps on
running.

—Susan Griffin, *Voices*

Menopause, the cessation of menstruation, is the second "change of life" that we go through as women.[1] In our society, with its emphasis on youth, this is an unappreciated, often maligned time; there is no bar mitzvah for menopause. As Ursula LeGuin said, "It seems a pity to have a built-in rite of passage and to dodge it, evade it, and pretend nothing has changed. That is to dodge and evade one's womanhood, to pretend one's like a man. Men, once initiated, never get the second chance. They never change again. That's their loss, not ours. Why borrow poverty?"[2] Despite the intrinsic appeal of LeGuin's position, there have been many forces at work pressing women to see themselves differently. It is no accident that the powerful male-dominated institutions of our society, particularly medicine, have functioned here as in so many other cases to define the ways in which women are different from men as deviant, diseased or, at the very least, undesirable.

The topic of menopause seems particularly to evoke such sentiments. This may derive from the conviction with which a nineteenth-century medical "authority" like the prominent gynecologist Charles D. Meigs could describe woman as "a moral, a sexual, a germiferous, gestative and parturient creature."[3] Another physician, Holbrook, pontificated that it was "as if the Almighty, in creating the female sex, had taken the uterus and built up a woman around it."[4] Historian Peter Stearns observed that in eighteenth- and nineteenth-century Europe physicians thought women decayed at menopause.[5] Victorian physicians invariably characterized it as the "Rubicon" in a woman's life, and medical popularizers of the day "blamed the frequency and seriousness of disease during this period upon the 'indiscretions' of earlier life."[6] Kellogg remarked that the woman who transgressed nature's laws will find menopause "a veritable Pandora's box of ills, and may well look forward to it with apprehension and foreboding."[7]

Since then, physicians' attitudes have not changed much. However, it might be surprising to some not familiar with the pervasive sexism in current gynecological writing,[8] or with the traditional anti-female ambiance in medical education,[9] to learn that at a conference on menopause and aging sponsored by the U.S. Department of Health, Education and Welfare—a conference uncontaminated by the presence of a single female participant—Johns Hopkins' obstetrician-gynecologist, Howard Jones, characterized menopausal women as "a caricature of their younger selves at their emotional worst!"[10,11] To quote Mary Brown Parlee, "It sometimes seems as if the only thing worse than being subjected to the raging hormonal influences of the menstrual cycle is to have these influences subside."[12]

Physicians are willing to prescribe "for the menopausal symptoms that bother him [the husband] most,"[13] even though the drugs may be of questionable value and can have harmful side-effects. Estrogen is the one most frequently prescribed. It has been touted as effective in controlling menopausal symptoms from general ones such as depression to specific ones such as hot flashes (which are nervous system responses triggered by lower estrogen levels). A gynecologist named Wilson (whose research, not coincidentally, was sponsored by a drug company) warned women that to stay "Feminine Forever," they should take estrogen as long as they live, because menopause is a "deficiency disease."[14,15] This redefinition of a natural event such as menopause into a disease is an example of the increasing medicalization of normal events in our lives. Childbirth is another example of an event—once considered a natural part of women's lives—that the medical establishment now treats as pathological.[16] This process is both a cause and an effect of the enormous power American physicians have to define and manipulate our reality.

Actually, Wilson's skill as salesman far exceeded his accomplishments as a medical researcher.[17] Over 300 articles promoting estrogen have appeared in popular magazines in the intervening years. Yet researchers started reporting increases in uterine and breast cancer in women taking estrogen as far back as the 1940s,[18] and almost twenty years ago, animal studies started showing that estrogen can induce cancer in estrogen-dependent organs (breast, uterus, cervix, and vagina).[19] More recently, well-designed studies have shown that the risk of developing cancer of the lining of the uterus is four and one-half to fourteen times greater for women who take estrogen than for women who don't, and that the longer a woman takes estrogen, the greater her risk becomes.[20,21] Another study showed a possible link between taking estrogen and developing breast cancer.[22]

Ayerst, the drug company that has been grossing about seventy million dollars a year from Premarin, the estrogen most frequently prescribed to menopausal women, took all this in its stride. Immediately after publication of the 1975 studies, Ayerst sent physicians a letter (which did not even mention the cancer studies) recommending "business as usual." Alexander Schmidt, then Food and Drug Administration Commissioner, called this act "irresponsible."[23] Later, Ayerst hired the public relations firm of Hill and Knowlton, which specializes in companies with image problems, to help them deal with their increasingly bad press. In a letter dated December 17, 1976, a Hill and Knowlton vice president recommended to Ayerst an impressively complete and cynical media campaign "to protect and enhance the identity of estrogen replacement therapy."[24]

Even such high-powered planning may not be able to save Ayerst's business entirely, though. On July 22, 1977, the Food and Drug Administration proposed a regulation that estrogen drugs shall contain patient package inserts detailing what estrogen does and does not do. The model insert that the FDA prepared for comment is unusually candid and direct: "You may have heard that taking estrogen for long periods (years) after the menopause will keep your skin soft and supple and keep you feeling young. There is no evidence that this is so, however, and such long-term treatment carries important risks." Further, "Sometimes women experience nervous symptoms or depression during menopause. There is no evidence that estrogens are effective for such symptoms and they should not be used to treat them. . . ."[25] The model insert also contains several other hard-hitting statements.

Before they become effective, federal regulations must be published to allow time for public review and comment. After a set period has elapsed, a final version of the regulation is published and becomes law in sixty or ninety days unless someone petitions the relevant agency and/or federal court to prevent its acceptance. Not surprisingly the drug companies, represented by the Pharmaceutical Manufacturer's Association, have both petitioned the FDA and sued in federal court to block the estrogen regulation. What is less expected and more distressing is that the American College of Obstetricians and Gynecologists joined in the suit. Subsequently, the American Pharmaceutical Association, the American Medical Association and the National Association of Chain Drug Stores have also lent their support to the court action.

There's something for everyone in menopause: patients for physicians, profits for drug companies, and cancer for women. The medical and pharmaceutical groups are claiming that including the information brochure in the package will interfere with the traditional doctor-patient-pharmacist relationship (which does seem possible—though perhaps in ways that will be of benefit to the patient!). In a predictably paternalistic tone, the American Pharmaceutical Association claims, "The officially composed leaflet is far from understandable—to many patients, it will be utterly incomprehensible. And much of the information mandated for the leaflet is not only in no way pertinent to *the proper concerns of the patient* (emphasis added), once therapy has been determined, but it is wholly unsuitable for lay persons without medical or scientific training."[26] (If you feel as if your intelligence has just been insulted, you are in good company.) Finally, the medical and pharmaceutical groups claim that the FDA has acted without legislative authority.

The following excerpt of an article entitled "DOCTORS' GROUPS FILE THREE SUITS: FDA's estrogen ruling fought" that appeared in *The Boston Globe* on 9 October 1977, perhaps best reveals the nature and extent of the medical establishment's opposition to the FDA action:

"Indeed," said Dr. H.J. Barnum Jr., owner of a public relations firm in New York, in a written affidavit, "it is hard to imagine a class of patients more susceptible to adverse psychological reactions than the menopausal female, the very target of this leaflet, and nearly every aspect of the [pamphlet] appears calculated to arouse an emotional reaction in the patient."

"The best the [pamphlet] will accomplish," added Detroit Gynecologist T. N. Evans in his affidavit, "is a massive scare which the medical evidence indicates is wholly unwarranted." Evans added that the patient could even experience such an erosion of confidence "that she would not bother contacting her doctor at all, but simply discontinue consultation with the physician in the mistaken[!] belief that he lacks competence or adequate concern for her."

On the other hand, the Detroit doctor said, so many women might call their doctors after reading the pamphlet that this would put "an additional strain on an already overtaxed health care system."

Another issue in the suits, somewhat more in the background, is the fear of both drug companies and doctors that women who read the pamphlets may be more inclined to sue for malpractice, claiming that uterine cancer or some other serious disorder was caused by estrogen pills or their misuse.

As this is being written (October 18, 1977), the FDA regulation has become law and all packages of estrogen are now required to contain a patient information brochure. Though the medical and pharmaceutical groups have lost their attempt to get a temporary injunction against the enforcement of the regulation, they continue to press their suit seeking a permanent injunction. Since the judge who ruled in the temporary injunction case cited the substantial evidence of risk of cancer and overuse, it seems unlikely that the permanent injunction will be won. (Indeed, the claim that the FDA is acting beyond its legislative authority is belied by the fact that there have been patient package inserts in oral contraceptives since 1970).[27] Nevertheless, these actions of the medical and pharmaceutical groups dramatize the sexism and general inhumanity of the male dominated, profit-oriented U.S. medical system. A "deficiency disease" was invented to serve a drug that could "cure" it, despite the suspicion that the drug caused cancer in women.[28] That the suspicion had been voiced for so many years before anyone chose to investigate it is yet another example of how unimportant the well-being of women is to the men who control research and the drug companies who fund much of it. And the unwillingness of physicians, pharmacists, and drug companies to give women the information now available about estrogens demonstrates once again that the powerful will not give up any of their power (to say nothing of their financial gain), even after it has been clearly shown that they are using that power to harm women.[29]

We also have the male medical establishment to thank for the paucity of information about menopause. (It is difficult to imagine such ignorance about an event in the life of every man.) Sonja and John McKinlay surveyed what little literature there is about menopause and found it wanting.[30] Much of it is based only on physicians' "clinical experience," which is notoriously selective and unreliable. Where more objective research has been attempted, it has frequently involved retrospective data (which introduces all the unreliability of memory), unclarified cultural differences in recognizing and reporting symptoms, and the use of non-uniform definitions of menopause and of its symptoms.

What little we know about how *most* women (that is, those who do not end up in doctors' offices) experience menopause comes, not surprisingly, from research done by women.[31] Sonja McKinlay and Margot Jeffreys surveyed over six hundred women between the ages of 45 and 54 in and around London.[32] Most women go through menopause around 50. Three-quarters of the menopausal women were experiencing hot flashes. One-quarter of the post-menopausal women continued to have hot flashes for five years or more. The six other symptoms that were specifically inquired about—headaches, dizzy spells, palpitations, sleeplessness, depression, and weight increase—were each reported by one-third to one-half the women. While having these symptoms was not related to whether a woman was pre-, post-, or currently menopausal,[33] a woman who reported any symptom was likely to report more than one. Neither employment outside of the home nor the work load within it were related to the experience of those problems. Despite the fact that three-quarters of the women who had hot flashes found them embarrassing or uncomfortable, only about one-fifth sought medical treatment.

Since professionals have offered women so little information about the menopause, women's self-help groups have done some of their own research. Two groups, one in Seattle and one in Boston, used mail-in questionnaires in an attempt to survey women's physical and emotional experiences of the menopause.

The Seattle group, calling itself "Women in Midstream," had originally set out to investigate what the experience of menopause was like for women who were middle-aged before estrogen replacement therapy was available. Accordingly, they sent one thousand questionnaires to nursing homes, but received only seventy replies. These older women were also unwilling to talk about the subject in face-to-face interviews. This experience is of interest because it shows the extent to which women are socialized to regard normal bodily processes and life experiences with shame and to hide them from public scrutiny. It also suggests that most, if not all, earlier studies of menopause may well suffer from the respondents' unwillingness to reveal to researchers the full extent of their actual feelings and experiences. It may well be that only now, when support is available in our culture for women to share these formerly private areas of their lives with other women, can we really learn about these experiences in a systematic way.

Clearly, some women are now eager to share knowledge about menopause. The "Women in Midstream" group has received more than seven hundred completed questionnaires from highly motivated women who wrote or called to request the questionnaire after they had learned of it in the newspaper or on the radio. Unfortunately, sufficient womanpower and resources have so far been available to analyze only 250.

Because of the method by which they obtained their sample, the "Women in Midstream" researchers think that the respondents are probably largely middle and upper class and have had a relatively difficult experience with their menopause. (Even so, half the group described it as "easy" or "moderately easy!") About 60 per cent come from the state of Washington and the rest from elsewhere in the country. (Most of the sample came from a newspaper column directed to middle-aged women that originates in the *Des Moines Register* but is nationally syndicated in smaller towns and cities.) The vast majority of the women in the sample are white, married, have had about three children, but live with only their husband at this time. The women range in age from 28 to 73, with two-thirds of the group between 45 and 55. 60 percent are Protestant and 18 percent Catholic. Half of the women consider themselves "in the midst" of menopause and one-sixth "all through." Two-thirds of the women are working outside the home.

One of the most striking findings in this group is that three-quarters of the women had been prescribed hormone (i.e., estrogen) therapy. Yet no more than 60 per cent of the group had sought physicians' help for hot flashes and/or thinning and drying of the vaginal walls, the only two menopausal conditions for which there is some agreement that estrogen treatment is effective. (Because doctors have not changed their prescribing habits since publication of the studies that clearly showed the link between estrogen therapy and increased risk of cancer, we would not expect different results were the survey re-run today.) A staggering 55 per cent of the group were prescribed tranquilizers, causing one to wonder how much of this was in response to the women's needs and how much in response to those of their husbands and/or physicians,[34] especially since psychiatric therapy was recommended for less than 10 per cent of the women. Fifteen per cent of the women received dietary supplements. (Obviously, some women received more than one recommendation.) Only 11 per cent were told that they needed no treatment.

Only slightly more than half of the women reported satisfaction with their doctor's attitudes and found her or him helpful. Of those among the group who sought help from non-medical sources, three-fifths found these people helpful. More than 75 per cent of the women discussed their menopausal problems with female friends or relatives and nearly half of them found this helpful. More than 66 per cent of the group discussed their problems with husbands, male friends or relatives and, again, nearly half found it helpful to do so.

The "Women in Midstream" group feels that social supports are very important for women going through menopause. Accordingly, they asked their sample if they would be interested in talking with other women about the health and social problems of older women. More than half the women were definitely interested and another sixth said they might be. The same total number of women indicated definite or possible interest in individual discussions, though more were uncertain here. (The desirability of such discussions is certainly borne out by an incident Paula Weideger reports, in which a woman participating in a menopause consciousness-raising group was surprised to find that while she had originally experienced her hot flashes as uncomfortable, they had now become pleasurable!)[35]

The Boston Women's Health Book Collective attempted an even more ambitious survey than the Seattle women. They set out to study the attitudes toward and experiences of menopause by women of all ages and in all parts of the country. Mail questionnaires were sent to friends and relatives of the Collective's members as well as to all the clinics and counseling centers that ordered the 1973 edition of *Our Bodies, Ourselves.* Replies were received from almost five hundred women. Most of them live in large northeastern or middle-Atlantic cities or suburbs. Less than two-thirds are married (compared with four-fifths of the Seattle group), though another fifth are divorced or widowed. Less than half of the group are Protestant, one-quarter Jewish, and one-fifth Catholic. As a group, they have slightly smaller families than the Seattle sample. Just over a third of the total group was menopausal or post-menopausal. As was the case in the Seattle sample, two-thirds of this group was employed outside the home.

Just over three-fifths of the Boston menopausal women received estrogen replacement therapy. While this is less than the Seattle group, it is still surprisingly high for a group that probably is experiencing many fewer menopausal "problems" (given the ways the respective samples were gathered). Five-sixths of the women talked with friends about their menopausal experiences, while over two-thirds talked to husbands (just about as many as in the Seattle group, though the questions asked were not exactly the same). More than two-thirds talked to children or other relatives and one-sixth each to therapists and women's groups. Two-thirds of the women had friends going through menopause at the same time. Two-thirds (largely, though not exactly, the same group) also received emotional support.

Where the physical experience of menopause blends into the psychological, we enter an area of many assumptions and few facts. Earlier, we quoted some choice descriptions medical authorities have given of the menopausal woman. It is not surprising then that psychoanalysts, who are usually physicians, have also tried to use their "science" to keep women narrowly contained within the roles of wife and mother. Since psychoanalysts follow Sigmund Freud, they tend to describe human development in terms of his phallocentric perspective. Helene Deutsch, who unfortunately demonstrates the fact that women in male-dominated professions sometimes internalize that perspective (frequently as a survival technique), described the menopausal woman as having "reached her natural end—her partial death—as servant of the species."[36] In fact two men, one of them a psychoanalyst, have managed to provide a clear view of this muddy area. They have summed it up this way: "It becomes obvious that a type of confused, fuzzy, and prejudicial thinking has existed in the minds of psychologists, psychiatrists, and physicians in general toward the female. Whereas it is clearly recognized that male psychology must be differentiated from male biology, with an awareness of the effects of one upon the other, no such differentiation has been allowed for females. Males are not governed by biologic maturation processes related to aging. They can be abstract, external, and worldly, and are concerned with jobs and events of the day. For whatever reasons, little attention has been given to the obviously similar dichotomy in the psychology and biology of women. Emotions and cognition both have been viewed as being a part of, and controlled by, biology. Almost all research has proceeded from this obviously muddled hypothesis."[37]

Some fuzzy thinking about women derives from the work of the psychoanalyst, Theresa Benedek. In one article, Benedek commented that menopause brings a diminution of the part "of the integrative strength of the personality which is dependent upon the stimulation by gonadal hormones"![138] In another, she found that there was a drop in hormone levels in the pre-menstrual phase of the menstrual cycle, which also seemed to be marked by "narcissistic regressive preoccupations" (which is psychoanalytic-ese for a woman spending more of her energies thinking about herself than our culture thinks she should!). By an impressive deductive leap, Benedek concluded that the declining hormonal capacity of menopause leads to a diminished capacity to love.[39] The methodological problems in Benedek's research are extensive. Among other things, the sample includes only women undergoing psychotherapy.[40]

Another group of researchers who have attempted to understand women's experience of menopause are social psychologists. Bernice Neugarten and her students (also women, demonstrating again that anyone interested in menopause is likely to be female in gender even if not in perspective) used an attitudes checklist to study women's feelings about menopause.[41] They found that women who had experienced menopause were less likely to consider it a significant event than were pre-menopausal women. Also, upper-middle-class women in particular denied the significance of menopause in a woman's life. While it is good that these studies questioned a wider group of women than just those who turned up in a doctor's office, the studies were done in the early 1960s, and thus before the Women's Movement gave women permission to "speak bitterness"[42] about the limitations of their socially prescribed roles.[43]

Some of the interesting findings which come out of these studies result from subdividing the women interviewed. For instance, Lenore Levit found no difference in general in the amount of anxiety menopausal women suffered compared with post-menopausal women.[44] However, women who were very much invested in their role as mothers were more anxious during menopause than they were afterwards. Thus, the transition appears harder for women who are giving up a role they highly value. (This was true whether or not the children were living at home.) Similarly, whereas middle-class women did not differ in anxiety during or after menopause, working-class menopausal women were more anxious than those post-menopause, possibly because working-class life offers women fewer rewards besides children.

Ruth Kraines studied one hundred women, one-third in menopause, one-third pre-menopause, and one-third post-menopause.[45] Using interviews, self-report forms, and checklists containing many symptoms, some of which are traditionally associated with menopause, she did not find appreciable differences among the three groups in their assessment of their own physical state (although menopausal women did check more of the symptoms particularly associated with the menopause). She did find that women who were low in self-esteem and life satisfaction were most likely to have difficulty during menopause, and that the relationship appeared to be circular, *i.e.*, low self-esteem led to difficulty during menopause which in turn led to low self-esteem. Kraines also found a continuity between a woman's previous reactions to bodily experiences (such as health problems, menstruation, and pregnancy) and her reaction to menopause. She concludes that, contrary to the medical studies, menopause in itself is not experienced as a critical event by most women. She suggests that women who seek medical help differ from most middle-aged women in their physical and emotional reactions to stress.

Medical people tell us that a woman undergoes extensive hormonal changes at menopause and therefore requires (their kind of) help; at the same time, social psychological surveys suggest that most women do not regard menopause as a difficult stage of life. The problem with much of the available literature is that physicians have generalized from the relatively small group of women who seek medical or psychiatric help. And social psychologists, working at a time before there was support in the culture for a woman to express her dissatisfaction with the lack of alternatives open to her after her children are grown, may also have come up with a biased picture. How, then, can we figure out what is really happening to women and how many physiological and sociocultural factors influence the experience of menopause?

One way to tease out the sociocultural from the physiological is to look at cross-cultural studies. Both Nancy Dowty and one of us (Pauline Bart) have studied menopause in this manner. Nancy Dowty worked in Israel studying five sub-cultures which she arrayed on a continuum of modernization, from traditional Arab women at one end to European-born Israeli women at the other, with Jews from Turkey, Persia, and North Africa in between. She found no linear relationship between social change and difficulty during the menopause. The transitional women, midway between traditional life styles and modernization, suffered most: they had lost the privileges afforded traditional women, while not receiving those benefits that modernization confers upon women. They had the problems of both groups but the advantages of neither.

Pauline Bart, using the Human Relations Area Files as well as ethnographic monographs, found that certain structural arrangements and cultural values were associated with women's changed status after the childbearing-years.[47] These are summarized in the following table.

Table I. Characteristics of societies in which women's status changes at menopause.

Status Rises	Status Declines
Strong tie to family of orientation (origin) and relatives.	Marital tie stronger than tie to family of orientation (origin).
Extended family system.	Nuclear family system.
Reproduction important.	Sex an end in itself.
Strong mother-child relationship reciprocal in later life.	Weak maternal bond; adult-oriented culture.
Institutionalized grandmother role.	Non-institutionalized grandmother role; grandmother role not important.
Institutionalized mother-in-law role.	Non-institutionalized mother-in-law role; mother-in-law doesn't train daughter-in-law.
Extensive menstrual taboos.	Minimal menstrual taboos.
Age valued over youth.	Youth valued over age.

It seems reasonable to assume that an increase in status would increase the likelihood of feelings of well-being, so that even if physiological stresses are experienced at menopause, they are well buffered. This appears to be the case in kinship-dominated societies. When this system begins to break down, as it is now doing in some Third World countries, problems arise similar to those faced by some women in our culture. For example, one (Asian) Indian mother who brought up her children to live in the modern manner, that is, independently, felt very lonely and commented: "I sometimes feel, 'What is the use of my living now that I am no longer useful to them [her children]!' "[48]

It is easy to see why some women in our society find middle age stressful. Except for the mother-child bond, which in our society is strong but *non*-reciprocal, we fall on the right hand side of the table, with the cultures in which women's status drops in middle age. It is true that for women whose lives have not been child-centered and whose strong marital tie continues, or for those whose children set up their own residence near the mother, the transition to middle age may be buffered. However, for women who have emphasized the maternal role or the glamor role, middle age may be difficult. Our emphasis on youth (particularly in women) and the stipulation that mothers should not interfere in the lives of their married children (the mother-in-law syndrome) can make middle age stressful for women who have not had the opportunity to invest themselves in anything besides wife and mother roles. By examining the question in a cross-cultural perspective, however, we can observe the multiplicity of *possible* roles for middle aged women and appreciate the fact that middle age need not be a difficult time. Indeed, it *can* and *should* offer women its own unique rewards.

While the cross-cultural evidence strongly suggests that the phenomenon of menopausal depression is related not to physiological changes but to social and cultural structures and factors, a study of individual, depressed, menopausal women in our society could shed a great deal of light on how these factors operate in American culture. One of us[49] has studied the hospital records of over five hundred women between the ages of forty and fifty-nine who had had no previous hospitalization for mental illness. The records were drawn from five mental hospitals ranging from an upper-class private institution to two state hospitals, and were used to compare all the women diagnosed as depressed (whether neurotic, psychotic involutional, or manic depressive) with those who had received other diagnoses. Twenty intensive interviews were also conducted to round out the picture obtained from the records. The interviews included questionnaires used in studies of "normal" middle aged women, and a projective biography test consisting of sixteen drawings of women at different stages of their life cycle and in different roles.

Statistical analysis of the hospital records indicated that depression was associated with current role loss and even with the prospect of loss. Housewives were particularly vulnerable to the effects of losing roles such as those of wife or mother. Ethnicity is another relevant variable, with Jewish women showing the highest rate of depression. When all women having overinvolved or overprotective relationships with their children are compared with women who do not, however, the ethnic differences almost wash out. (Thus you do not have to be Jewish to be a Jewish mother, but it helps a little.) Overall, the highest rate of depression was found among housewives who had overprotective or overinvolved relationships with their children who were currently or soon leaving home. Thus the lack of meaningful roles and the consequent loss of self-esteem, rather than any hormonal changes, seemed largely to account for the incidence of menopausal depression.

This hypothesis received further support from the results of the interviews. All the women with children, when asked what they were most proud of, replied "my children." None mentioned any accomplishments of their own, except being a good mother. When asked to rank seven roles available to middle-aged women in order of importance, the mother role ("helping my children") was most frequently ranked first or second. When children leave home, however, the woman is frustrated in attempting to carry out this role she values so highly and suffers a consequent loss of self-esteem. It is precisely to the extent that a woman "buys" the traditional norms and seeks vicarious achievement and identity (which society has told her is the appropriate route to "true happiness" and "maturity") that she is vulnerable when her children leave. Moreover, many of these women then experience their life situation as unjust and meaningless because the implicit bargain they had struck with fate did not pay off. In the words of two of the women:

> I'm glad that God gave me . . . the privilege of being a mother
> . . . and I loved them [my children]. In fact, I wrapped my love
> so much around them . . . I'm grateful to my husband since if it
> wasn't for him, there wouldn't be the children. They were my
> whole life . . . My whole life was that because I had no life
> with my husband, the children should make me happy . . . but
> it never worked out.

> I felt that I trusted and they—they took advantage of me. I'm
> very sincere, but I wasn't wise. I loved, and loved strongly and
> trusted, but I wasn't wise. I—I deserved something, but I thought
> if I give to others, they'll give to me. How could they be different?
> . . . But, you see, those things hurt me very deeply, and when I had
> to feel that I don't want to be alone, and I'm going to be alone,
> and my children will go their way and get married—of which
> I'm wishing for it—and then I'll still be alone, and I got more and
> more alone, and more and more alone.

Statements such as these poignantly portray middle age as a time when reality overtakes women's dreams for the future and some women are confronted with the meaninglessness of their lives. Women have been taught to believe that they can achieve "true happiness" through self-abnegation and sacrifice for their husbands and their children, that to do anything for themselves is selfish. Some women are able to evade or overcome this script. Some continue to receive the vicarious gratification they sought—their husbands are still alive, well and attentive; their children have made "proper" marriages and/or embarked upon "proper" careers; grandchildren have arrived; and significant others congratulate them on a job well done. But for others, the story is different. They are bewildered by their children's life style, which rejects the values the mothers have worked so hard to attain, indeed not so much for themselves as for those very children. They cannot understand why their daughters do not want to have the children they were taught were their destiny, thus denying them the grandchildren they so joyfully anticipated and who would give new meaning to their lives. Their husbands may have left them for younger women to bolster their own waning egos and diminished potencies (on average, the second wife is younger than the current age of the first wife).[50] For these women to seek a younger mate would be thought ludicrous, since they are no longer considered sexual beings (though of course physiologically, unlike men, they are as capable of sex as they ever were).[51] And because so many of these events coincide with menopause, their effects are attributed to this physiological change.

Alice Rossi, analyzing recent census data, notes that maternity has become a very small part of the average adult woman's life: a woman who marries at twenty-two, has two children two years apart, and dies at seventy-four, on average will spend one-quarter of her adult life without a husband, two-fifths with a husband but no children under 18, one-third with spouse and at least one child under 18; but only one-eighth of her life in full-time maternal care of pre-school age children.[52] This projection dramatizes the inappropriateness of the standard script. It is important that women learn this message, and early, so they will not become casualties of the culture in middle age.

As feminists, we believe that societal problems cannot be dealt with on an individual basis: there are no individual solutions except for those few women who may slip through by chance or special privilege. (It is no accident that Bernice Neugarten's view—that middle age is a neutral and frequently almost positive experience—relies heavily on a study of middle-aged people who are elite professionals or in business and for whom terms like autonomy, predictability, and choices have meaning, particularly at this stage of the life cycle.[53] As the saying goes: "Rich or poor, as long as you have money!"). For most women's lives to change, sweeping economic and social reforms are essential. For the present, the only changes we can count on are those that can be brought about by the organized efforts of many women working together to structure alternatives for themselves and for others. While true long-range solutions would require changes in women's lives from very early ages (and many women are actively working at bringing these about), there is growing support for women already in their middle years. The National Organization for Women (NOW) has a task force on older women. In the Fall of 1980, a founding convention for a national Older Women's League was held.[54] There are also increasing numbers of rap groups for middle-aged women both here and in other countries.

For instance, there are menopause rap groups throughout Holland. Since few Dutch married women work outside their homes, the usual problems of transition are exacerbated by the lack of alternative sources of self-esteem. While the women originally join the groups to discuss problems around menopause, broader life style and life cycle issues emerge. In the supportive atmosphere of the groups, many women later are able to express sexual or marital dissatisfactions.[55]

One of the signs of the successful impact of the Women's Health Movement is the fact that gynecological self-examination, once a revolutionary cry of a small group, has begun to be a part of routine office practice among some of the more forward-looking gynecologists. As women demand more participation in, and control over, the various elements of their lives, we expect that the heavy taboos on bodily functions (and what woman has not been anxious lest she "stain through" during her menstrual cycle?) will decrease, so that the embarrassment caused by the hot flashes most menopausal women experience can be alleviated without the use of cancer-causing drugs.

There is much talk of the wider range of options available to middle-aged women. Group support can enable those women who have greater options to use their new freedom to change their life styles and fulfill some of their deferred dreams. However, options are limited by economic conditions, racism and previous educational opportunities. Only in a society in which racism, sexism and poverty are not endemic can all women live full lives. We shall end with a poem by former Ann Arbor City Councilwoman Kathy Kozachenko describing both the treatment women receive when they try to change their way of being-in-the-world and their ultimate triumph.

Mid-Point

She stored up the anger
for twenty-five years,
then she laid it on the table
like a casserole for dinner.

"I have stolen back
my life," she said.
"I have taken possession
of the rain and the sun
and the grasses," she said.

"You are talking
like a madwoman,"
he said.

"My hands are rocks,
my teeth are bullets,"
she said.

"You are
my wife,"
he said.

"My throat is an eagle
My breasts
are two white hurricanes," she said.

"Stop!" he said.
"Stop or I shall call
a doctor."

"My hair
is a hornet's nest,
my lips
are thin snakes
waiting for their victims."

He cooked his own dinners,
after that.

The doctors diagnosed it
common change-of-life.

She, too, diagnosed
it change of life.
And on leaving the hospital
she said to her woman-friend
"My cheeks
are the wings
of a young
virgin dove.
Kiss them."

Notes

1. We would like to thank the Boston Women's Health Book Collective and Women in Midstream for generously sharing their data with us.
2. Ursula LeGuin, "The Space Crone," *The CoEvolution Quarterly*, Summer 1976, p. 110.
3. Quoted in Carroll Smith-Rosenberg and Charles Rosenberg, "The Female Animal: Medical and Biological Views of Woman and her Role in Nineteenth-Century America," *The Journal of American History*, 60 (1973), 332–356.
4. *Ibid.*
5. Peter Stearns, "Interpreting the Medical Literature on Aging," *Newberry Library Family and Community History Colloquia: The Physician and Social History*, 30 October 1975.
6. John S. Haller, Jr. and Robin M. Haller, *The Physician and Sexuality in Victorian America* (Urbana: University of Illinois Press, 1974) p. 135.
7. *Ibid.*, p. 135.
8. Diana Scully and Pauline B. Bart, "A Funny Thing Happened on the Way to the Orifice: Women in Gynecology Textbooks," *American Journal of Sociology*, 78 (1973), 1045–1050.
9. Margaret A. Campbell, *"Why Would a Girl Go Into Medicine?"* (Old Westbury, N.Y.: The Feminist Press, 1973).
10. Howard W. Jones, Jr., E. J. Cohen, and Robrt B. Wilson, "Clinical Aspects of the Menopause," *Menopause and Aging*, John K. Ryan and D. C. Gibson, eds. (Washington, D.C.: U.S. Government Printing Office, 1971), p. 3.
11. Contrast this view with the feminist perspective offered by June Arnold in the novel *Sister Gin* (Plainfield, Vt.: Daughters, 1975), p. 189: "Bettina will be all right when she reaches menopause. She will be old again as soon as her body stops being under the moon's dominion. The child and the old don't go by clocks and don't know fear. Time took away the child and only time can give her back."
12. Mary Brown Parlee, "Psychological Aspects of the Climacteric in Women" (Paper delivered to the Eastern Psychological Association, New York, April 1976).
13. This is from a drug ad quoted in the Boston Women's Health Book Collective, *Our Bodies, Ourselves* (New York: Simon and Schuster, 1976), p. 327. More recently, under pressure from the Food and Drug Administration, the drug ads are markedly more responsible. There is no evidence, however, that the physicians' attitudes have changed.
14. Robert A. Wilson, *Feminine Forever* (New York: M. Evans and Co., 1966), p. 18.
15. In testimony before the Senate Health Subcommittee, January 21,

1976, then FDA Commissioner Alexander Schmidt said, "It is clear that such treatment [estrogen] must be viewed as providing symptomatic relief for an annoying problem and not as essential therapy for a disease, and that all such treatment is a phenomenon largely of modern Western medicine and medical affluence"—and physicians' desire to keep it that way we might add!

16. See Datha Brack's article in this collection for further discussion.
17. Anita Johnson, "The Risks of Sex Hormones as Drugs," *Women and Health*, 2:1 (1977), pp. 8–11.
18. S. B. Gusberg, "Precursors of Corpus Carcinoma Estrogens and Adenomatous Hyperplasia," *American Journal of Obstetrics and Gynecology*, 54:6 (1947), pp. 905–27.
19. W. V. Gardner, "Carcinoma of the Uterine Cervix and Upper Vagina; Induction under Experimental Conditions in Mice," *Annals of the New York Academy of Science*, 75 (1959), pp. 543–64.
20. Harry K. Ziel and William D. Finkle, "Increased Risk of Endometrial Carcinoma Among Users of Conjugated Estrogens," *New England Journal of Medicine*, 293:23 (1975), pp. 1167–70.
21. Donald C. Smith, Ross Prentice, Donovan J. Thompson, and Walter L. Herrmann, "Association of Exogenous Estrogen and Endometrial Carcinoma," *New England Journal of Medicine*, 293:23 (1975), pp. 1164–7.
22. Robert Hoover, Laman A. Gray, Philip Cole, and Brian MacMahon, "Menopausal Estrogen and Breast Cancer," *New England Journal of Medicine*, 295:8 (1976), p. 501.
23. Sharon Lieberman, "But You'll Make Such a Feminine Corpse ...," *Majority Report*, 3–4, 19 February–4 March, 1977.
24. "New Discovery: Public Relations Cures Cancer," *Majority Report*, 9–10, 8–18 February, 1977. This letter came to light because an employee in one of the firms felt it important that women know about it, and made the letter available to this feminist newspaper in New York.
25. Morton Mintz, "FDA Requiring Direct Warning to Users of Estrogen," *Washington Post*, 21 July 1977.
26. *F-D-C Reports*, 10–12, 5 September 1977, p. 12 (A newsletter published by the U.S. Food and Drug Administration).
27. And that also was accomplished only by organized pressure from feminist groups.
28. Anita Johnson ("The Risks ...") says that the suspicion dates back to the 1890s but does not indicate her source.
29. A spot check of physicians after publication of the 1975 studies showed that they had not changed their practices with regard to estrogen prescription. One San Francisco physician compared menopause to diabetes and went on to say: "Most women suffer some symptoms whether they are aware of them or not, so I prescribe estrogens

for virtually all menopausal women for an indefinite period" (*New York Times*, 5 December 1975). A symptom is an outward sign of an underlying problem. That this physician chooses to give carcinogens to women to deal with problems they are not experiencing is the ultimate in medical arrogance and represents medicine's attempt to take over not only our bodies but even our sense of reality. Medical sociologist John McKinlay also has reported that physicians are still prescribing estrogens as frequently as before (10 June 1977).

30. Sonja M. McKinlay and John B. McKinlay, "Selected Studies of the Menopause: An Annotated Bibliography," *Journal of Biosocial Science*, 5 (1973), pp. 533–55.

31. Similarly, one of us (Pauline Bart) found that whether an anthropological study included information about menopause in the culture being studied depended upon whether any of the anthropologists doing the study were women!

32. Sonja M. McKinley and Margot Jeffreys, "The Menopausal Syndrome," *British Journal of Preventive and Social Medicine*, 28:2 (1974), pp. 108–115.

33. "Menopausal" means in the first twelve months following final cessation of menses.

34. Jane E. Prather and Linda S. Fidell, "Sex Differences in the Content and Style of Medical Advertising," *Social Science and Medicine*, 9 (1975), pp. 23–26. Robert Seidenberg, "Drug Advertising and Perception of Mental Illness," *Mental Hygiene*, 55 (1971), pp. 21–31.

35. Paula Weideger, *Menstruation and Menopause* (Revised and Expanded) (New York: Dell, 1977), p. 235.

36. Helene Deutsch, *The Psychology of Women, Volume II* (New York: Grune and Stratton, 1945), p. 459.

37. Howard J. Osofsky and Robert Seidenberg, "Is Female Menopausal Depression Inevitable?" *American Journal of Obstetrics and Gynecology*, 36 (1970), pp. 611–15.

38. Therese Benedek, "Climacterium: A Developmental Phase," *Psychoanalytic Quarterly*, 19:1 (1950), pp. 1–27.

39. Therese Benedek and Boris B. Rubenstein, "Psychosexual Functions in Women," *Psychosomatic Medicine* (New York: Ronald Press, 1952).

40. For critiques and alternative perspectives, *see* Randi Daimon Koeske, "Premenstrual Emotionality: Is Biology Destiny?" *Women and Health*, 1:3 (1976), pp. 11–14; Mary Brown Parlee, "The Premenstrual Syndrome," *Psychological Bulletin*, 80:6 (1973), pp. 454–65; and K. Jean Lennane and R. John Lennane, "Alleged Psychogenic Disorders in Women—A Possible Manifestation of Sexual Prejudice," *New England Journal of Medicine*, 288 (1973), pp. 288–92.

41. Bernice L. Neugarten, Vivian Wood, Ruth J. Kraines, and B. Loomis,

"Women's Attitudes Toward Menopause," in *Middle Age and Aging*, Bernice L. Neugarten, ed. (Chicago: The University of Chicago Press, 1968).

42. The term comes from the Chinese revolutionary groups which gathered to talk about the bad old days in order to purge themselves of the feudal mentality so that they could create a new society. Women's consciousness-raising groups function similarly to help women free themselves from their constricting role socialization.

43. Vivian Wood, one of the authors of that article, has since personally expressed doubt about the reliability of her findings due to the change in mores.

44. Lenore Levit, *Anxiety and the Menopause: A Study of Normal Women* (Doctoral Dissertation, University of Chicago, 1963).

45. Ruth J. Kraines, *The Menopause and Evaluations of the Self: A Study of Middle-aged Women* (Doctoral Dissertation, University of Chicago, 1963).

46. Nancy Dowty, "To Be a Woman in Israel" *School Review*, 80 (1972), pp. 319–332.

47. Pauline B. Bart, "Why Women's Status Changes in Middle Age," *Sociological Symposium*, 3 (1969), pp. 1–18.

48. Cormack cited in Pauline B. Bart, "Why Women's Status Changes . . . ," p. 13.

49. Pauline B. Bart, *Depression in Middle-Aged Women: Some Socio-cultural Factors* (Doctoral Dissertation, University of California at Los Angeles, 1967). *Dissertation Abstracts* 28:4752-B, (University Microfilms No. 68-7452), 1968; see also Bart, "Mother Portnoy's Complaints," *Trans-Action*, 8:1–2 (1970), pp. 69–74; and "Depression in Middle-Aged Women," *Women in Sexist Society*, Vivian Gornick and B. K. Moran, eds. (New York: Basic Books, 1971).

50. Inge P. Bell, "The Double Standard," *Trans-Action*, 8:1–2 (1970), pp. 76–81.

51. Zoe Moss, "It Hurts to be Alive and Obsolete: The Aging Woman," *Sisterhood is Powerful*, Robin Morgan, ed. (New York: Vintage, 1970).

52. Alice Rossi, "Family Development in a Changing World," *American Journal of Psychiatry*, 128 (1972), pp. 1057–80.

53. Bernice L. Neugarten and Nancy Datan, "The Middle Years," *American Handbook of Psychiatry* (2nd ed.), Vol. 1, Salvator Arieti, ed. (New York: Basic Books, 1974).

54. Information is available from the Older Women's League, 3800 N. Harrison Street, Oakland, CA 94611.

55. Another interesting function these groups have performed is the treatment of agoraphobia (fear of open spaces), a widespread problem in Holland. They seem to have succeeded in a number of cases in which years of professional treatment have failed. It is interest-

ing to note that some clinicians regard agoraphobia as the attempt of a person who considers herself powerless to seize some kind of power. Apparently these groups have been able to supply their members with another, more direct, sort of personal power.

[Marlyn Grossman *is a psychologist in Chicago where she was a founding member of Women in Crisis Can Act (W.I.C.C.A.), a feminist hot line, and the Chicago Abused Woman Coalition. She worked with Pauline Bart on her research on women who were attacked and avoided being raped. She does therapy with women and children from a feminist perspective and is a member of the Association for Women in Psychology.*]

[Pauline Bart *is a radical feminist sociologist at the University of Illinois Medical Center. After 12 years of being a trapped suburban housewife, she returned to U.C.L.A. to write her dissertation on depression in middle-aged women (better known as "Portnoy's Mother's Complaint"). While teaching in the usual marginal positions reserved for women at U.S.C. and Berkeley, she presented their first course on women in the Spring of 1969. Most of her work has focussed on the interface between sex roles and health issues— including, with Diana Scully, "A Funny Thing Happened on the Way to the Orifice: Women in Gynecology Textbooks," and with Linda Frankel,* The Student Sociologist's Handbook. *She is a founding mother and active participant in Sociologists for Women in Society, and co-founder and first chair of the Section on the Sociology of Sex Roles of the American Sociological Association. She is currently studying both rape victims and women who successfully avoided rape so that women may be given data-based advice.*].

Datha Clapper Brack

Displaced—The Midwife by the Male Physician

From earliest times, women have assisted other women in childbirth. This was true for preliterate peoples, agricultural societies, and our own Western culture prior to the Industrial Revolution. Even today, considerably more babies throughout the world are delivered by midwives than by physicians. Then how did it happen that in our society the female midwife was replaced by the male obstetrician?

Before this could come about, two fundamental conditions had to be met. First, men had to be defined as socially acceptable persons to attend childbirth, and second, normal childbirth, traditionally regarded as a "natural" event, had to be redefined as one that called for professional medical assistance. However, although these conditions allowed men to enter and practice midwifery, they did not require them to dominate the occupation; on the European Continent and in Great Britain, medically trained midwives continued to practice, and today they deliver sixty to eighty per cent of the babies there.[1] But in the United States, by the twentieth century, the midwife had all but disappeared. During the 1960s, nearly ninety-nine per cent of American babies were delivered by physicians, more than ninety-three per cent of whom were male.[2] When a joint study group of the International Confederation of Midwives and the International Federation of Gynecology and Obstetrics compiled a report of maternity care throughout the world in 1966, the writers found it necessary to treat the United States as a "special case" in the tables "because of its tendency not to recognize midwifery as an independent profession."[3]

It is likely that professionalization of medicine allowed male physicians to break the midwife's traditional hold of the occupation. As medicine developed in Europe from a learned calling to a consulting profession, the prestige of the profession allowed the physicians to make claims to superior competence as childbirth attendants, and co-opt the developing medical technology and knowledge applicable to childbirth. In the process of developing their expertise, they redefined the

*Figure XVIII. Pre-Columbian bowl from the Mimbres Valley in New Mexico.
(Peabody Museum, Harvard University)
Figure XIX. Women's Laboratory at M.I.T.
(M.I.T. Historical Collections)*

nature of childbirth and reorganized childbirth care to suit their own professional needs.

However, it was *as a professional group* that physicians in the United States subsequently were able to protect their own interests and dominate the field. Powerful medical associations gave them access to the power of the state, through which they were able to control licensing legislation, restrict the midwife's sphere of activity, and impose legal sanctions against her. Male physicians blocked women not only from delivering babies as midwives, but also from delivering them as practicing physicians. The exclusionary practices of medical associations and the use of informal professional networks worked both to bar women from studying medicine and to reduce their effectiveness as practitioners when they were qualified.[4]

To a certain extent, these pressures existed on both sides of the Atlantic, but additional social and cultural factors combined in the United States to give physicians an advantage. First, there had been no established tradition for training midwives in the United States such as had developed in Europe alongside the profession of medicine. This European tradition gave professional midwives more access to developing medical knowledge and more legal protection than their American counterparts had. Second, medicine itself was a newly imported profession in America, one that was having difficulty staking its claims against those of lay practitioners of all kinds in the nineteenth century. Its practitioners were therefore more jealous of what they saw as their territory than were members of the older, established profession in Europe.[5] Third, women were more likely to be restricted from entering prestigious professions in the United States. And finally, perhaps most importantly, at several critical periods, physicians were in greater supply here than in Europe. It was during these periods that attempts were made to restrict delivery of babies to the medical profession. At times when physicians were in short supply, midwives were more likely to be tolerated and suggestions for establishing and improving schools for their training more likely to be entertained.

The Social Organization of Childbirth Care

It is useful to think of a profession as an occupation which has assumed a dominant position in a division of labor, so that it gains control over the substance of its own work. Unlike most occupations, it is autonomous or self-directing. In developing its own "professional" approach, the profession changes the definition and shape of problems as experienced and interpreted by the layman. The layman's problem is recreated as it is managed—a new social reality is created.[6]

It was necessary for medical men to change the definition of childbirth in order to manage it, because it had been institutionalized as a social event rather than an illness, and one that specifically demanded female participation. During childbearing, a woman passes through a sequence of exclusively female social statuses—pregnant woman, laboring woman, nursing mother—and for each of these, she has to learn role behavior expected by her own culture. In preliterate societies, experienced older women were the holders of wisdom about childbearing. They provided role models for behavior, taught the new mother what she needed to know, and gave emotional and social support during the birth event.[7]

When midwifery became established as an occupation in traditional society, it is not surprising that it continued to be woman's business. Midwives generally came from the social class of the women they served and therefore shared the same cultural expectations for childbearing behavior. They provided counselling and personal support not only at the birth event, but also before and during labor, and through the post partum period. Among common people, as a rule, the midwife moved in with the family when labor was imminent, attended the birth, stayed afterward to nurse the mother and child, did light housekeeping until the mother was able, and gave advice on infant care and breastfeeding. Even today's highly professional midwife gives care that is characterized as "sympathetic," "personal," "counselling," "total."[8] A recent article on midwives in the *New York Times* was captioned, "Delivery by an Old Friend."

A professional man could hardly have served the same functions. As the physician's role developed, it ideally demanded emotional detachment and services limited to the specific condition being treated.[9] Consequently, as obstetrics became a specialty, labor and delivery were isolated from the rest of childbirth care and defined as medical and surgical events. Development of modern hospitals and transportation made it feasible for childbirth to be moved out of the home away from its previous family and social context, and placed in the physicians' domain where they had access to the increasingly complex technology of obstetrics. In the United States in 1968, 98.5% of all deliveries were performed in hospitals.[10]

Furthermore, childbirth is spontaneous and unscheduled. Any particular delivery has its own timetable for day, time, and length of labor.

This presented no problem for the midwife, who was prepared to settle down and wait. But physicians were able to perform efficiently only as the profession developed a technology that allowed them more control over time and place. A study by Rosengren and De Vault shows how anesthesia is used in the modern delivery room to hurry or slow deliveries. The authors add: "The use of forceps is also a means by which the tempo is maintained in the delivery room, and they are so often used that the procedure is regarded [by physicians] as normal."[11]

A third characteristic of childbirth is that most deliveries are uncomplicated.[12] Although percentages of complicated births can vary depending upon a number of social and cultural factors, approximately 95% of deliveries—even in modern society—require little or no technical assistance other than catching the baby and tying the umbilical cord, chores the mother herself may perform quite casually in less sophisticated cultures. (And as police manuals point out, they are chores any patrolman can handle.) The origins of names for childbirth attendant suggest that the role is mainly one of waiting and supplying support. "Midwife" is derived from the Old English "with woman;" "obstetrician" from the Latin "to stand by;" "accoucher" from the French "to put to bed." The midwife waited for childbirth to take its natural course, intervening only in cases of abnormality; the physician developed care that stressed intervention rather than waiting. Freidson has pointed out that it is "the task—indeed the mission—of the medical men to find illness," that the profession is "first of all prone to see illness and the need for treatment more than it is prone to see health and normality."[13] In the United States, the profession came to define all childbirths as not only potentially complicated, but also as in need of "aid"; as a condition that ideally should always be hospitalized, medicated, anesthetized, and require an episiotomy.[14]

Finally, childbirth is associated with a woman's sexuality. Pregnancy begins with coitus, it alters the balance of female hormones, and the infant may nurse at the mother's breast—an organ invested in Western culture with strong sexual meaning. Furthermore, a woman's genitals are exposed and touched during examination and delivery—acts that in any other social context would be invested with erotic meaning. Mead and Newton point out that sexually allied emotions may be aroused at the time of delivery in both the attendant and the parturient, and that although this is often recognized in less sophisticated societies, it is taboo in America.[15]

While the sexual connotation of childbirth posed no problems for women attending women, it presented a considerable barrier to men.

Tracts published against men entering midwifery often expressed outrage at the impropriety of having any man other than the woman's own husband present while she was giving birth. This attitude remained a prime obstacle to the immigrant woman's acceptance of the American doctor as late as the 1920s. The social organization of modern obstetrical care denies or hides the sexual implications of childbirth, and thus minimizes or eliminates role conflict for the male physician. The removal of the woman from a familiar environment; the emotional neutrality of the doctor-patient relationship; the rituals of aseptic technique and surgical preparation; the routine use of medication and forceps; and finally, the routine use of anesthesia, which removes the woman from participating in the birth act—all place a medical rather than a social or sexual connotation on childbirth. It is, perhaps, a measure of the "success" of the medical profession in de-sexualizing childbirth that the claims of some prepared childbirth advocates to sensual and orgasmic experiences during delivery can be viewed with amusement and disbelief, and easily dismissed.

Men in Midwifery—Conflict and Controversy

By the 16th and 17th centuries in Europe, the midwife held a recognized position in society and was sometimes well-educated and well-paid. She acquired her knowledge by formal apprenticeship to other midwives, was licensed to practice by the bishop, and successful midwives were considered little, if at all, inferior to doctors. Early textbooks were written by and for women midwives, and were used in the early universities.[16]

When men entered midwifery in the 17th century, they did so as university-trained physicians, and practiced among upper-class women who were accustomed to being attended in illness by men from their own class. The early professions (law, medicine, theology) had developed in the early university; the university in turn had developed within the powerful church, the seat of learning during the 12th and 13th centuries. The university was supported and attended by members of the aristocracy, and the professions became as strongly male-dominated as the patriarchal elite classes and the church from which they developed. By the 17th century, when men first entered midwifery, medicine was an established male profession in Europe, and the power structure of the time legitimized the physician's claim to expertise.[17]

But still they encountered obstinate barriers, for social institutions, laws, and customs all legitimized the woman midwife's control over delivering babies. For example, physicians were required to obtain special licenses to practice midwifery in London, and as late as 1762, when the first instruction in midwifery was given at the University of Edinburgh Medical College, it was confined solely to female midwives.[18] A midwife, Mrs. Elizabeth Nihill, gives the following account of the opposition physicians met when they attempted to practice midwifery in a public hospital in Paris. The men had learned their skills from midwives:

> and yet have many of those very men-practitioners, influenced by that self-interest which has such power in all human affairs, revolted against their mistresses in the art, and their benefactresses. They have, at various times, commenced law suits about the Hotel-Dieu at Paris, in order to get lyings-in there committed to them: but the administrators, the persons of a just sense of things, together with the parliament of the town, ever attentive to decency . . . have constantly opposed and frustrated the pretensions of these innovators. These again thus disappointed, were forced to content themselves with practising upon some women of quality under the favor and protection of some of the old ladies of the court of Louis XIV who had their reasons for the propagation of this fashion.[19]

Then, Nihill explains, physicians began to "run down" the midwives while "exalting themselves," and the "novelty prevailed," eventually spreading to the provinces.

Among common people, men entered midwifery through the authority of the guild. Sarah Stone, an English midwife, states that "almost every young man who has served his apprenticeship to a Barber-Surgeon immediately sets up for a man-midwife; although as ignorant, and indeed much ignoranter, than the meanest woman of the Profession."[20]

Although the authority of the professions and the guilds made it possible for men to gain a foothold in midwifery, it was the obstetrical forceps, developed early in the 18th century, which encouraged them to enter the occupation in large numbers and challenge the midwife's dominance. Using forceps, a birth attendant could extract the infant during labor, gain control over the timing of delivery, and thus operate more efficiently.[21]

The ascendancy of men-midwives precipitated a controversy that raged in popular and medical literature for the better part of two centuries. It persisted until medicine had firmly established its monopoly over healing and its control over obstetrics.[22] The use of instruments—

principally obstetrical forceps—was a pivotal issue around which claims of superior competency and charges of abuses turned. The "natural right" of women to attend women in childbirth and the "immorality" of having a man attend delivery also figured prominently in the debates. And finally, the profit motive—a persistent theme in any story of occupational displacement—pervaded the controversy.

At first there were no clearly drawn battle lines between midwives and male practitioners. Midwives defended men-midwives when they were competent,[23] and there were male physicians who saw women as more capable in the occupation than men.[24] Forceps and other instruments were used by trained midwives as well as men practitioners. The difference was that midwives still waited for a delivery to take its normal course and used instruments only in extreme cases. The men were accused both by midwives and by other physicians of using them indiscriminately to hasten the delivery and make a profit, damaging mothers and infants in the process.[25] But other physicians defended the use of forceps as an advance, referring to them as "noble and beneficient" instruments that had "rescued many lives."[26] Early obstetricians were presented as brave men who were pioneering in improvements while they withstood the attacks of the midwife who saw her livelihood slipping away from her.

Pamphlets debating the morality of the issues bore such colorful titles as the following (both published mid-19th century):

> *Medical Morals. Designed to show the Pernicious Social and Moral Influence of the Present System of Medical Practice, and The Importance of Establishing Female Medical Colleges.*[27]

> *Man Midwifery Exposed; or The Danger and Immorality of Employing Men in Midwifery Proved; and the Remedy for the Evil Found.*[28]

And finally, the vaginal examination practiced by physicians came under attack both by medical men who spoke of the danger of introducing infection and by the pamphleteers for whom the sexual implications were frequently just as worrisome. For example:

> Where then is the excuse for the indecencies and outrages of man-midwifery? Why is it that, without plea of necessity our wives are exposed to the shame and pollution of examinations which are *invariable*, and manipulations such as a pure-minded and sensitive woman must blush to think of—such as must excite

the indignation of every man who regards the person of his wife as sacred?[29]

However, by early 20th century, as medical knowledge became more advanced, the controversy had shifted to other issues: to responsibility for the high maternal and infant death rates, and the problem of inadequate training both for midwives and for physicians.

The Dominance of the Medical Profession over Childbirth

In the case of medicine a significant monopoly could not occur until a secure and practical technology of work developed.[30]

At the beginning of the 19th century, the claim of the medical profession to practice midwifery was considered legitimate regardless of the controversy surrounding the physician's management of delivery, but the midwife was still an independent practitioner. By the end of the century, medicine's preeminence among the healing occupations had been established, and its dominant position over the occupation of delivering babies was assured. While in England and the rest of Europe this dominance meant that the medical profession controlled most decisions about the licensing of midwives, and about which medical and surgical procedures licensed midwives could legitimately perform, in the United States it meant a slowly developing monopoly of the occupation.

Generally, in Europe it was assumed that childbirth would be normal and therefore the responsibility of trained midwives, unless there were indications that a delivery might be complicated. The midwife's role specifically included the obligation to recognize indications of abnormality, and turn such patients over to the physician. In actual practice she was trained to handle emergencies, and it was implicitly recognized that she would do so "on her own responsibility" if a doctor's assistance were not available on time. In the United States, on the other hand, all deliveries came to be seen as potentially complicated and treated as such. They were therefore the province of the physician. As the 20th century progressed, the tendency was to treat all deliveries as surgical events: i.e., ideally they were performed in a surgical operating room of a hospital, with the "patient" surgically prepared (enema, surgical shave, no food), and with surgical interventions (medication, anesthesia, routine episiotomy).

A number of factors contributed to medicine's domination of childbirth management. One was the increasing sophistication of the medical and surgical knowledge applied to the complications of childbirth, much of which was progressively restricted by law from use by midwives. Anesthesia, medications, laboratory techniques for predicting complications, all were made more available to physicians than to midwives. A second factor was the dramatic increase in large public hospitals, a development that reflected the widespread social change at that time. The Industrial Revolution brought increasing urbanization, with its concomitant poverty and crowded ghettos of the urban slums. Women as well as men worked long hours under brutal conditions and were poorly nourished. Sewage systems and sanitation were inadequate, and in the summer poor refrigeration made the contamination of food likely. Under these conditions, disease rates for the entire population were high, and maternal and infant death rates soared. The situation stimulated widespread concern over public health and resulted in the building of hospitals with both public and private funds. In the United States, in the fifty years between 1873 and 1923, the number of hospitals increased from 149 to 6,762.[31] It was in these hospitals that medicine developed its expertise in obstetrics.

Finally, during the 19th century, the profession of medicine organized powerful medical associations that grew to exert strong influence over directions of health care. These associations gave the

Figure XIX. Women's Laboratory at M.I.T.
(M.I.T. Historical Collections)

profession the power to dominate obstetrics as a medical specialty and to define its scope, and in the United States, the leverage to monopolize it.

Occupational Displacement in the United States

Although it is likely that in this country midwives still delivered over 50 per cent of the babies early in the 20th century, they were an unorganized, unrepresented group of women who had little social prestige or power.[32] By and large midwives served lower-class immigrant women of their own ethnic groups whose cultural expectations led them to seek midwives; indigenous "granny" midwives served women in out-of-the-way places such as the Kentucky hills and the deep south. These patients would have been unprofitable for a private physician. While most state laws were requiring medical men to be licensed, and, at least by the second decade, medical associations were setting more rigid standards for training and licensing, there was little uniformity and control over qualifications to practice midwifery. Only six states required a midwife to be trained; only twelve plus Washington, D.C., required her to pass an examination. In seven, the regulations were inadequate, and in fourteen, there were no laws at all governing her training, registration and practice. Since she had no medical degree, laws in most states prohibited her from giving medication or performing even minor surgery. Furthermore, while medical school standards were growing more exacting, few quality schools for midwives had survived from the past century.[33]

By comparison, in Great Britain and elsewhere in Europe, midwifery was a profession that was still recognized and respected. In most cases, midwives were required by law to be trained and licensed, their practice was regulated, and as professionals, they were allowed to collect professional fees for services. Their services were provided to the poor at public expense. Excellent schools with sound medical training were available to them.

England's situation had been developing similarly to that in the United States, until the Midwives' Act was passed by parliament in 1902.[34] The act provided for the establishment of a Central Midwives' Board, required all midwives to be trained, and empowered the Board to regulate their training and supervision.

The question remains why the situation developed one way on one side of the Atlantic and another way in the United States. One factor may have been a difference in cultural history. Midwifery had at one time been the occupation of women of education and prestige in England, trained and licensed by the state, so that the profession had both tradition and official sanction to give it legitimacy. The reverse was the case in the U.S.

A second factor may have been a difference in cultural attitudes about women. Cynthia Epstein has pointed out that the Victorian "idealization" of women that depicted them as frail, in need of protection, and not very capable (middle-class, not working-class, women), had a greater hold in this country than elsewhere, and was one of the factors that prevented women from entering professional life.[35] This attitude pervades the literature of the midwife controversy throughout the 19th century, and was used as an argument for discouraging women from studying medicine as well as from practicing midwifery.[36] The following example from a pamphlet published in 1820 by a Boston physician is typical:

> I do not intend to imply any intellectual inferiority or incompetency in the sex. My objections are founded rather on the nature of their moral qualities, than on the power of their minds, and upon those very qualities which render them, in their appropriate sphere, the pride, the ornament, and the blessing of mankind . . . .
> [W]omen are distinguished for their passive fortitude . . . they have not that power of action, or active power of the mind which is essential to the practice of surgery. . . . [I]t is obvious that we cannot instruct women as we do men in the science of medicine.[37]

A third important factor is that physicians were in greater supply here than they were in Europe throughout most of the 19th century, and in greater supply in the years that were crucial to the midwife's survival—that is, until 1920—than they have been in recent years. This would have given economic competition more weight here than in Europe. The medical historian Shyrock notes:

> the ratio of doctors to population was far higher here than in Europe [in the period from 1860–1909], and competition was already keen when women began to enter it [medicine] . . . . [T]here was no little suspicion that doctors opposed women for fear of economic competition, the usual motive for suppressing a minority element. Occasionally this matter came out in the open, as when an article in the Boston Medical Journal noted as early as 1853 that competition was becoming serious: women were already cutting in on profits in obstetrical cases. In the same

city, Dr. Gregory, who suffered no inhibitions, accused the doctors of desiring a male monopoly for the market.[38]

In 1898, Henry Jacques Garrigues, an obstetrical surgeon to the Maternity Hospital in New York, wrote:

> Although an evil, midwives are, however, in most countries a necessity in view of the fact that physicians would be unable to find the time to do the work. This is not so here, where there is a superabundance of medical men.[39]

However, in 1921, when physicians were in much shorter supply, another physician describing a new program for inspecting and supervising midwives in Philadelphia made this interesting apology:

> It seems almost incredible that anyone, not a confirmed idealist, cognizant of conditions as they exist, will fail to realize without further argument that the most vital and compelling argument for considering the midwife a necessity today is that without her aid, it would be impossible to care for the cases of childbirth in any large city or other locality in which a large proportion of the population is composed of poor people. . . .
>
> We . . . resent the charge that we hold a brief for the midwife. On the contrary, we assert that she is an anachronism and a menace under the conditions at present operative, and that as soon as any practicable (emphasize practicable) scheme is evolved, that we will be the first to applaud her retirement to the oblivion which is her due.[40]

If indeed competition played a role in the midwife's disappearance, it would be logical to assume that powerful professional associations gave the physician an advantage in the eventual outcome. The role they played in blocking women from practicing medicine is described in Mary Roth Walsh's article. Just one example is added here:

> In 1859 the Medical Society of the County of Philadelphia passed "resolutions of excommunication" against every physician who should "teach in a medical school for women" and every one who should "consult with a woman physician or with a man teaching a woman medical student." In Massachusetts, after qualified women physicians were given state certificates to practice, the Massachusetts Medical Society forbade them membership, thus refusing to admit the legality of diplomas already sanctioned by the highest authority.[41]

The previously quoted Garrigues, a member of the prestigious New York Academy of Medicine, gives an example of the same sort of professional pressure being brought to bear on the activities of midwives. He states that in 1884 a bill was introduced into the state legislature to grant a charter to a certain college of midwives in the City:

> Dr. Albert Warden and I were sent to Albany as delegates to confer with persons of influence on the matter, and the bill was eventually killed.[42]

In 1898, he had himself offered a resolution at the Academy of Medicine, "To take immediate steps to pass a law confining the practice of midwifery to qualified medical practitioners. The resolution was passed January 1898."[43]

Childbirth Care Today

Today, in the United States, there is renewed interest in training midwives, and graduate programs in midwifery are appearing in the more prestigious schools of nursing.[44] This trend coincides with a health care delivery crisis that is perceived and defined as a shortage of medical personnel,[45] and has given rise to the development of new programs to train physician's assistants, primary care nurses, and other paraprofessionals to lighten the physician's load. However, since the United States has more physicians per capita than many other developed countries with better health statistics, it seems likely that the "shortage" is due to distribution of physicians, an increased patient demand for services, as well as a continued reluctance of physicians to serve less affluent members of the population, and not to a lack of numbers.

Interestingly, women being trained to deliver babies are called *nurse*-midwives today, a clue to their status in the United States as paraprofessionals rather than independent practitioners. It appears likely that by and large the medical profession will continue to define and limit the scope of the nurse-midwife's field, and that she will perform under implicit if not explicit supervision of physicians.

Today, childbirth care is changing in several different directions. On the one hand, a slowly developing Birth Reform Movement, which has its roots in the work of Grantly Dick-Read over a generation ago, is bringing a safer, more personal, woman-and-infant oriented delivery to increasing numbers of women. On the other hand, concerned women and men, both lay and professional, have been working through such organizations as the International Childbirth Education Association, and the American Society for psychoprophylaxis in Obstetrics, to break the medical profession's definition of childbirth as pathological, and present it once again as a normal human event.[46] They have pioneered in educating women for unanesthetized (prepared) childbirth, and are bringing about some changes in the organization of hospital care. However, these organizations essentially work within the system to reform it. They do not address the issue of present-day medical and male dominance of childbirth, which is perhaps more subtle than it was previously. Women are encouraged to participate in their own deliveries, but obstetricians still retain final authority to decide on intervention. And another male, the husband, is added, although it is rare indeed for a woman to be permitted another woman of her own choosing to be with her in labor. While it is important for a father to regain access to the birth of his own children and to be able to support the mother in labor, prepared childbirth methods frequently give him an authoritarian role he did not have before. In the Lamaze method, for example, he is taught to monitor the labor and instruct the woman in breathing. In the method taught by the American Academy of Husband-Coached Childbirth, he is given even more responsibility. Often he is cast in partnership with hospital personnel. In a popular film on Lamaze instruction, the father is shown conferring with the physician to decide whether to "permit" labor to continue or to use intervention to hasten it.

Leboyer's book *Birth Without Violence,* which has found wide appeal in its advocacy for gentler handling of babies after childbirth, provides another example of physician control. In describing his method for "non-violent" childbirth, the author ignores violence done to women by modern technological birthing. However, he describes in colorful language the imagined agony of the baby during labor. Not only the uterus but the woman herself is cast in the role of the one doing the violence—"the enemy" who "crushes and twists [the baby] in a refinement of cruelty"—a "monster" who "stands between the child and life."[47] After delivery, Leboyer disregards the mother and the mother-baby bond as he takes center stage, bathes the baby in warm water and caresses it with his gentle hands. It may be observed that his method seems "new" only in places where modern obstetrics has moved delivery into brightly lit, noisy, surgical operating rooms. In many other places (as in the past) babies are given warm baths, are gently handled by midwives and others, and are caressed by their mothers.

In discussing these attempts to reform hospital care, Anne Seiden has remarked that currently there is a "profound ambivalence about 'letting' the laboring woman control the situation, using companions of her choice including medical personnel as sources of support, consultation and technical expertise on an informed consent basis," and that there is "the potential indeed for a new and more sophisticated kind of infantilization of the pregnant or newly delivered woman."[48]

In the 1970s more radical movements have been seeking alternatives to the system, rather than reform. The Women's Health Movement is raising women's consciousness to the sexist orientation and poor quality of obstetric/gynecological care offered them. It is encouraging women to recognize the damage done them and their children by irresponsible hospital procedures; to learn about the functioning of their own bodies; and to assume control of their own health care, including childbirth.[49] In some areas, lay midwives and a few physicians who have rejected traditional hospital practices are assisting women in home deliveries.[50] However, these changes, as well as the more conservative reforms mentioned above, are affecting a minority of childbearing women, principally a self-selected population from the educated middle class.

Meanwhile, the majority of hospital deliveries by obstetrical specialists is moving toward *more* rather than less intervention. More sophisticated medication, anesthetics, and technology are used routinely, not only for the 5 per cent of deliveries that are complicated and in which intervention may save lives, but also for the 95 per cent in which it would be unnecessary, if sound prenatal and birthing care were available.[51] The number of deliveries by Caesarian section is increasing, running as high as 30 or 40 per cent in some areas. The risk to both mother and infant when intervention is used in otherwise normal deliveries has been well documented.[52] The higher rates of birth injuries, cerebral palsy, minimal brain damage, and infant death occurring in the United States than in countries where intervention is not the rule, leave little doubt that sophisticated technological birthing takes a grim toll. Unfortunately, this is still standard childbirth care for the great majority of American women, and certainly for the poor.

It remains to be seen what the Birth Reform Movement will mean for these women in the long run, and what part the "new" nurse-midwife may play in the story. Those who are working for change testify that it comes slowly and against great resistance from the medical profession.

Notes

An earlier version of this essay has appeared in *Women and Health,* vol. 1, no. 6 (Nov./Dec./, 1976). Available from SUNY, College at Old Westbury, Westbury, L.I., NY 11568.

1. International Federation of Gynaecology and Obstetrics and the International Confederation of Midwives, *Maternity Care in the World: International Survey of Midwifery Practice and Training,* Report of a joint study group (Oxford: Pergamon, 1966), p. 13.
2. U.S. Department of Health, Education and Welfare, *Vital Statistics of the United States, Volume I—Natality* (Washington, D.C.: Gov. Printing Office, 1968, Table I-S, pp. 1–20.
3. International Federation of Gynaecology and Obstetrics, p. 3.
4. For a more detailed discussion *see* Mary Roth Walsh's essay in this collection, as well as her recent book, *Doctors Wanted: No Women Need Apply* (New Haven: Yale University Press, 1977).
5. European physicians, of course, had encountered the same problem establishing a claim superior to that of lay healers. But they could draw on the prestige of the Church, University and social class. For a discussion, *see* Barbara Ehrenreich and Deirdre English, *Witches, Midwives and Nurses: A History of Women Healers* (Old Westbury, NY: The Feminist Press, 1971).
6. Eliot Freidson, *Profession of Medicine* (New York: Dodd, Mead, 1970), p. 17.
7. Niles Newton and Margaret Mead, "Cultural Patterns of Perinatal Behavior," *Childbearing: Its Social and Psychological Aspects,* Stephen A. Richardson and Alan Guttmacher, eds. (Baltimore: Williams and Wilkins Co., 1967), pp. 193–194.
8. *See Education for Nurse Midwifery* (New York: The Maternity Center Association, 1967).
9. For a discussion of the physician's role *see* Talcott Parsons, *The Social System* (New York: Free Press, 1951), pp. 428–97.
10. U.S. Department of Health, Education and Welfare, *Vital Statistics,* pp. 1–20. Moving childbirth to the hospital has *not* meant safer deliveries for women and babies. By comparison, in the Netherlands in the late 1960s, over half of all births occurred at home with the assistance of a midwife and a maternity aide [Doris Haire, *The Cultural Warping of Childbirth* (Milwaukee: The International Childbirth Education Association, 1972), p. 15.] Yet in that country, the infant mortality rate for 1970 was 12.7 compared to 19.8 in the United States, which had higher infant death rates than fourteen other developed countries (United Nations Statistical Office).
11. William R. Rosengren and Spencer De Vault, "The Sociology of Time and Space in an Obstetrical Hospital," *The Hospital in Modern*

Society, Eliot Freidson, ed. (New York: Free Press, 1963), pp. 266–92.

12. Doris Haire, p. 12. For a recent comparative study showing lower rates of complications in home births compared to obstetrical hospital managed deliveries *see* Lewis E. Mehl, M.D., "Options in Maternity Care," *Women and Health,* September/October 1977, pp. 29–43. For a discussion of increased pathology in childbirth as a consequence of technological intervention in obstetrical management, *see* Frederick M. Ettner, M.D., "Hospital Technology Breeds Pathology," pp. 17–23, in the same issue of *Women and Health.*

13. Eliot Freidson, *Professional Dominance* (New York: Atherton, 1970), p. 277.

14. Episiotomy is the surgical cutting and suturing of the birth canal, ostensibly to prevent its "tearing" during delivery. For a discussion of the widespread prevalence of this and other surgical procedures in United States obstetrics, see Rosengren and De Vault cited above. For a discussion of the greater risk to mother and baby of surgical intervention in normal childbirth, *see* Haire; Suzanne Arms, *Immaculate Deception* (New York: Houghton Mifflin, 1975); and Nancy Stoller Shaw, *Forced Labor: Maternity Care in the United States* (New York: Pergamon Press, 1974).

15. Newton and Mead, *Childbearing,* pp. 171–72.

16. A. M. Carr-Saunders and P. A. Wilson, *The Professions* (Oxford: Oxford University Press, 1933), pp. 121–25.

17. Ibid. *See also* Ehrenreich and English's discussion of the Church's persecution of women healers, and its protection of physicians.

18. Herbert K. Thoms, *Chapters in American Obstetrics* (Springfield, Ill.: Charles C. Thomas Press), reprinted from *Yale Journal of Biology and Medicine* (1933), pp. 11–13.

19. Elizabeth Nihill, Professed Midwife, *A Treatise on the Art of Midwifery* (London: A. Morley, 1760), pp. 5–6.

20. Sarah Stone, *A Complete Practice of Midwifery* (London: T. Cooper, 1737), p. 16. *See also* the discussion of men-midwives and the barber surgeon guilds in Carr-Saunders and Wilson, p. 122.

21. Thoms, *Chapters,* p. 13.

22. *Ibid.,* p. 13.

23. *See* for example Nihill, *Treatise,* p. 9, and Stone, *Complete Practice,* p. 12.

24. For example, a London physician named John Stevens published a pamphlet in 1830 with the title *An Important Address to Wives and Mothers on the Dangers and Immorality of Man-Midwifery.* In it, he says: "We see in the works on midwifery, especially those which profess to trace the art to its origin, that surprise is expressed how *few* accidents would happen in former times when none but females assisted during parturition. The City of London Hospital

affords a most powerful proof of what female intelligence can perform" (pp. 29–30).

25. Thoms, *Chapters*, p 21. *See also* Samuel Bard, M.D., *A Compendium on the Theory and Practice of Midwifery* (1807). Dr. Bard was the president of the College of Physicians and Surgeons in the University of the State of New York, and his text, which went into several editions, was the first text on obstetrics published here.

26. For example *see* William Felix Mengert, M.D., "The Origin of the Male Midwife," *The Annals of Medical History* 4, 5 (1932): pp. 453–65. This article is a reprint of a work originally published in mid-19th century.

27. George Gregory. Published in New York, 1853.

28. John Stevens, M.D. Published in London, 1830.

29. Pamphlet, 1851.

30. Freidson, *Dominance*, p. 21.

31. George Rosen, "The Hospital: Historical Sociology of a Community Institution," *The Hospital in Modern Society*, Eliot Friedson, ed. (New York: Free Press, 1963), p. 25.

32. United States Bureau of the Census, *Statistical Abstract of the United States* (Washington DC, 1970), p. 45. *See also* Carolyn Conant van Blarcomb, "Midwives in America," *American Journal of Public Health* 8, 4 (1914): 197–207.

33. *Ibid.*, p. 198.

34. For a discussion of the issues, and a report of Parliamentary hearings leading to the passage of the act, *see* British Parliamentary Papers, *Report of the Select Committee on Midwives Registration Together with the Proceedings of the Committee*, 17 June 1892. Ordered by the House of Commons General Health Session 1890–1894.

35. Cynthia Epstein, *Woman's Place* (Berkeley: University of California Press, 1971), p. 41.

36. *See* Mary Roth Walsh's paper in this collection, as well as her book, *Doctors Wanted: No Women Need Apply*.

37. A Physician, *Remarks on the Employment of Females as Practitioners in Midwifery* (Boston, 1820).

38. Richard Harrison Shyrock, *Medicine in America, Historical Essays* (Baltimore: Johns Hopkins Press, 1966), p. 187.

39. Henry Jacques Garrigues, "Midwives," *Medical News* (Philadelphia), 19 February 1898, p. 98.

40. A. B. Nicholson, M.D., "The Midwife, an Anachronism of the Twentieth Century," Lecture delivered in Philadelphia, 1921, pp. 10–11 (Reprint in New York Academy of Medicine Library).

41. Anna Garlin Spencer, "Woman's Share in Social Culture," *Feminism: The Essential Historical Writings*, Miriam Schneir, ed. (New York: Vintage, 1972), pp. 269–85. Quote from p. 282.

42. Garrigues, *Midwives*, p. 11.

43. *Ibid.*, p. 12.
44. John Kosa, "Women and Medicine in a Changing World," *The Professional Woman*, Athena Theodore, ed. (Cambridge: Schenkman, 1971), pp. 709–20. *See also, Education for Nurse Midwifery.*
45. Patricia L. Kendall and George G. Reader, "Contributions of Sociology to Medicine" *Handbook of Medical Sociology*, 2nd ed., Howard E. Freeman, Sol Levine, and Leo G. Reeder, eds. (Englewood Cliffs: Prentice-Hall, 1972), p. 9.
46. Anne M. Seiden, "The Birth Reform Movement: Strengths and Some Limitations" (Paper delivered at the Third Annual Conference on Psychosomatic Obstetrics and Gynecology, held at Temple University, Philadelphia, Pa., February 2, 1975).
47. Frederick Leboyer, *Birth Without Violence* (New York: Knopf, 1975).
48. Anne M. Seiden, *Birth Reform*, p. 5.
49. *Proceedings of the First Childbirth Conference* (Stamford, Ct.: New Moon Communications, 1973).
50. Raven Lang, ed., *The Birth Book* (Ben Lamond, CA: Genesis Press, 1972); and Suzanne Arms, *Immaculate Deception.*
51. Anne M. Seiden, "The Sense of Mastery in the Childbirth Experience," *Primary Care* 3, 4 (1976), 717–25.
52. Doris Haire, *Cultural Warping*, p. 12.

[Datha Clapper Brack *is Assistant Professor of Sociology at Bergen Community College, Bergen, New Jersey, where she teaches courses on Sociology of Sex Roles and on Society and Women's Health. She received her Bachelor and Master degrees in Sociology from Columbia University, and is currently completing her dissertation in Medical Sociology at C.U.N.Y. While specializing in women's health issues, she is presently conducting research on social forces influencing women to choose breastfeeding and to breastfeed successfully. Some of her preliminary research on this topic has been published in* Nursing Outlook. *She is a member of Bergen Community College's Affirmative Action Committee, and the national and local chapters of Sociologists for Women in Society. She writes: "Bearing and rearing six children in suburbia probably did more to raise my feminist consciousness and push me to study women's health issues than any other single experience."*]

Beverly Smith

Black Women's Health: Notes for a Course

The health of Black women is a subject of major importance for those of us who are committed to learning, teaching, and writing about our sisters. By exploring this topic, we discover some of the ways in which living in this oppressive society breaks us physically and mentally and also how we have struggled and survived.

The ideas and resources in this syllabus have been developed as part of my involvement in the Feminist Movement and the Women's Health Movement. My participation has included leading workshops on Black and Third World Women's health issues at a number of conferences and working in the Combahee River Collective—a Black feminist organization, and in the Boston Committee to End Sterilization Abuse. My experiences teaching courses on Women and Health and on Black Americans and the Health Care System have also contributed to the syllabus. The inclusion of imaginative literature results from the success and pleasure I have had in using such materials in my courses. I have found that Black women writers, in their efforts to tell the truth about our lives, necessarily deal with health. Their works are extraordinarily valuable, much more so than most articles in medical journals, because they are vivid, whole, and accessible to all types of students.

A course similar to the one designed here would be valuable for a variety of students. Those who are preparing to become health workers or who are already working in health care might find it especially useful. I think the students for whom it would be most meaningful would be Black and other Third World women. I have found that one of the most exciting aspects of teaching courses on Women's Health has been experiencing how we, as women, can interact with and augment the material being studied. Black women would be able to bring the richness of their lives to such a course. The course focuses almost entirely on Black women because this is the subject with which I am most familiar. I have found that other Third World women also see such material as very relevant to their own experiences and needs. I think the addition of other resources on Third World women in the United States and abroad would greatly increase this course's appeal and usefulness. White women also seem to enjoy and benefit from discussion of these issues.

Figure XX.
(Photo: G. Dunkel)

The course might be offered as an alternative to the more standard (and usually much more white) women's health courses given by Women's Studies Programs or by other college departments. It should also fit comfortably in Black Studies Departments. Another possibility would be to take such a course out of the usual academic setting and into the various communities to which it relates. Shelters for battered women, drug and alcoholism treatment programs that serve Third World women, women's centers, and community centers in Black neighborhoods are all potential locations. There are many exciting possibilities. I can envision a version of this course being offered at a neighborhood health center with both workers and consumers participating.

Readings are often listed under more than one heading, for several reasons. One is to provide ideas about the different ways in which they may be used and combined. Another is that the distinctions among topics are not rigid. There is a good deal of overlap and, in approaching these materials, one should be open to the connections that can be made among them. This interrelationship may be indicative of the holistic way in which feminists view health and all of life. For example, the strict distinction made between mind and body in Western, male-dominated medicine is a distortion of reality. "It Happened on My Birthday," written by a young Black woman who was sterilized, though it is listed under "Reproductive Health," could just as appropriately be included in the resources on mental health. Similarly, in discussing violence against Black women, the devastating emotional effects must be considered along with the physical.

The greatest shortcoming in the following materials is the frequent absence of awareness that Black women's health is affected by sexism, racism, and class position. In many cases, instead of insights into this complexity we have the bare bones of acts or the frustrating hints and guesses of a partial analysis—for example, one that comprehends racism but not sexism, heterosexism, or class oppression. There is not a single article which provides a comprehensive analysis of all these factors as they relate to health. I see the following, though it is in the form of a syllabus, as an initial attempt at gathering the concepts that would be necessary for such an analysis.

The Politics of Black Women's Health

The poor health of Black women can be documented in a number of ways. We die sooner than white women; our death rates from pregnancy and childbirth are several times higher; and our death rates from a number of diseases are also higher. These facts clearly indicate that our political position as Black women affects our health. The effects of racism, sexism, and class oppression make us less healthy and also deprive us of decent health care. Several of the articles describe how, as Black women, we are often singled out for especially brutal "treatment"—for example, forced sterilization and experimentation—while others deal with the unavailability of care of any kind. A careful consideration of the issues raised by these readings is essential background for the rest of the course.

Alexander, Daryl, "A Montgomery Tragedy: The Relf Family Refused to Be the Nameless Victims of Involuntary Sterilization," *Essence* (September 1973), pp. 42 *et seq.*

"Crimes in the Clinic: A Report on Boston City Hospital," *The Second Wave* 2, no. 3 (1973), pp. 17-20. This short article provides many examples of the abuse of Black, Latina, and white women in a large public hospital. It is one of the best articles available on the subject. *The Second Wave*'s address is 20 Sacramento Street, Cambridge, MA 02138.

Poor Black Women, including "Birth Control Pills and Black Children," a statement by the Black Unity Party (Peekskill, N.Y); "The Sisters Reply" by Patricia Haden *et al.*; and "Poor Black Women" by Patricia Robinson (Boston: New England Free Press, n.d.). "The Sisters Reply" is also reprinted as "Statement on Birth Control" by the Black Women's Liberation Group, Mount Vernon, NY, in *Sisterhood Is Powerful: Anthology of Writings from the Women's Liberation Movement,* Robin Morgan, ed.

This pamphlet may be ordered from the New England Free Press, 60 Union Square, Somerville, MA 02143 for 15¢ per copy. Orders of five or more are given a 12½ percent (1/8) discount. Orders for which payment is made in advance do not have to be accompanied by postage; however, the payment of postage is greatly appreciated.

Reedy, Juanita, "Diary of a Prison Birth," *Majority Report* (May 31, 1975), pp. 1, 3.

Shakur, Assata (Joanne Chesimard), "Birth Journal" (unpublished ms., typewritten, n.d.).

These two grimly similar articles detail the tortures endured by Black women who gave birth in prison. Their descriptions of their resistance to what was happening to them are inspiring. Unfortunately, the article by Assata Shakur is very difficult to obtain. I do not think it was every published, although it was sent to *Ms. Magazine*, which, not surprisingly, rejected it. Copies of the manuscript can be obtained by sending a self-addressed envelope to Ruth Hubbard, Biological Laboratories, Harvard University, Cambridge, MA 02138.

"Sterilization: Relevance for Black Women," Black Women's Community
Development Foundation, *Mental and Physical Health Problems of
Black Women* (Washington, D.C., 1975), pp. 19-26.

This article is from the only book I know that focuses exclusively on
the health of Black women. It is a report on a conference sponsored by
the Foundation in March 1974. Topics covered are hypertension,
cancers and fibroids, sterilization, and suicide and depression.
Obviously, it is far from comprehensive, but its most serious flaw is
that it is neither feminist nor socialist in its analysis, thus, even though
it is antiracist, it is unsatisfactory. Nevertheless, it is an extremely
valuable resource, and many of its articles are included in this syllabus.
It may be ordered from the Foundation, 1028 Connecticut Avenue,
N.W. (Suite 1010), Washington, DC 20036. The price of $9.95 includes
postage.

Walker, Alice, "Strong Horse Tea," *In Love and Trouble: Stories of Black
Women* (New York: Harcourt Brace Jovanovich, 1973), pp. 88-98.

A poor, Black woman wants a "real doctor" for her gravely ill baby.
This is a devastating story about how racism and poverty maim and
destroy human life.

Wright, Sara E., *This Child's Gonna Live* (New York: Dell, 1969). Set on the
Maryland eastern shore, this beautifully poetic novel tells the harrow-
ing and haunting story of Mariah Upshur and her family, whose fundamental struggles include poverty, sickness, and lack of health care.

Reproductive Health

Black women, like all women, have always attempted to control reproduction. These attempts have been made against terrible odds—forced breeding
during slavery; the selling away of the children of slaves; experimentation on
Black women, which began during slavery and continues today; forced
sterilization; the unavailability of safe, legal abortions to most poor women
and the continued threat to all women of losing this right. An aspect of this
struggle for control has been the development of traditional healing. Excellent resource people for this topic are rural Black midwives. Articles on
prenatal and maternal health that provide a more clinical treatment might
also be assigned.

Bradley, Valerie Jo, "It Happened on My Birthday," *Mental and Physical
Health Problems of Black Women* (Washington, DC, 1975),
pp. 103-110.

The Chicago Maternity Center Story, Kartemquin/Haymarket Films, P. O. Box 1665, Evanston, IL, (312) 869-0602. Purchase: $400; rental: $75 to hospitals and universities, $60 to high schools and churches, $40 to community groups. 60 minutes, black and white. This film documents the fight of Black, Latina, and white women to save the Center's home birth service.

Lerner, Gerda, ed., *Black Women in White America: A Documentary History:* "A Mother Is Sold Away from Her Children," pp. 10-12; and "A Woman's Fate," pp. 45-53 (New York: Vintage Books, 1972).

Poor Black Women (Boston: New England Free Press, n.d.).

Reedy, Juanita, "Diary of a Prison Birth," *Majority Report* (May 31, 1975), pp. 1, 3.

Sanders, Marion K., "The Right Not to Be Born," *Harper's Magazine* (April 1970), pp. 92-99.

The first part of this article describes the experience of a Black woman who was denied an abortion after being exposed to German measles and who consequently gave birth to a severely retarded daughter. It is a very moving and enraging article.

Shakur, Assata (Joanne Chesimard), "Birth Journal" (unpublished ms., typewritten, n.d.).

"Sterilization: Relevance for Black Women," Black Women's Community Development Foundation, *Mental and Physical Health Problems of Black Women* (Washington, DC, 1975), pp. 19-26.

Swartz, Donald P., "The Harlem Hospital Center Experience, *The Abortion Experience: Psychological and Medical Impact*, Howard J. Osofsky and Joy D. Osofsky, eds. (Hagerstown, MD: Harper and Row, 1973), pp. 94-121.

Diseases of Black Women

While there are probably no diseases that are exclusive to Black women, there are some conditions which we are more likely to develop than other groups. These diseases include high blood pressure, uterine fibroids, and lupus. It is essential to consider the implications of such conditions for contraception, pregnancy, and childbearing. It is important, too, to note the different impact of diseases on Black women. For example, while fewer Black women get breast cancer, more die from it. It is probable that this higher death rate results from later detection and the lower quality of care we receive. Additional clinical information can be found by checking recent medical journals.

Lang, Frances, "The Sickle Cell and the Pill" (unpublished ms., typewritten, n.d.).

Copies are available from Women's Community Health Center, Inc., 639 Massachusetts Avenue, Cambridge, MA 02139. A donation of 50¢ plus the cost of postage should be sent with requests for the article.

Loebl, Suzanne, "'SLE': Another Black Disease," *Essence* (September 1973), pp. 50-51.

This article describes a support group for women who have systemic lupus erythematosus, a chronic disease which has a higher incidence among Black women than among any other group.

"Cancers and Fibroids: Relevance for Black Women," Black Women's Community Development Foundation, *Mental and Physical Health Problems of Black Women* (Washington, DC, 1975), pp. 11-17.

"Hypertension: Relevance for Black Women," Black Women's Community Development Foundation, *Mental and Physical Health Problems of Black Women* (Washington, DC, 1975), pp. 1-9.

Mental Health

The minds of Black women have been battered along with our bodies. One of the great, mostly untold, stories about Black women and men is how the centuries of oppression we have endured in this country have damaged our psyches. One theme that recurs in several of the following readings is the extreme narrowness of Black women's lives. Our possibilities are constricted by the system's severely limited notions of what we should be. We are officially defined by others and not by ourselves. Yet self-definition is the source of our strength.

Cox, Ida, "Wild Women Blues," *The World Split Open: Four Centuries of Women Poets in England and America, 1552-1950*, Louise Bernikow, ed. (New York: Vintage Books, 1974), pp. 278-279.

Hughes, Langston, "The Gun," *Something in Common and Other Stories* (New York: Hill and Wang, 1963), pp. 154-161.

This story describes some of the tragic ways in which racism and sexism intersect in the life of an isolated domestic worker.

Morrison, Toni, *The Bluest Eye* (New York: Holt, Rinehart and Winston, 1970).

A major focus of this superlative novel is the destructive effect of white beauty standards on the psyches of Black women.

Slater, Jack, "Suicide: A Growing Menace to Black Women," *Ebony* (September 1973), pp. 152 *et seq.*

"Suicide and Depression: Relevance for Black Women," Black Women's Community Development Foundation, *Mental and Physical Health Problems of Black Women* (Washington, DC, 1975), pp. 27-35.

Toupin, Elizabeth Ann, and Zella, Luria, "Some Cultural Differences in Response to Co-ed Housing: A Case Report," Black Women's Community Development Foundation, *Mental and Physical Health Problems of Black Women* (Washington, DC, 1975), pp. 126-135.

A Black male student declares in this article, "I have nothing against girls, now; I just don't want to live next door to them." This is a fascinating and discouraging article both for what it reveals about Black women's position in society and because the authors are oblivious to the sexism inherent in the limited options of Black women.

Walker, Alice, "Her Sweet Jerome," *In Love and Trouble: Stories of Black Women* (New York: Harcourt Brace Jovanovich, 1973), pp. 24-34.

Walker, Alice, "'Really, *Doesn't* Crime Pay?,'" *In Love and Trouble: Stories of Black Women* (New York: Harcourt Brace Jovanovich, 1973), pp. 10-23.

These stories about a hairdresser and a middle-class "housewife" who wants to be a writer both treat the damaging psychological effects conventional marriage can have on Black women.

Black Women Health Workers

The Black women health workers who comprise a large part of the health labor force can be valuable resources for exploring this topic. Guest speakers who work at different levels in the health hierarchy—for example, a nurse's aide, a dietary worker, a laboratory technician, a nurse, or a doctor—can be invited to talk with the class. Members of the class might also be able to share their experiences working in health care.

Brown, Carol A. "Women Workers in the Health Service Industry," *International Journal of Health Services* 5, no. 2 (1975), pp. 173-184.

This article provides a good overview of the topic and includes some interesting observations on the implications of racism as well as sexism for union organizing.

Ferris, Louanne, *I'm Done Crying*, Beth Day, ed. (New York: New American Library, 1970).

In this autobiographical work, the author describes her experiences working in a large city hospital. She is first hired as a dietary worker and eventually becomes a nurse. The book is valuable for its treatment of hospital working conditions and for the information it provides on the inferior health care available to poor, Black people.

I Am Somebody, Contemporary/McGraw-Hill, Princeton Road, Hightstown, NJ 08520, (609) 448-1700; or 1714 Stockton Street, San Francisco, CA 94133, (415) 362-3115. 28 minutes, color, 1970.

This film records the organizing of an 1199 Local in a Southern city. Most of the hospital workers are Black women.

Women Health Workers, Women's Health Collective of Philadelphia. This slide presentation is available from the Slide Tape Collective, 36 Lee Street, Cambridge, MA 02139, (617) 492-2949. Purchase: $100; rental: $40 regular price, $12 women's, worker, and community groups. 180 slides in carousels, script, and tape.

The slides focus on the Chicago Maternity Center.

Sexuality

The pervasiveness of racism and sexism in Black women's lives is revealed in the following readings on sexuality. Included in this section are articles that focus on sexual identity and on the emotional and social aspects of intimate relationships. The fact that our position as Black women affects something so seemingly private and individual as the ways in which we express our sexual feelings is a striking illustration of the basic feminist belief that the personal is political. In too many cases we have been deprived of the right to act in ways that are sexually authentic for us, in part because once again we have been defined by others. For example, white men for centuries have justified their sexual abuse of Black women by claiming that we are licentious, always "ready" for any sexual encounter. A desperate reaction to this slanderous myth is the attempt some of us have made to conform to the strictest versions of patriarchal "morality." Those of us who are lesbians have had to face the profound homophobia (hatred and fear of lesbians and homosexuals) of both Blacks and whites. Implicit in our communities' attitudes toward Black lesbians is the notion that they have transgressed both sexual *and* racial norms. Despite all of the forces with which we must contend, Black women have a strong tradition of sexual self-determination. The blues lyrics written and sung by Black women are examples of this tradition, as are many of the works listed below.

Bambara, Toni Cade, "My Man, Bovanne," *Gorilla, My Love* (New York: Pocket Books, 1973), pp. 13-20.

An older woman asserts her right to pursue a relationship with a man over her children's objections.

Hurston, Zora Neale, *Their Eyes Were Watching God* (Urbana, IL: Univ. of Illinois Press, 1978).

This novel, first published in 1937, provides a rare description of an egalitarian relationship between Janie and her lover Tea Cake.

Lorde, Audre, "Scratching the Surface: Some Notes on Barriers to Women and Loving," *The Black Scholar* (April 1978), pp. 31-35.

Lorde analyzes attitudes toward Black lesbians in the Black community and provides a stunning description of woman-identified women in Africa.

"The Myth of the 'Bad' Black Woman," *Black Women in White America: A Documentary History*, Gerda Lerner, ed. (New York: Vintage, 1972), pp. 163-171.

"Sharon," *Our Bodies, Ourselves*, The Boston Women's Health Book Collective, eds. 2nd ed. (New York: Simon and Schuster, 1976), pp. 84-85.

Thoughts on what it means to be a Black woman and a lesbian.

Suncircle, Pat, "A Day's Growth," *Christopher Street* (February 1977), pp. 23-27.

This short story describes a day in the life of a Black lesbian teenager. It is especially effective because the characters and events are rooted in familiar Afro-American cultural experiences.

Toupin, Elizabeth Ann, and Zella, Luria, "Some Cultural Differences in Response to Co-ed Housing: A Case Report," Black Women's Community Development Foundation, *Mental and Physical Health Problems of Black Women* (Washington, DC, 1975), pp. 126-135.

Tyson, Joanne and Richard, "Sex and the Black Woman: They Are Now Seeking Advice," *Ebony* (August 1977), pp. 103 *et seq.*

Although this article contains some useful observations, the efforts of Black heterosexual women to create less oppressive relationships with men are not acknowledged and the existence of Black lesbians is totally ignored.

Walker, Alice, "The Child Who Favored Daughter," *In Love and Trouble: Stories of Black Women* (New York: Harcourt Brace Jovanovich, 1973), pp. 35–46.

A young woman, "a black-eyed Susan . . . a slight, pretty flower . . . [who] pledge[s] no allegiance to banners of any man," is destroyed because her loving defies the bounds dictated by her race and class.

Violence Against Black Women

Violence against Black people has been synonymous with our experience in this country. The very fact that we are here is the result of the abduction of our ancestors from their homes in Africa. After the nightmarish Middle Passage, our kin found themselves enslaved in one of the most physically and psychologically brutal systems ever devised.

Black women have never been exempted from racist violence because we are women. The physical abuse of Black women is unique, however, because of its sexual and racial dimensions. The tradition of rape of Black women by white men and of forced sterilization points up two ways in which we are terrorized simultaneously as women and as Black people. For us to comprehend violence against Black women, we must realize that all violence against all women is related to our position under patriarchy. It is also essential to acknowledge, in approaching this topic, that Black men as well as white men violate and attack us.

I would like to comment here on the importance of Alice Walker's works in this syllabus. She is a writer dedicated to illuminating the complexities of our lives. Her unflinching portrayals of violence against Black women illustrate this commitment.

The readings that follow are grouped into several subcategories:

Battering:

Hurston, Zora Neale, "Sweat," *I Love Myself When I Am Laughing . . . And Then Again When I Am Looking Mean and Impressive: A Zora Neale Hurston Reader*, Alice Walker, ed. (Old Westbury, N.Y.: The Feminist Press, 1979), pp. 197–207.

In this short story, first published in 1926, Hurston creates a classic character, a "traditional" Black woman who finally stands up to her emotionally and physically battering husband.

Petry, Ann, "Like a Winding Sheet," *Miss Muriel and Other Stories* (Boston: Houghton Mifflin, Co., 1971), pp. 198-210.

A Black man who is powerless in his miserable, menial job strikes at the only target available to him, his wife. This story and the following work by Walker provide insights into how racism and capitalism foster violence by Black men against Black women.

Walker, Alice, *The Third Life of Grange Copeland* (New York: Harcourt Brace Jovanovich, 1970), pp. 43-123.

Walker chronicles a man's incredibly cruel and systematic destruction of his wife.

Rape:

Davis, Angela, "The Dialectics of Rape," *Ms. Magazine* (June 1975), pp. 74 *et seq.*

The author relates the case of JoAnne Little to the history of rape and racism in the United States.

Friedman, Deb, "Rape, Racism—and Reality," *FAAR and NCN Newsletter* (July/August 1978), pp. 17-26.

In this well-written article, Friedman includes a historical summary of the issues and discusses Black women's critiques of Susan Brownmiller's *Against Our Will*. The publication in which this article appeared has changed its name to *Aegis: Magazine on Ending Violence Against Women* (c/o FAAR, P. O. Box 21033, Washington, DC 20009).

"The Rape of Black Women as a Weapon of Terror," *Black Women in White America: A Documentary History*, Gerda Lerner, ed. (New York: Vintage Books, 1972), pp. 172-193.

Violence Against Children:

Angelou, Maya, *I Know Why the Caged Bird Sings* (New York: Bantam Books, 1969), pp. 58-74.

This autobiographical work includes an extremely moving account of the sexual violation of a child. It is especially valuable because it is told from the child's perspective.

Hosken, Fran P. "Female Circumcision and Fertility in Africa," *Women and Health: Issues in Women's Health Care* (November/December 1976), pp. 3-11.

Morrison, Toni, *The Bluest Eye* (New York: Holt, Rinehart, and Winston, 1970).

shange, ntozake. "is not so gd to be born a girl (1)," *The Black Scholar* (May/June 1979), p. 28.

Shange writes in stark, simple language about the varieties of violence against children. This poem was also published in *Sojourner: The New England Journal of News, Opinions, and the Arts* in February 1979.

Walker, Alice, "The Child Who Favored Daughter," *In Love and Trouble: Stories of Black Women* (New York: Harcourt Brace Jovanovich, 1973), pp. 35-46.

General Analysis and Description:

Combahee River Collective, "Eleven Black Women—Why Did They Die?" (Pamphlet, 1979).

This Black feminist analysis of violence against Black women was written in response to the murders that occurred in Boston during the winter and spring of 1979. Copies can be obtained by writing to the Combahee River Collective, c/o AASC, P. O. Box 1, Cambridge, MA 02139. An earlier version, "Six Black Women—Why Did They Die?," was published in the May/June 1979 issue of *Aegis: Magazine on Ending Violence Against Women.*

shange, ntozake, "with no immediate cause," *nappy edges* (New York: St. Martin's Press, 1978), pp. 114-117.

[Beverly Smith *is the founder and director of the Black Women Artists Film Series in Boston. She has published in* Conditions, Azalea, *and* Common Lives, Lesbian Lives *and is a contributor to the anthology,* This Bridge Called My Back: Writings by Radical Women of Color *(Persephone Press). She lives in Cambridge, Massachusetts.*]

Figure XXI. Boston University Medical Center, class of 1886.
(Boston University School of Medicine; collection of Mary R. Walsh)

Mary Roth Walsh

The Quirls of a Woman's Brain

That there was a substantial number of women physicians in the nineteenth century comes as a surprise to many. For example, Dr. Charles
Phelps recently wrote in the *Journal of Medical Education* that the increased number of women medical students "represents a rather remarkable change considering the fact that there were no women in medicine
only a hundred years ago." In reality, the late nineteenth century witnessed a sharp upsurge in the numbers of women physicians. By 1890,
only forty-three years after the admission of the first woman, Elizabeth
Blackwell, to medical school, there were 4,557 women doctors in the
United States. In the same year, 18 percent of Boston's physicians were
women, a remarkable number when one considers that in 1976, women
accounted for only 11.7 percent of Boston's medical population.[1]

Women's success in gaining access to medical schools in the late
nineteenth century seemed to guarantee that the number of female
physicians would continue to mount. In 1893, women made up 10
percent or more of the student enrollment at eighteen regular medical
schools across the nation. In the same year, women accounted for 19
percent of the University of Michigan Medical School and 31 percent
of Kansas Medical College. These figures, of course, must be balanced
against cities like New York, Philadelphia, and Chicago, where all
of the regular medical schools were still sex segregated, clustering the
women students in women's medical colleges. In other cities, women
found themselves welcome only at the irregular medical colleges, particularly the homeopathic schools. (See Table I for the names of some
of the long established regular medical schools that admitted 10 percent
or more women in 1893–94.)

Moreover, the evidence of women doctors' success in the nineteenth
century is not limited to their numbers. Articles in the press measured
their success with traditional financial yardsticks as well. *The Boston
Daily Advertiser* claimed that scores of women doctors, who counted
among their patients the city's "most cultivated, influential, and high

Table 1

Regular Medical Schools with 10 percent enrollment of female students in 1893–94

College and Location	1893–1894 Total Enroll.	% Women
1. Univ. Southern Calif.	39	15.38
2. Cooper Med. College (Calif.)	228	12.28
3. Univ. of Calif.	109	11.01
4. Med. Dept., Univ. Col.	42	19.05
5. Denver Medical Coll. (Col.)	36	25.00
6. Gross Medical Coll. (Col.)	72	35.85
7. National University (D.C.)	88	12.50
8. Council Bluffs Med. Coll. (Iowa)	12	25.00
9. Kansas Medical College	44	31.11
10. Coll. Phy. and Surgeons (Mass.)	135	21.48
11. Tufts University (Mass.)	80	28.75
12. Johns Hopkins Univ. (Maryland)	83	15.66
13. Univ. of Michigan	375	18.93
14. University of Buffalo	186	11.83
15. Syracuse Univ. (New York)	49	14.29
16. National Normal Univ. (Ohio)	26	13.33
17. Toledo Medical College (Ohio)	38	10.53
18. Univ. of Oregon	29	17.24

These figures are drawn from figures given in The U.S. Commissioners of Education Report.

born women," had incomes of five figures. Another article in the *New York Herald Tribune* singled out the success of Boston women physicians and called attention to the "surprising number of Back Bay offices luxuriant in appointment of tasteful furniture, paintings, and bric-a-brac belonging to women who add M.D. to their names." The *Boston Post* cited several Boston University Medical School graduates "whose practice is lucrative" and whose "professional services are in demand in some of the best families of the city."[2]

Novelists of the day quickly took note of this new phenomenon and women physicians made their appearance in William Dean Howells' *Dr. Breen's Practice* (1881), Sarah Orne Jewett's *A Country Doctor* (1884), Elizabeth Stuart Phelps' *Dr. Zay* (1886), Henry James's *The Bostonians* (1886), and Annie Nathan Meyer's *Helen Brent, M.D.* (1891).[3]

A great deal of the success of nineteenth-century women physicians was due to the support of the feminist movement.[4] Women raised funds to help other women gain a medical education. They also helped finance hospitals such as the New York Infirmary and the New England Hospital for Women and Children, institutions with all-female staffs run by women for women.

These medical bastions not only symbolized female solidarity, they became centers where women physicians could expand their medical base. As we shall see shortly, those who wished to keep women out of medicine often based their arguments on the physical inferiority of women. Female physicians in women's hospitals were able to counter these attacks with research that demonstrated their scientific as well as human capacity for practicing medicine.

Another major source of encouragement was the feminist press. Lucy Stone's *Woman's Journal* regularly attacked the barriers to female careers in medicine and urged its readers to patronize female physicians. A dramatic illustration of the commitment of the feminist movement to the professional advancement of women was its push to open the elite medical schools to women. Although more and more medical colleges began to admit women in the last quarter of the nineteenth century, a

Figure XXII. Dr. Anna Howard Shaw and other suffragists.
(Collection of Mary R. Walsh)

number of women believed that matriculation at a prestigious school would insure their success. Their strategy was simple: to find an elite medical school in need of funds and offer a gift of money in return for the beneficiary's promise to admit women. In 1873, Dr. Mary Putnam Jacobi, one of America's leading physicians, commented on her own purchasing of privileges while a student at the École de Medicine in Paris: "It is astonishing how many invincible objections on the score of feasibility, modesty, propriety, and prejudice will melt away before the charmed touch of a few thousand dollars."[5]

But the women soon discovered that it would take more than a few thousand dollars to buy their way into a major medical school. In 1878, Harvard Medical School rejected feminist Marion Hovey's offer of $10,000, if the school would agree to educate women on equal terms with men. Three years later, a group of women physicians in Boston raised $50,000 in pledges from women in several cities. At first, Harvard accepted the offer, but quickly reversed itself when an enraged medical school faculty threatened to resign if women were admitted as students.[6]

Undaunted by the collapse of the Harvard venture, women shifted their campaign to an even more attractive possibility, Johns Hopkins Medical School. The financially beleaguered Baltimore school had long been in the planning stage, but it had been unable to open because of lack of funds. Convinced that Johns Hopkins offered the best hope, a number of women collected $500,000 by 1892. They demanded that Johns Hopkins admit women on the same terms as men, *and* require an A.B. degree for admission, a prerequisite that no other medical school in the country imposed. The administration had no choice but to accept the women's terms and, in 1893, three out of the twenty-one entering students in the first class were women.

The victory over Johns Hopkins appeared to signal the end of women's second-class status in medicine. The *Nation* echoed the hopes of many when it enthusiastically proclaimed that John Hopkins could be the turning point for women physicians. *Century* magazine marked the decision by publishing a series of letters of approbation from a diverse group of public figures that included M. Carey Thomas, the president of Bryn Mawr College, Dr. Mary Putnam Jacobi, Josephine Lowell, and Cardinal Gibbons. What is remarkable, the *Nation* noted, is that not a single one of these writers had "a word to say on the question of whether the quirls of a woman's brain have any peculiarities which necessarily unfit her from profiting from the most advanced medical instruction."[7]

The "quirls of a woman's brain" had been an issue, however, for the last half of the nineteenth century. Moreover, the *Nation* was naive in assuming that the Johns Hopkins capitulation marked the final surrender of those who saw woman's biology as an insurmountable barrier to professional achievement.

The efforts of women to secure equal opportunity in medicine had, of course, encountered determined opposition from the male medical establishment. The arguments of the more articulate medical men formed not only a major obstacle to women's advancement in medicine, but also furnished the nineteenth-century antifeminists with a rich source of ammunition. And while historians have recently begun to examine the scientific and medical rhetoric about appropriate sexual spheres, the highly charged situations that spawned the rhetoric have been ignored.

During the last half of the nineteenth century, the age-old question of woman's place became a central issue. Women were knocking on doors marked "men only" and demanding to be admitted. Medical men found themselves confronted with some of the same problems that their friends in other fields faced, but felt themselves particularly vulnerable to a female onslaught. In colonial times, women had dominated midwifery,[8] and if women were to enter any profession, their special "talent" for nurturing seemed to dictate a career in medicine.

Unfortunately, American women began to press their efforts to be admitted to medical schools at a time when physicians were concerned about the depressed state of the profession, a condition that many believed stemmed from an oversupply of doctors. Moreover, women made up a large portion of many a medical man's practice; what would happen if women heeded the feminist press's call to support their sisters who were struggling to establish themselves? One solution was to prove that a woman's nature, far from being an asset in a medical career, was an insurmountable liability. Nowhere in the profession was there a greater urgency to promote this idea than among those men who specialized in gynecology and obstetrics, the areas where women physicians posed the greatest threat.

What resulted was the most extensive emotional debate on women's biology in American history. In 1866, the *Boston Medical and Surgical Journal* declared that the issue of women physicians was creating chaos in the profession and now was the time to bring the "unsettled question" into the open. As if waiting in the hospital wings, Dr. Horatio Storer quickly responded to the challenge.

Storer was a complex figure in Boston medicine. The son of David Humphreys Storer, Head of the Obstetrical Department at Harvard, Horatio had pioneered in separating the study of gynecology from obstetrics, and was one of the founders and, later, president of the

Figure XXIII. Dr. Horatio Robinson Storer.
(Countway Library of Harvard Medical School; collection of Mary R. Walsh)

Gynecological Society of Boston. In 1863, he had joined the New England Hospital for Women and Children as head surgeon, the only male to serve on the hospital staff throughout the nineteenth century.

Storer's response to the *Boston Medical & Surgical Journal* came a few months after the challenge was printed, in his letter of resignation from the staff of the Women's Hospital. The letter, which focused on women's unsuitability for a career in medicine, was all that the medical journal could have hoped for.[9] And what better evidence could there be? Here was a man who had observed female physicians, not from afar, but alongside them in the wards and operating rooms of their own hospital. Furthermore, he had been assisted for two years in his private practice in Boston by Dr. Anita Tyng, whom he described as "one of the very best woman physicians . . . as I suppose there is at present in the country . . . [whose] natural tastes and inclinations . . . fit her, more than I should have supposed any woman could have become fitted, for the anxiety, the nervous strain, and shocks of the practice of surgery." Nevertheless, she was dismissed by Storer as an "exception," for women were "naturally" lacking the courage and the daring to pursue the dangerous and difficult decisions involved in gynecological surgery.

But what else could one expect? How could women act freely and confidently when they were the captives of their own biology? Here, Storer turned to the subject to which every medical opponent of women was irresistibly drawn in the nineteenth century—the female reproductive system. Storer asked rhetorically, who could trust the great questions of life and death to women whose equilibrium varied from "month to month and week to week . . . up and down?" It was not to women as physicians, *per se,* that Storer objected, for he claimed they made "most agreeable and charming attendants." What Storer found objectionable was "their often infirmity during which neither life nor limb submitted to them would be as safe as at other times." It was clear from Storer's description of menstruation as "periodical infirmity . . . mental influences . . . temporary insanity," that he believed women to be monthly cripples, certainly more in need of medical aid than able to furnish it. But if women remained in their proper sphere, all would be right with the world. Storer's views of the proper sphere for that crippled class of individuals coincided with those of another widely quoted physician of the era, Dr. Charles Meigs, who wrote in his textbook on obstetrics that woman "has a head almost too small for intellect but just big enough for love."[10]

Storer's opinion was very persuasive to those who wished to be persuaded. As a leading physician, a man of obvious good will who had worked in a women's hospital and so, in a sense, had risked his reputation to conduct a scientific experiment in the opposition's laboratory, Storer's pronouncement that the experiment had proven to be a failure was not taken lightly. Yet the question must be asked, were the experiment and Storer's findings as objective as Storer intimated?

For Storer, 1866 had certainly been a difficult year. His disagreement with senior members on the Harvard Medical Faculty had led to his dismissal in the spring of that year from the position he had held as assistant in obstetrics, and had ended permanently his connection with the Harvard Medical School. And tucked away in his letter of resignation to the New England Hospital was a brief reference to his immediate reason for leaving: his objection to a new requirement that surgeons must consult with their colleagues before performing high risk surgery. On 13 August 1866, the Board of Directors of the New England Hospital had joined forces to prevent further danger to women patients and put forth the following declaration:

> Whereas: The Confidence of the Public in the Management of the Hospital rests not only on the character of the Medical attendants, having its immediate charge, but also on the high reputation of the consulting physicians and surgeons, and, Whereas: We cannot allow them to be responsible for cases over which they have no control—
> Resolved: That in all unusual or difficult cases in medicine, or where a capital operation in surgery is proposed, the attending and Resident Physicians and Surgeons shall hold mutual consultations, and if any one of them shall doubt as to the propriety of the proposed treatment or operation one or more of the consulting physicians or surgeons shall be invited to examine and decide upon the case.[11]

The Board's action was in response to the fact that all three patient deaths during the previous year had occurred in the surgical wards of the hospital, after what Dr. Lucy Sewall described as "hazardous operations." As Dr. Mary Putnam Jacobi, who had been an intern at the hospital while Storer was on the staff, later recalled, the results of Storer's operations often failed to match the boldness of his plans. And in 1866, the hospital's action came as part of a double blow, following Storer's recent dismissal from Harvard. In addition, the requirement that he clear his operations with female physicians put Storer in what he regarded as a subordinate position. Of even greater significance was the fact that these restrictions blocked what had been the hospital's major attraction for Storer—a free hand to develop his skills as a gynecological surgeon.[12]

In 1863, when Storer had first joined the hospital, none of the other hospitals in Boston allowed abdominal gynecological surgery. In fact, the entire field of gynecology was treated with great suspicion by the conservative Boston medical establishment. Storer had even been warned that the profession in New England would never tolerate in its ranks an "avowed gynecologist." Storer's son, Malcolm, who also became a physician, later attributed the prejudice regarding gynecology to the low status associated with the treatment of women's diseases: "In the ears of conservative men, the very name of diseases of women savored strongly of quackery; and it was the honest belief of many a doctor of the old school that the preservation of a man's personal morality was highly dubious if he was constantly engaged in treating the female genitals . . ."[13]

Moreover, the few experiments in gynecological surgery in Boston had all proven to be failures. Six women had been operated on for ovarian tumors between 1830 and 1858 at Massachusetts General Hospital; but, after all six women died, gynecological surgery was not permitted inside the hospital until asepsis (modern sterilization procedure) was fully established in the 1880s. The other major Boston hospital, Boston City, had similar strictures against such operations, a fact not surprising since the medical leadership in Boston was a closely knit group that oversaw all hospital affairs. That the women's hospital in Boston permitted surgery out of concern for women patients who might be saved by an operation is testimony to the women doctor's confidence in their own procedures. Other evidence indicates that many leading male physicians in the city shared their faith in the hospital's procedures.[14]

Although Storer was later able to continue his work in the newly opened, but less prestigious, operating rooms of the Carney and St. Elizabeth's hospitals, the increasingly conservative medical atmosphere in Boston limited the extent of his experimentation and publications. Nevertheless, Storer's later operations continued to exhibit much of the "boldness" to which the New England Hospital had objected. Male surgeons have historically taken a cavalier view of operations on the female reproductive system, in sharp contrast to their reluctance to experiment on the male organs. Thus, a later physician turned historian could describe a three-hour gynecological operation by Storer in 1868 as "the greatest feat" in his career. Similarly, Storer's assistant in the operation depicted it as "the most heroic of the bold precedures as yet resorted to." Both observers, however, glossed over the fact that the woman died![15]

What made Storer's lost or limited opportunities especially frustrating was the knowledge that free-wheeling gynecological surgery was being performed elsewhere in the nation. During the late nineteenth century, Storer was easily outdistanced by his daring medical rival in New York, Dr. J. Marion Sims, who performed his operations in crowded amphitheatres. In one marathon display of his surgical virtuosity, Sims performed a series of varied operations for four successive days, capped off by an entertainment dinner for the large audience of distinguished American doctors.

Furthermore, whereas Storer had been thwarted by the New England Hospital women, Sims had vanquished the wealthy society women who had founded the New York Hospital for Women in 1856. Sims had been appointed head surgeon with the expectation that he would engage a woman as his assistant, hopefully Dr. Emily Blackwell, who had just returned from study abroad and was eminently qualified. Sims at first resisted, but when the women insisted, he derisively appointed a female acquaintance who had been serving at the hospital as matron and general superintendent and who had no medical training whatsoever. Six months later, the board of lady managers of the hospital backed down and a man was appointed to assist Sims in his surgery, the selection being based not on medical qualifications, but on the fact that the man had married a young Southern friend of Sims' acquaintance.[16]

Sims had apparently recognized that female physicians might serve as a check on his aspirations. Unhampered by the type of resistance Storer had encountered at the New England Hospital, Sims performed an incredible variety of gynecological operations in the following years. Having previously performed thirty experimental operations on a slave woman named Anarchia, he now performed a similar number on an Irish woman, Mary Smith, in New York. His slashing scalpel dazzled the medical world, and earned him the reputation of "one of the immortals" in gynecological surgery. Among the medical students at Harvard, he was recognized as possessing "divinity."[17]

Physicians such as Storer and Sims, who were engaged in gynecology and obstetrics, were especially apprehensive about women in medicine. Because of the sexual anxiety of Victorian culture, the increasing numbers of women physicians threatened a revival of the old charges of immorality and insensitivity laid against men when male midwives first began to practice. In the seventeenth and eighteenth centuries, the "man-midwives" had to struggle to gain acceptance from their female patients and were much criticized by woman-midwives and looked down upon by doctors. That this criticism spilled over to the field of gynecology in the nineteenth century is demonstrated by the fact that at the first meeting of the Gynecological Society of Boston, founded by Storer in 1869, a resolution was passed confirming that the male physician who treated female patients exclusively was not affected by lust and sensuality. The Society declared "all impurity of thought and even the mental appreciation of a difference in sex is lost by the physician."[18]

Storer's career demonstrates that a good deal of the pseudo-scientific opposition to women in medicine stemmed from what can hardly be described as dispassionate sources. Far from the picture that they sought to project of men passing judgments from a vantage point above the battle, the view of male physicians was often shaped by their own special conflicts with women in medicine. We do know that when Horatio Storer was finally forced to withdraw from active medical practice in

1872, because of a near fatal infection from an accident in one of his hazardous operations, the opponents of women in medicine lost a vigorous spokesman.

It is testimony to the growing intensity of the struggle that his place was taken by an even more formidable figure, one who became a national spokesman for the anti-feminists in America, Dr. Edward Clarke. Clarke had succeeded the distinguished Dr. Jacob Bigelow in 1855 as professor of *materia medica* at the Harvard Medical School, but resigned from the Harvard faculty in 1872 and became a member of Harvard's Board of Overseers. Although Clarke was in no way a colleague of Storer's— he had, in fact, been one of those on the faculty responsible for the gynecologist's dismissal from Harvard in 1866—the two men shared similar views on the need for a masculine and "scientific" examination of the woman question.

Like Storer, Clarke had been a liberal for several years on the question of women's advancement. Early on, in 1869, Clarke had felt obligated to respond to the growing animosity shown by the male medical students in Philadelphia who had driven the women students out of the classroom with a shower of tobacco quids and tinfoil.[19] In an article published in the *Boston Medical and Surgical Journal,* he described the Philadelphia incident as "unfortunate" because of both the "unenviable notoriety" brought upon the young men involved and the fact that "nothing advances any cause so much as the martyrdom or persecution of its disciples. In this way the Philadelphia medical class have given an unexpected impetus to the cause they opposed." And indeed, this is exactly what happened. That very same year, almost directly linked to the incident, the Philadelphia County Medical Society proposed a resolution for the admission of women. Clarke apparently hoped to ward off the brash activities of immature young medical students, but also any premature action by his medical brethren. He hoped that the publication of his remarks, originally presented to the graduating class the previous spring at Harvard, would contribute to a discussion of the subject of women physicians "till a satisfactory solution is reached."[20] There is little doubt that Clarke hoped that without any catalytic incidents such as had occurred in Philadelphia, the discussion would proceed at a leisurely pace.

In his essay he reminded his audience, who probably needed no reminding, that women were knocking hard at the physician's door. Pointing out that whatever woman can do, she has a right to do and eventually *will* do, Clarke stated that *a priori,* she had "the same right to every function and opportunity which our planet offers, that man has." Neither did Clarke believe that there was anything in medicine that was improper for women to study—"for science . . . may ennoble, it can never degrade man, woman, or angel." But, he warned, the real question was not one of right, but of capability. If woman were capable, no law, argument, or ridicule would prevent her success. Therefore, Clarke urged, neither the medical profession nor the community should stand in her way: "Let the experiment . . . be fairly made . . . [and] in 50 years we shall get the answer." Clarke volunteered to predict that women could master the science of medicine.[21]

The address was certainly a welcome contrast to Storer's repeated declaration that the experiment had in fact already been tried and proven a failure. There were, however, a few items in Clarke's talk that should have set off warning signals in the feminist camp. Although he had stated that women could master medicine, he did not define what he meant by this. Nor did he explain what he meant when he declared that, with the exception of a few areas, he felt women would not become successful medical practitioners. But what difference did such comments make? All the women physicians wanted was the fair test Clarke had sanctioned. They were convinced that they would pass with flying colors.

More menacing was Clarke's insistence that the test should take place in separate classrooms. Although Clarke felt that there was nothing improper in the study of medicine by women, he seemed to believe that something mysterious and dangerous would materialize when the pursuit took place in a coeducational context. After all, a bath was a necessary and even purifying process for all, claimed Clarke, but he warned that it did not follow that the two sexes needed to bathe "at the same time and in the same tub." No narrow-minded reactionary, Clark welcomed the suffrage for women but "God forbid that I should ever see men and women aiding each other to display with the scalpel the secrets of the reproductive system; . . . or charmingly discuss together the labyrinthine ways of syphilis."[22]

But the most ominous note was struck by Clarke when he outlined the types of questions that should be included in the test of woman's fitness to be a physician. The heart of the matter was the issue of whether or not woman's nature would enable her to advance in medicine. He claimed the answer in the final analysis would be found in woman's physiology—"the facts which physicians can best supply." Nevertheless, in spite of its promises of trouble, Clarke's address did not touch off any public criticism from the women, who apparently chose to concentrate on the positive notes in his talk.

Women, quite clearly, were both eager and able to participate in Dr. Clarke's 50-year experiment. In fact, they were too eager as it turned out

—at least, as far as Clarke himself was concerned. It is difficult to explain his sudden reversal and the disappearance of his liberality with regard to the advancement of women. It may have been due to the fact that women were now pressing Harvard itself to open its doors. And Clarke, one assumes, was elected to the Board of Overseers to guard Harvard's male sanctity. It was one thing to talk nobly of a half-century's test of time; it was something quite different to look out and see contestants preparing to storm the gates of one's own home. Clarke, who had received his M.D. at the University of Pennsylvania, loved Harvard with the zeal of a convert.

He spoke to all these issues—the immediacy of the women's petitions to enter Harvard, the proper role of women, and the sacredness of an all-male educational environment—at the New England Women's Club of Boston in December, 1872. Although his own medical specialty was otology, the study of the ear, he picked up the challenge he had thrown out in 1869 and proceeded to discuss the relationship between women's education and their physiology. Not surprisingly, he found women limited by their biology. Clarke was surprised by the furor that erupted after his talk in Boston and decided to review his statements carefully with an eye to publishing them in a more comprehensive form.

The result was *Sex in Education; or, A Fair Chance for the Girls,* published in the fall of 1873. Perhaps no other single book on the limitations of the female system evoked such a wave of controversy. Within thirteen years, Clarke's book went through seventeen editions. As far away as Ann Arbor, Michigan, it was reported that everyone was reading Dr. Clarke's book. A local bookseller there claimed sales of two hundred copies in a single day, chortling, "the book bids fair to nip coeducation in the bud." But neither the number of printings nor their geographical distribution reflects the full impact of the book. Years later, M. Carey Thomas, the first President of Bryn Mawr College, recalled that "we did not know when we began whether women's health could stand the strain of education. We were haunted in those days, by the clanging chains of that gloomy little specter, Dr. Edward H. Clarke's *Sex in Education.*"[23]

Whereas Clarke in his 1869 address at Harvard had originally opposed coeducation only in the sensitive area of medical instruction, in his 1872 speech, he reworked and expanded his thesis to encompass all post-puberty education. The most dangerous threat, Clarke believed, came from the mistake of educating females as if they were males. Clarke argued that unlike the male whose development into manhood he viewed as a continuous growth process, the female at puberty experienced a sudden and unique spurt during the development of her reproductive system. If this did not occur at puberty, or if some outside force interfered, this "delicate and extensive mechanism within the organism,— a house within a house, an engine within an engine" would fail to develop. Since he believed the uterus was connected to the central nervous system, energy expended in one area (i.e. the growth of one's mind) was necessarily removed from another (i.e. the growth of one's uterus).

In *Sex in Education,* Clarke cited the dramatic case of Miss D_____ who entered Vassar College at the age of 14, a normal and healthy girl. Within a year, menstruation began and Miss D_____ continued to follow what Clarke described as the normal regimen of a male student. The results were predictable. Fainting was followed by painful menstruation, but she persisted in her studies until at last she graduated with "fair honors and a poor physique." In the following year, the young woman was "tortured" for two or three days out of every month and left weak and miserable for several more days. Then, the flow stopped altogether and Miss D_____ became pale, nervous, hysterical and complained of constant headaches. On examining the girl, Clarke found evidence of arrested development of the reproductive system. Confirmatory proof was found in his examination of her breasts, "where the milliner had supplied the organs Nature should have grown."[24]

In the pages that followed, Clarke went on to describe six similar cases in which the women all experienced "those grievous maladies which torture a woman's earthly existence: leucorrhoea, amenorrhea, dysmenorrhoea, chronic and acute ovaritis, prolapsus uteri, hysteria, neuralgia, and the like." And, if this were not enough, Clarke painted the end results of female education: "monstrous brains and puny bodies; abnormally active cerebration and abnormally weak digestion; flowing thought and constipated bowels. . . ."[25] The wonder is not that Dr. Clarke's book loomed so large in the thinking of women like M. Carey Thomas, but that they dared to entertain any thought of education at all.

The implications Clarke drew went far beyond the field of education; they extended into the sphere of population problems. The increasing number of educated women would mean that within 50 years "the wives who are to be mothers in our republic must be drawn from trans-Atlantic homes."[26] Clarke's study dovetailed neatly with the growing concern over the shrinking size of the American family—especially among the genteel classes. As early as 1850, the Massachusetts Census returns indicated that the foreign-born had a considerably higher birth rate than most white native Americans, a situation that would later give rise to fears of race suicide.[27]

Once again, women found themselves blocked by someone who had originally appeared to be a friend. Five years after urging a fair test of women's capabilities, Clarke declared his dread of seeing "the costly experiment" tried. His only solution was to provide women with *"a special and appropriate education, that shall produce a just and harmonious development of every part."*[28] Clarke was vague about the particulars of this special education, except to recommend that girls spend only two-thirds the time boys spent on their studies, and that the girls be given time off during their menstrual periods.

The seriousness with which Clarke's opponents treated his work can be measured by the extent of their response, which included at least four books published in 1874 alone: *Sex and Education; No Sex in Education; Woman's Education and Woman's Health;* and *The Education of American Girls.* Each critic recognized the dangers if Clarke's book remained unchallenged. As Mrs. E. B. Duffey, editor of *No Sex in Education,* noted, Clarke's covert plan "has been a crafty one and his line of attack masterly. He knows if he succeeds . . . and convinces the world that woman is a 'sexual' creature alone, subject to and ruled by 'periodic tides,' the battle is won for those who oppose the advancement of women."[29]

The opposition hammered away at each of Clarke's points. His critics agreed that his study would fail any scientific test. As Thomas Wentworth Higginson pointed out, to take seven cases out of a physician's notebook, assuring the readers that there were a good many more, was simply not good enough. Furthermore, Clarke neglected to present seven "representative" males for comparison, but simply assumed that boys could withstand any educational pressures to which they were exposed. Similarly, the resident physician at Vassar questioned Clarke's integrity in selecting the seven women, noting that the case of Miss D_____ was not even possible since no girl as young as 14 had ever enrolled at the college. She claimed that an error of such proportion could not help but shake one's confidence in the book's other cases, and, indeed, in its very thesis.[30]

Critics generally agreed that whatever difficulties were experienced by female college students were environmentally induced. Julia Ward Howe asserted that, if anything, a woman's education should be more like a man's, in that she should be given equal amounts of exercise and fresh air rather than be confined to the home after school to perform her domestic duties. Furthermore, she argued, rather than suffering from the pressures of keeping up with the male students, women were, in fact, victims of the constant reminder that for them education "does not matter."[31]

Clarke's vague recommendation for a special educational program for women was rejected as so impractical that it would lead to little or no education if it were implemented. Mrs. Duffey pointed out that Clarke's plan could only work if each student were subject to a uniform menstrual period. "But each girl has her own time; and if each were excused from attendance and study during this time, there could be neither system nor regularity in the classes." And what of the teacher, who probably also was a woman? "She too requires her regular furlough, and then what are the scholars to do?" Duffey wondered whether Clarke would extend his argument to the home so that wives could leave children uncared for and dinners uncooked for three or four days each month. "I think a concerted action among women in this direction," Duffey wrote, "would bring men who are inclined to agree with the doctor to their senses sooner than anything else."[32]

Most of the critics were unwilling to go as far as a general women's menstrual strike, and simply called for scientific studies to test Clarke's thesis. The first response to their cries came from the Harvard Medical School in 1874, announcing that one of the two topics to be considered for its annual Boylston Medical Prize competition was, "Do women require mental and bodily rest during menstruation and to what extent?" Since the applicants' names were not revealed to the committee, it was possible for a woman to be judged fairly in the competition, and Cambridge friends urged Dr. Mary Putnam Jacobi of New York to apply.[33]

Dr. Jacobi had previously examined the question in "Mental Action and Physical Health," in Anna Brackett's *The Education of American Girls,* one of the four books responding to Clarke in 1874. The article had been written with the general public in mind, and Jacobi had concentrated on causes other than education that might explain female "disabilities," such as "competition, haste, cramming, close confinement, long hours, and unhealthy sedentary habits." For the Boylston competition, which she eventually won, Jacobi sent out 1,000 questionnaires concerning the relationship between rest and general health.[34] She received 286 responses to such questions as how far young women walked; the presence or absence of uterine disease; and the degree and intensity of mental activity in and after school. Tests were also run on a smaller number of women at the New York Infirmary where scientific measurements could be taken on biological responses during the menses and at other times.

Her scientific findings were clearly at odds with the Clarke thesis. More than half, 54 percent, of the women did not experience any menstrual difficulties whatsoever; and, since most of those who did experience difficulties suffered only moderate pain, Jacobi felt there was nothing in the nature of menstruation to imply the necessity, or even

Figure XXIV. Dr. Mary Putnam Jacobi.
(Collection of Mary R. Walsh)

the desirability, of periodic rest for the vast majority of women. In fact, proper physical exercise, along with better nutrition, could do a great deal in Jacobi's view to prevent the development of menstrual pain. Jacobi struck a modern note in her comments when she declared that most women would be better off continuing their normal work patterns rather than taking to their beds to rest. Equally important was her research finding that mental activity was not physically dangerous. In those cases of severe menstrual pain, the women inevitably had some anatomical difficulty causing the disability.

Dr. Jacobi's monograph was followed by a number of studies that appeared to demonstrate clearly that higher education was not injurious to the health of American women. One of the first acts of the Associated Collegiate Alumnae, formed in 1882, was to commission an examination of the health of college women. Questions were drawn up with the help of a group of physicians and mailed to 1290 women college graduates in the United States. The 705 responses were analyzed by the Massachusetts Bureau of Labor Statistics. The Bureau found that 78 percent of the respondents were in excellent health; 5 percent were classified as in fair health; and 17 percent were in poor health. If anything, college women appeared to enjoy better health than the national average, though this is not surprising in view of their economic backgrounds. Similarly, L.H. Marvel, in an article in the journal, *Education,* found that college life, far from being deleterious to woman's physical adjustment, "had resulted in a stronger physique and a more perfect womanhood." In addition, Marvel's statistics demonstrated that the mortality rate for graduates of Mount Holyoke was substantially lower than that for men's schools such as Amherst, Bowdoin, Harvard and Yale.[35]

The debates on women's biology confirm Ruth Hubbard's point that "what questions we ask about the world, in what way and to what end, depends on who is asking, when, and where."[36] The doctors who had a vested interest in proving women's unsuitability for medical careers concentrated their attack on the female reproductive system. Convinced of the correctness of their position, the medical spokesmen repeated each others' prejudices rather than getting themselves involved in any serious scientific investigation of the subject. This course often led to some interesting contradictions. For example, the male medical establishment, though opposing the entry of women into the profession as doctors, welcomed them as nurses. Somehow, the nurses' uniform was a successful antidote to the biological limitations that had been the curse of women doctors. Thus the *Boston Medical and Surgical Journal,* which had objected to women doctors because their delicate health would prevent them from enduring the strain of constant house calls, would later praise the nurses for their constant round of visits to the sick in the slums of the city.[37]

Medical women and their friends, on the other hand, made every effort to support their position through careful investigation. The result was the first series of scientific studies to dispel the myth of female inferiority. But women physicians did more than simply develop arguments in favor of women's entry into medicine. Their research led to a wide ranging investigation of factors related to women's health. I have located 145 scientific articles by women physicians in the period 1872–90 dealing with such subjects as feminine hysteria, hysterectomies, menstrual difficulties, midwifery, and female insanity.

A number of female physicians also played a role in opposing the excessive surgical experimentation on women in the nineteenth century. Dr. Lillian Towslee, in 1903, had charged the medical profession with having gone "mad in the direction of gynecological tinkering, womb prodding and probing." Towslee noted that women were "rarely permitted to have an ache or a pain referable to any other part of the anatomy." Though she was writing long before the recent feminist call for a move away from surgery to solve gynecological difficulties, Towslee concluded her discussion with the same recommendation.[38]

It seems reasonable to assume that if the percentage of women physicians had continued to increase, medical treatment and research on women would have been significantly different. Instead, women experienced a sharp retrenchment in the form of medical school admissions quotas after the mid 1890s. As I have demonstrated elsewhere, this retrenchment preceded the famous Flexner report and appears to have begun just at the time when women physicians were feeling secure as a result of their successful victory over Johns Hopkins.[39] While the retrenchment did not mean that women were entirely shut out of medical schools, they were made to feel unwelcome there and quotas were established, admitting only about five percent women students for almost fifty years. Such small numbers of women doctors have not been able to have a significant impact on medical research or practices. Women's health, therefore, has been almost entirely in the hands (and heads) of male doctors. The few women who entered the profession not only were trained largely by men, but also were forced to "measure up" to the standards laid down by the (male) leaders of the profession in order to survive as doctors.

In 1970, Dr. Frances Norris testified before a Congressional subcommittee that the federal government's failure to investigate sex discrimination in medicine was in large part responsible for the poor quality of medical care given women in this country.[40] Not only has the paucity of women physicians in the American medical community limited the freedom of women to choose physicians of their own sex; it has also determined the priorities of medical research itself. There is a growing awareness of the relationship between the discrimination women face as medical students and physicians, and as patients—both contributing to the present reliance on radical mastectomies to treat breast cancer; the excessive frequency of hysterectomies; the lack of concern for the hazards of the various methods of birth control and of hormone treatments; and the generally deficient health care for women.

The dramatic increase in the number of women entering medical school during the past few years, up more than 700 percent since 1960, marks the first real progress for medical women since the late nineteenth century. It would, of course, be unfair to expect the new crop of medical women to transform the whole of American medicine. However, the pioneering work of their sisters almost a century ago provides a model for what can be accomplished by research done from a woman's point of view and focused on issues of concern to women. Yet the failure of nineteenth-century women physicians to sustain their progress serves as a warning to those who argue that the battle of women in medicine has now been won.

Notes

1. These and other statistics cited later in this essay are based on handcounts of female names on lists from medical colleges which were broken down by gender. Mary Roth Walsh, *Doctors Wanted: No Women Need Apply* (New Haven: Yale University Press, 1977) contains detailed tables on women physicians and medical students on pp. 185, 186 and 193.
2. Newspaper dates are 1894, 1886 and 1881, respectively.
3. For a detailed discussion of these novels, *see* Mary Roth Walsh, "Images of Women Doctors in Popular Fiction: A Comparison of the 19th and 20th Centuries," *Journal of American Culture,* 1:2 (1978), pp. 276–284.
4. Mary Roth Walsh, "Feminism: A Support System for Women Physicians," deals with this phenomenon in detail. *See* the *Journal of the American Medical Women's Association,* 31 (1976): pp. 247–250; also, Chapter 3 of *Doctors Wanted . . . ,* "A Feminist Show-place."
5. Mary E. Putnam Jacobi, M.D., "Social Aspects of the Readmission of Women into the Medical Profession" *Papers and Letters Presented at the First Women's Congress of the Association for the Advancement of Women, October, 1873* (New York: Association for the Advancement of Women, 1874), p. 177.
6. Marion Hovey to Harvard Medical School, March 21, 1878, Harvard Medical School Dean's Records; Agnes Vietor, *A Woman's Quest: The Life of Marie Zakrzewska,* (New York: D. Appleton and Co., 1924); *Doctors Wanted . . . ,* p. 173.
7. *Nation,* (12 February 1891), p. 131; *Century* (May 1891), pp. 632–36.
8. A bibliography of major publications on this topic is in *Doctors Wanted . . .* pp. 4–8; fn. 7–18. *See also* Jean Donnison, *Midwives and Medical Men: A History of Inter-Professional Rivalries and Women's Rights* (New York: Schocken Books, 1977). The most recent interpretation of the reasons for American women's loss of dominance in midwifery is Datha Clapper Brack's essay in this collection.
9. *Boston Medical and Surgical Journal,* 75 (1866), pp. 191–92.
10. C. D. Meigs, "Lecture on Some of the Distinctive Characteristics of the Female, delivered before the class of the Jefferson Medical College, January 1847" (Philadelphia, 1847), p. 67.
11. *Annual Reports of the New England Hospital* (1866), pp. 10–11.
12. For more detailed information on Storer's life history, see the references in *Doctors Wanted . . .* pp. 113–115, fn. 16 and 17.
13. Malcolm Storer, "The Teaching of Obstetrics and Gynecology at

Harvard," *Harvard Medical Alumni Association,* 9 (1903), pp. 439–40.

14. *See* circular, dated 1864 with letter from John H. Stephenson endorsed by Drs. Horatio Storer, Walter Channing, C. P. Putnam, S. Cabot, and Henry I. Bowditch in New England Hospital Collection, Schlesinger Library, Radcliffe College.

15. Frederick C. Irving, *Safe Deliverance* (Boston: Houghton-Mifflin Co., 1942), pp. 114–116.

16. Vietor, *A Woman's Quest . . . ,* p. 226.

17. G. J. Barker-Benfield, *The Horrors of the Half-Known Life* (New York: Harper and Row, 1976), pp. 91–119; and Seale Harris with F. H. Brown, *Woman's Surgeon* (New York: Macmillan, 1950), pp. 235–272.

18. Horatio R. Storer, "Report of the First Regular Meeting of the Gynecological Society of Boston, January 22, 1869," *Journal of the Gynecological Society of Boston* (July 1869), p. 14.

19. *Boston Medical Surgical Journal,* 81 (1869), p. 345; Hiram Corson, *Brief History of Proceedings in the Medical Society of Pennsylvania to Procure Recognition of Women Physicians* (Norristown, 1894); *Evening Bulletin* (Philadelphia, November 8, 1869); "Women as Physicians," *Philadelphia Medical and Surgical Reporter* (April 1867), p. 2; *The Press* (Philadelphia, November 18, 1869); Clara Marshall, *The Woman's Medical College of Pennsylvania* (Philadelphia, 1897), pp. 17ff.

20. *Boston Medical and Surgical Journal,* 81 (1869), p. 345.

21. *Ibid.,* pp. 345–346.

22. *Ibid.,* p. 346.

23. Edward H. Clarke, *Sex in Education; or, A Fair Chance for the Girls* (Boston: James R. Osgood and Co., 1873); M. Carey Thomas, "Present Tendencies in Women's College and University Education," *Educational Review,* 25 (1908), p. 68; Dorothy Gies McGuigan, *A Dangerous Experiment: 100 Years of Women at the University of Michigan* (Ann Arbor: Center for Continuing Education of Women, 1970), p. 38; Lilian Welsh, M.D., *Reminiscences of Thirty Years in Baltimore* (Baltimore, 1925).

24. *Sex in Education,* pp. 81–82.

25. *Ibid.,* pp. 23, 41.

26. *Ibid.,* p. 63.

27. *See* Oscar Handlin, "The Horror," in *Race and Nationality in American Life* (New York: Doubleday Anchor Books, 1957), pp. 111–132.

28. *Sex in Education,* p. 140.

29. Eliza Bisbee Duffey, *No Sex in Education: or, an equal chance for both girls and boys* (Syracuse, 1874), p. 117.

30. Julia Ward Howe (ed.), *Sex and Education. A Reply to Dr. Clarke's "Sex in Education"* (Boston, 1874), pp. 191–92.

31. *Ibid.,* pp. 27–28.
32. Duffey, *No Sex in Education* . . . , pp. 115–16, 97.
33. Alice C. Baker to Mary Putnam Jacobi, Cambridge, Massachusetts, 7 November 1874, folder 18, Jacobi papers, Schlesinger Library, Radcliffe College.
34. This survey was published in book form: Mary Putnam Jacobi, *The Question of Rest for Women During Menstruation* (New York: G. P. Putnam, 1877).
35. Louis H. Marvel, "Why Does College Life Affect the Health of Women?" *Education,* 3 (1883), p. 501.
36. Ruth Hubbard, "When Women Fill Men's Roles . . .," *Trends in Biochemical Sciences,* 1 (1976), pp. N 52–53.
37. *Boston Medical and Surgical Journal,* 76 (1867), p. 217. *See* Walsh, *Doctors Wanted* . . . , Chapter 4, for a fuller discussion of this whole issue.
38. Lillian G. Towslee, M.D. *Women's Medical Journal,* 13 (1903), p. 121.
39. A detailed discussion of this retrenchment process appears in *Doctors Wanted* . . . , pp. 186–206.
40. *Hearings before the Special Subcommittee on Education of the Committee on Education and Labor, House of Representatives, 91st Congress, 2nd Session on Section 805 of H.R. 16098, Part 1* (Washington, 1970), p. 511.

[Mary Roth Walsh *is Professor of Psychology at the University of Lowell in Massachusetts. Her book,* Doctors Wanted: No Women Need Apply *(1977 and 1979), published by Yale University Press, describes and analyzes barriers to female achievement in medicine over the years 1835–1975. She has actively worked on the problems of women in medicine since 1973 when she became a research manager at the Harvard Medical School for the Committee on the Status of Women. Since 1977 she has led leadership training sessions for women physicians, lectured nationally on their behalf, and has served as Director of Academic Curriculum for a Women's Educational Equity Act project. Her article, "The Rediscovery of the Need for a Feminist Medical Education,"* Harvard Educational Review, *Vol. 49, No. 4, 1979 summarizes some of this more recent work. Currently she teaches both the psychology of women and social psychology and is actively involved in research projects on women, social policy, and psychology.*]

Figure XXV. Men inspect Biology's Woman
(M.I.T. Historical Collection)

Naomi Weisstein

Adventures of a Woman in Science

I am an experimental psychologist, doing research in vision. The profession has for a long time considered this activity, on the part of one
of my sex, to be an outrageous violation of the social order and against
all the laws of nature. Yet at the time I entered graduate school in the
early sixties, I was unaware of this. I was remarkably naive. Stupid,
you might say. Anybody can be president, no? So, anybody can be a
scientist. Weisstein in Wonderland. I had to discover that what I wanted
to do constituted unseemly social deviance. It was a discovery I was
not prepared for: Weisstein is dragged, kicking and screaming, out of
Wonderland and into Plunderland. Or Blunderland, at the very least.

What made me want to become a scientist in the first place? The
trouble may have started with *Microbe Hunters,*[1] de Kruif's book about
the early bacteriologists. I remember reading about Leeuwenhoek's discovery of organisms too small to be seen with the naked eye. When he
told the Royal Society about this, most of them thought he was crazy.
He told them he wasn't. The "wretched beasties" were there, he insisted;
one could see them unmistakably through the lenses he had so carefully
made. It was very important to me that he could reply that he had
his evidence: evidence became a hero of mine.

It may have been then that I decided that *I* was going to become a
scientist, too. I was going to explore the world and discover its wonders.
I was going to understand the brain in a better and more complete
way than it had been understood before. If anyone questioned me, I
would have my evidence. Evidence and reason: my heroes and my
guides. I might add that my sense of ecstatic exploration when reading
Microbe Hunters has never left me through all the years I have struggled
to be a scientist.

As I mentioned, I was not prepared for the discovery that women
were not welcome in science, primarily because nobody had told me.
In fact, I was supported in thinking—even encouraged to think—that my

aspirations were perfectly legitimate. I graduated from the Bronx High School of Science in New York City where gender did not enter into intellectual pursuits; the place was a nightmare for everybody.[2] We were all, boys and girls alike, equal contestants; all of us were competing for that thousandth of a percentage point in our grade average that would allow entry into one of those high-class, out-of-town schools, where we could go, get smart, and lose our New York accents.

I ended up at Wellesley and this further retarded my discovery that women were supposed to be stupid and incompetent: the women faculty members at Wellesley were brilliant. I later learned that they were at Wellesley because the schools that had graduated them—the "very best" schools where you were taught to do the "very best" research—couldn't or did not care to place them in similar schools where they could continue their research. So they are our brilliant unknowns: unable to do research because they labor under enormous teaching loads; unable to obtain the minimal support necessary for scholarship; unable to obtain the foundations for productive research: graduate students, facilities, communication with colleagues.

While I was still ignorant about the lot of women in the academy, others at Wellesley were not. Deans from an earlier, more conscious, feminist era would tell me that I was lucky to be at a women's college where I could discover what I was good at and do it. They told me that women in a man's world were in for a rough time. They told me to watch out when I went on to graduate school. They said that men would not like my competing with them. I did not listen to the deans, however; or, when I did listen, I thought what they were telling me might have been true in the nineteenth century, but not in the late fifties.

So my discovery that women were not welcome in psychology began when I arrived at Harvard, on the first day of class. That day, the entering graduate students had been invited to lunch with one of the star professors in the department. After lunch, he leaned back in his chair, lit his pipe, began to puff, and announced: "Women don't belong in graduate school."

The male graduate students, as if by prearranged signal, then leaned back in their chairs, puffed on their newly bought pipes, nodded, and assented: "Yeah."

"Yeah," said the male graduate students. "No man is going to want you. No man wants a woman who is more intelligent than he is. Of course, that's not a real possibility, but just in case. You are out of your *natural* roles; you are no longer feminine."

My mouth dropped open, and my big blue eyes (they have since changed back to brown) went wide as saucers. An initiation ceremony,

I thought. Very funny. Tomorrow, for sure, the male graduate students will get theirs.

But the male graduate students never were told that they didn't belong. They rapidly became trusted junior partners in the great research firms at Harvard. They were carefully nurtured, groomed, and run. Before long, they would take up the white man's burden and expand the empire. But for me and the other women in my class it was different. We were shut out of these plans; we were *shown* we didn't belong. For instance, even though I was first in my class, when I wanted to do my dissertation research, I couldn't get access to the necessary equipment. The excuse was that I might break the equipment. This was certainly true. The equipment was eminently breakable. The male graduate students working with it broke it every week; I didn't expect to be any different.

I was determined to collect my data. Indeed, I *had* to collect my data. (Leeuwenhoek had his lenses. Weisstein would get her data.) I had to see how the experiment I proposed would turn out. If Harvard wouldn't let me use its equipment, maybe Yale would. I moved to New Haven, collected my data at Yale, returned to Harvard, and was awarded my Ph.D. in 1964. Afterward, I could not get an academic job. I had graduated Phi Beta Kappa from Wellesley; had obtained my Ph.D. in psychology at Harvard in two and one half years, ranked first in my graduate class, and I couldn't get a job. Yet most universities were expanding in 1964, and jobs were everywhere. But at the places where I was being considered for jobs, they were asking me questions like, "How can a little girl like you teach a great big class of men?" At that time, still unaware of how serious the situation was, I replied, "Beats me. I guess I must have talent." At another school, a famous faculty liberal challenged me with, "Who did your research for you?" He then put what I assume was a fatherly hand on my knee, and said in a tone of deep concern, "You ought to get married."

Meanwhile, I was hanging on by a National Science Foundation postdoctoral fellowship in mathematical biology at the University of Chicago, attempting to do some research. Prior to my second postdoctoral year, the University of Chicago began negotiations with me for something like a real job: an instructorship jointly in the undergraduate college and the psychology department. The negotiations appeared to be proceeding in good faith, so I wrote to Washington and informed them that I would not be taking my second postdoctoral year. Then, ten days before classes began, when that option as well as any others I might have taken had been closed, the person responsible for the negotiations called to tell me that, because of a nepotism rule—my husband taught history at the University of Chicago—I would not be hired as a regular faculty member. If I wanted to, I could be appointed lecturer, teaching general education courses in the college; there was no possibility of an appointment in psychology. The lectureship paid very little for a lot of work, and I would be teaching material unconnected with my research. Furthermore, a university rule stipulated that lecturers (because their position in the university was so insecure) could not apply for research grants. He concluded by asking me whether I was willing to take the job: ten days before the beginning of classes, he asked me whether I was willing to take the only option still available to me.

I took the job, and "sat in," so to speak, in the office of another dean, until he waived the restriction on applying for research grants. Acknowledging my presence, he told a colleague: "This is Naomi Weisstein. She hates men."

I had simply been telling him that women are considered unproductive precisely because universities do their best to keep women unproductive through such procedures as the selective application of the nepotism rule. I had also asked him whether I could read through the provisions of the rule. He replied that the nepotism rule was informal, not a written statute—flexibility being necessary in its application. Later, a nepotism committee, set up partly in response to my protest, agreed that the rule should stay precisely as it was; that it was a good idea, should not be written out, and should be applied selectively.

Lecturers at major universities are generally women. They are generally married to men who teach at these major universities. And they generally labor under conditions which seem almost designed to show them that they don't belong. In many places, they are not granted faculty library privileges; in my case, I had to get a note from the secretary each time I wanted to take a book out for an extended period. Lecturers' classrooms are continually changed; at least once a month, I would go to my assigned classroom only to find a note pinned to the door instructing me and my class to go elsewhere: down the hall, across the campus, out to Gary, Indiana.

In the winter of my first year, notices were distributed to all those teaching the courses I was teaching, announcing a meeting to discuss the following year's syllabus. I didn't receive the notice. As I was to learn shortly, this is the customary way a profession that prides itself on its civility and genteel traditions indicates to lecturers and other "nuisance personnel" that they're fired: they are simply not informed about what's going on. I inquired further. Yes, my research and teaching had been "evaluated" (after five months: surely enough time), and they had decided to "let me go" (a brilliant euphemism). Of course, the decision had nothing to do with my questioning the nepotism rules and explaining to deans why women are thought unproductive. I convinced them to "let me stay" another year. I don't know to this day why they changed their minds. Perhaps they changed their minds because it looked like I was going to receive the research grant for which I had applied, bringing in money not only for me, but for the university as well. A little while later, Loyola University in Chicago offered me a job.

So I left the University of Chicago. I was awarded the research grant and found the Psychology Department at Loyola at first very supportive. The chairman, Ron Walker, was especially helpful and enlightened about women at a time when few academic men were. I was on my way, right? Not exactly. There is a big difference between a place like Loyola and a place with a heavy commitment to research—a large state university, for example—a difference that no amount of good will on the part of an individual chairman can cancel out. The Psychology Department was one of the few active departments at Loyola. The other kinds of support one needs to do experimental psychology—machine and electrical shops, physics and electrical engineering departments, technicians, a large computer—were either not available or were available at that time only in primitive form.

When you are a woman at an "unknown" place, you are considered out of the running. It was hard for me to keep my career from "shriveling like a raisin" (as an erstwhile colleague predicted it would). I was completely isolated. I did not have access to the normal channels of communication, debate, and exchange in the profession—those informal networks where you get the news, the comment and the criticism, the latest reports of what is going on. I sent my manuscripts to various people for comment and criticism before sending them off to journals; few replied. I asked others working in my field to send me their pre-publication drafts; even fewer responded. Nobody outside Loyola informed me about special meetings in my area of psychology, and few inside Loyola knew about them. Given the snobbery rife in academic circles (which has eased lately since jobs are much harder to find and thus even "outstanding" young male graduates from the "best" schools may now be found at places formerly beneath their condescension), my being at Loyola almost automatically disqualified me from the serious attention of professional colleagues.

The "inner reaches" of the profession, from which I had been exiled, are not just metaphorical and intangible. For instance, I am aware of two secret societies of experimental psychologists in which fifty or so of the "really excellent" young scientists get together regularly to make themselves better scientists. The ostensible purpose of these societies is to allow these "best and brightest" young psychologists to get together to discuss and criticize each other's work; they also function, of course, to define who is excellent and who is not, and to help those defined as excellent to remain so by providing them with information to which "outsiders" in the profession will not have access until much later (if at all).

But the intangibles are there as well. Women are subjected to treatment men hardly ever experience. Let me give you a stunning example. I wrote up an experiment with results I thought were fascinating, and sent the paper to a journal editor whose interests I knew to be close to what was reported in my paper. The editor replied that there were some control conditions that should be run and some methodological loose ends; so they couldn't publish the paper. Fair enough. He went on to say that they had much better equipment over there, and they would like to test my idea themselves. Would I mind? I wrote back, and told them I thought it was a bit unusual, asked if they were suggesting a collaboration, and concluded by saying that I would be most happy to visit with them and collaborate on my experiment. The editor replied with a nasty letter explaining to me that by suggesting that they test my idea themselves, they had merely been trying to help me. If I didn't want their help in this way, they certainly didn't want mine; that is, they had had no intention of suggesting a collaboration.

In other words, what they meant by "did I mind" was: Did I mind if they took my idea and did the experiment themselves? As we know, taking someone else's idea and pretending it's your own is not at all an uncommon occurrence in science. The striking thing about this exchange was, however, that the editor was arrogant enough, and assumed that I would be submissive enough, for him to openly ask me whether I would agree to this arrangement. Would I mind? No, of course not. Women are joyful altruists. We are happy to give of ourselves. After all, how many good ideas do you get in your lifetime? One? Two? Why not *give* them away?

Generally, the justification for treating women in such disgraceful ways is simply that they are women. Let me give another example. I was promised the use of a small digital laboratory computer, which was to be purchased on a grant. The funds from the grant would become available if a certain job position entailing administration of this grant could be filled. I was part of the group which considered the candidates and which recommended appointing a particular individual. During the discussion of future directions of the individual's work, it was agreed that he would, of course, share the computer with me. He was hired, bought the computer, and refused me access to it. I offered to put in money for peripherals which would make the system faster and easier for both of us to work with, but this didn't sway him. As justification for his conduct, the man confessed to the chairman that he simply couldn't share the computer with me: he had difficulty working with women. To back this up, he indicated that he'd been "burned twice." Although the chairman had previously been very helpful and not bothered in the least about women, he accepted that statement as an explanation. Difficulty in working with women was not a problem this man should work out. It was *my* problem. Colleagues thought no worse of him for this problem; it might even have raised him in their estimation. He obtained tenure quickly, and retains an influential voice in the department. Yet if a woman comes to *any* chairman of *any* department and confesses that she has difficulty working with *men,* she is thought pathological.

What this meant for me at the time was that my research was in jeopardy. There were experimental conditions I needed to run that simply could not be done without a computer. So there I was, doing research with stone-age equipment, trying to get by with wonder-woman reflexes and a flashlight, while a few floors below, my colleague was happily operating "his" computer. It's as if we women are in a totally rigged race. A lot of men are driving souped-up, low-slung racing cars, and we're running as fast as we can in tennis shoes we managed to salvage from a local garage sale.

Perhaps the most painful of the appalling working conditions for women in science is the peculiar kind of social-sexual assault women sustain. Let me illustrate with a letter to *Chemical and Engineering News* from a research chemist named McGauley:

> There are differences between men and women ... just one of these differences is a decided gap in leadership potential and ability ... this is no reflection upon intelligence, experience, or sincerity. Evolution made it that way .... Then consider the problems that can arise if the potential employee, Dr. Y (a woman) [*sic*: he could at least get his chromosomes straight] will be expected to take an occasional business trip with Dr. X. ... Could it be that the guys in shipping and receiving will not take too kindly to the lone Miss Y?[3]

Now what is being said here, very simply, and to paraphrase the Bible, is that women are trouble. And by trouble, McGauley means sexual trouble. Moreover, somehow, someway, it is our fault. *We* are provoking the guys in shipping and receiving. Women—no matter who the women are or what they have in mind—are universally assigned by men, first, to sexual categories. Then, women are accused by men of taking their minds away from work. When feminists say that women are treated as sex objects, we are compressing into a single, perhaps rhetorical phrase, an enormous area of discomfort, pain, harassment, and humiliation.

This harassment is especially clear at conventions. Scientific meetings, conferences, and conventions are harassing and humiliating for women because women, by and large, cannot have male colleagues. Conversations, social relations, invitations to lunch, and the like are generally viewed as sexual, not professional, encounters if a woman participates in them. It does not cross many men's minds that a woman's motivation may be entirely professional.

I have been at too many professional meetings where the "joke" slide was a woman's body, dressed or undressed. A woman in a bikini is a favorite with past and perhaps present presidents of psychological associations. Hake showed such a slide in his presidential address to the Midwestern Psychological Association, and Harlow, past president of the American Psychological Association, has a whole set of such slides, which he shows at the various colloquia to which he is invited. This business of making jokes at women's bodies constitutes a primary social-sexual assault. The ensuing raucous laughter expresses the shared understanding of what is assumed to be women's primary function to which we can always be reduced. Showing pictures of nude and sexy women insults us: it puts us in our place. You may think you are a scientist it is saying, but what you really are is an object for our pleasure and amusement. Don't forget it.

I could continue recounting the horrors, as could almost any woman who is in science or who has ever been in science. But I want now to turn to the question of whether or not the commonly held assumptions about women are true. If they are, this would in no way justify the profession's shameful treatment of women, but it might lead us to different conclusions about how to remedy the situation.

I began the inquiry into women's "nature" while I was in graduate school. I wanted to investigate the basis on which the learned men in my field had pronounced me and my female colleagues unfit for graduate study.

I found that the views of the experts reflected, in a surprisingly transparent way, the crudest cultural stereotypes. Erik Erikson wrote:[4]

Young women often ask me whether they can 'have an identity before they know whom they will marry and for whom they will make a home'

He explained (somewhat elegiacally) that:

Much of a young woman's identity is already defined in her kind of attractiveness and in the selectivity of her search for the man (or men) by whom she wishes to be sought. . . .

Mature womanly fulfillment, for Erikson, rested on this "fact":

[that a woman's] . . . somatic design harbors an 'inner space' destined to bear the offspring of chosen men, and with it, a biological, psychological, and ethical commitment to take care of human infancy.

Bruno Bettelheim, speaking at a symposium on American women in science and engineering, commented:

We must start with the realization that, much as women want to be good scientists and/or engineers, they want first and foremost to be womanly companions of men and to be mothers.[5]

And Joseph Rheingold, a psychiatrist at the Harvard Medical School, tied the reluctance of women to give in to their true natures to society's problems:

Woman is nurturance . . . anatomy decrees the life of a woman . . . When women grow up without dread of their biological functions and without subversion by feminist doctrine, and therefore enter upon motherhood with a sense of fulfillment and altruistic sentiment, we shall attain the goal of a good life and a secure world in which to live it.[6]

So the learned men in my field were saying essentially the same thing as the male graduate students, but it still did not sound right to me. Since scientists (supposedly) do not assess the truth or falsity of propositions on the basis of who said them (they look instead for evidence), I decided to look for the evidence on which these eminent men had based their theories. I determined that there is no evidence behind their fantasies of the servitude and childish dependence of women. On the contrary, the idea of human possibility which rests on the accident of sex at conception has strangled and deflected psychology so that it is still relatively useless in describing, explaining, or predicting human behavior. This is true for men as well as women. It becomes especially pernicious when the theories are not only wrong but are proscriptive as well.

I have elsewhere gone through the arguments showing the near uselessness of these kinds of theories of human nature;[7] here, let me just provide the briefest of summaries.

The basic reason that this kind of psychology (i.e., personality theory, and, for the most part, theory from psychotherapists and psychoanalysts) tells us next to nothing about human nature, male as well as female, is that it has been looking in the wrong place. It has assumed that what people do comes from a fixed, rigid, inside directive: sex organs, or fixed cognitive traits, or what happened until, but no later than, the age of 5. This assumption has been shown to fail again and again in tests; a person will be assessed by psychologists as possessing a particular constellation of personality traits and then, when different criteria are applied, or someone different is asked to judge, or, more importantly, when that person is in a different kind of social situation, s/he will be thought to exhibit a completely different set of traits.[8]

One might argue, then, that personality is a somewhat subtle and elusive thing, and that it would be difficult to obtain a set of measures that would distinguish personality types. This is a reasonable argument. But even when one looks at what one would expect to be gross differences between a certified schizophrenic, say, and a normal,[9] or between a male homosexual and a male heterosexual,[10] or between a "male" personality and a "female" personality,[11] one finds that the same judges who claim to be able to differentiate human personalities, simply cannot distinguish one from the other. In one study,[12] for example, judges who were supposed to be experts at this kind of thing could not tell, on the basis of clinical tests and interviews (in which one is allowed to ask questions such as, "When did you first notice that you had grown antlers?"), which one of a group of people had been classified as schizophrenics and which as normals. Even stranger, some weeks later, when the same judges were asked to judge the same people, in many cases, they reversed their own judgments. Judges (again, allegedly clinical experts), attempting to distinguish between homosexuals and heterosexuals on the basis of what is assumed to be differences in their personalities, have done no better; nor was my graduate class at Harvard able correctly to distinguish stories written by males from those written by females, even though we had just completed a month and a half's study of the differences between men and women. In short, if judges cannot agree on whether a person belongs in a certain personality category, even when those categories are assumed to be as different from each other as normal/crazy, male/female, and straight/gay; if the measurement depends on who is doing the measuring, and on what time of day it is being done, then theories that are based on these personality categories are useless.

The other "test" that has frequently been cited as a way of confirming such theories is the test of therapy. Since most of the theorizing about men and women (and normals and schizophrenics, and homosexuals and heterosexuals) has been done by clinical psychologists who cite as evidence for their theories "years of intensive clinical experience," one test of their understanding of human personality might be their effectiveness in helping people solve these "problems." Of course, one might question what is going on in these years of intensive clinical practice when clinicians cannot even agree on descriptions; that is, they cannot agree on their categories. But suppose one countered that clinical psychologists really do have an understanding of the depths of human personality, and, although any two clinical psychologists may not be able to agree on a "verbal" level on what categories they are using, nevertheless (so this argument would go), they are operating at an intuitive level which "works" (i.e., they help their patients change their behavior). The fact is that, to the limited extent that therapy may change behavior (if at all), it doesn't matter *which* therapy is used: in general, no one therapy is reported to be any *more* effective than any other, even when the same symptoms are being treated.[13] Since theories upon which different therapies are based are different, and in some cases, conflict with each other, the extent to which a particular therapy may work cannot be taken to lend credence to that particular theory.

What all this means for women is that personality theory has given us no idea of what our true "natures" are; whether we were intended from the start to be scientists and engineers and were thwarted by a society that has other plans for us, or whether we were intended, as claimed by some of the learned men in the field, only to be mothers. There are a number of arguments based on selected primates[14] that also purport to show that females are suited only for motherhood (hopefully with a sense of fulfillment and altruistic sentiment). These arguments are even more specious than those from the clinical tradition, as I have discussed elsewhere.[15]

But while personality psychology and clinical psychologists have failed miserably at providing any statements we can trust about women's "true nature," or about anyone's "true nature," the evidence is accumulating from a different area of psychology, *social* psychology: what humans do and when they will do it is highly predictable. What people do and who they believe themselves to be will, in general, be a function of what the people around them expect them to be, and what the overall situation in which they are acting implies that they are. Let me describe three experiments that have made this fact clear.

The Experimenter Bias Experiments

These studies[16] have shown that if one group of experimenters has one hypothesis about what they expect to find, and another group of experimenters has the opposite hypothesis, both will obtain results in accord with their differing hypotheses. And this is not because the experimenters lie or cheat or falsify data. In the studies cited, the experimenters are closely observed, and they are made outwardly to behave in identical fashion. The message about their different expectations is somehow picked up by their subjects through nonverbal cues, head nods, ways of communicating expectations that we do not yet know about. The moral here is that, even in carefully controlled conditions, when we are dealing with humans (and in some cases rats),[17] the hypotheses we start with will influence the behavior of the organism we are studying. It is obvious how important this would be when assessing the validity of psychological studies of the differences between men and women.

Inner Physiological State Versus Social Context

Subjects[18] were injected with adrenalin, a hormone that tends to make people "speedy"; when placed in a room with another person (a confederate of the experimenter) who acted euphoric, the subject became euphoric. Conversely, if a subject was placed in a room with another person who acted angry, the subject became angry. These data seem to indicate that the far more important determinant to how people will act is not their physiological state, but the social context in which they are acting. Thus, no matter how many physiological differences we may find between men and women, we must be very cautious in assigning any fixed behavioral correlates to the physiological states. The point is made even more strongly, perhaps, in studies of hermaphrodites in whom the genetic, gonadal, hormonal sex, the internal reproductive organs, and the ambiguous appearance of the external genitalia were identical. It was shown that one will consider one's self male or female depending simply on whether one was defined and raised as a male or a female:

There is no more convincing evidence of the power of social interaction on gender-identity differentiation than in the case of congenital hermaphrodites who are of the same diagnosis and similar degree of hermaphroditism but are differently assigned and with a different postnatal medical and life history.[19]

The Obedience Experiments

In Milgram's experiments,[20] a subject is told that s/he is administering a learning experiment, and that s/he is to deal out shocks each time another "subject" (who is in fact a confederate of the experimenter) answers incorrectly. The equipment appears to provide graduated shocks ranging upwards from 15 V through 450 V; for each of four consecutive voltages, there are verbal descriptions such as mild shock; danger; severe shock; and finally, for the 435 V and 450 V switches a red XXX marked over the switches. Each time the confederate answers incorrectly, the subject is supposed to increase the voltage. As the voltage increases, the confederate begins to cry out in pain and demands that the experiment be stopped, finally refusing to answer at all. When all responses are stopped, the experimenter instructs the subject to continue increasing the voltage. For each shock administered, the confederate shrieks in agony. Under these conditions, about 62 per cent of the subjects administered shocks that they believed to be possibly lethal.

No tested individual differences among subjects predicted who would continue to obey, and who would break off the experiment. When forty psychiatrists predicted how many of a group of 100 subjects would go on to give the lethal shock, their predictions were orders of magnitude below the actual percentages; most expected only one or two of the subjects to obey to the end.

But even though psychiatrists have no idea how people will behave in this situation, and even though individual differences do not predict who will and will not obey, it is easy to predict when subjects will be obedient and when they will be defiant. In a variant of Milgram's experiment, two confederates were present in addition to the "victim," working along with the subject in administering electric shocks. When the two confederates refused to continue with the experiment, only 10 per cent of the subjects continued to the maximum voltage. This is critical for personality theory. It says that behavior is predicated largely on the social situation, not solely on the individual's history.

To summarize: if subjects under quite innocuous and noncoercive social conditions can be made to kill other subjects, and under other types of social conditions will positively refuse to do so; if subjects can react to a state of physiological arousal by becoming euphoric because there is someone else around who is euphoric, or angry because there is someone else around who is angry; if subjects will act a certain way because experimenters expect them to act in that way, and another group of subjects will act in a different way because experimenters expect them to act in that different way; then it appears obvious that a study of human behavior requires first and foremost a study of the social contexts within which people move, the expectations as to how they will behave, and the authority that tells them who they are and what they are supposed to do.

The relevance to males and females is obvious. We do not know what immutable differences in behavior, nature, ability, or possibility exist between men and women. We know that they have different genitalia and at different times in their lives, different sex hormone levels. Perhaps there are some unchangeable differences; probably there are a number of irrelevant differences. But all these differences are likely to be trivial compared to the enormous influence of social context. And it is clear that, until social expectations for men and women are equal and just; until equal respect is provided for both men and women, our answers to the question of immutable differences, of "true" nature, of who should be the scientist and who should be the secretary, will simply reflect our prejudices.

I want to stop now and ask: What conclusions can we draw from my experience? What does it all add up to?

Perhaps we should conclude that persistence wins out. Or that life is hard, but cheerful struggle and a "sense of humor" can make it bearable. Or perhaps we should search back through my family, and find my domineering mother and passive father, or my domineering father and passive mother, to explain my persistence. Perhaps . . . but all these conclusions are beside the point. The point is that none of us should have to face this kind of offense. The main point is that we must change this man's world and this man's science.

How will other women do better? One of the dangers of this kind of narrative is that it may validate the punishment as it singles out the few survivors. The lesson appears to be that those (and only those) with extraordinary strength will survive. This is not the way I see it. Many have had extraordinary strength and have *not* survived. We know of some of them, but by definition we will of course never know of most.

Much of the explanation for my own professional survival has to do with the emergence and growth of the women's movement. I am an experimental psychologist, a scientist. I am also a feminist. I am a feminist because I have seen my life and the lives of women I know harassed, dismissed, damaged, destroyed. I am a feminist because without others I can do little to stop the outrage. Without a political and social movement of which I am a part, without feminism, my determination and persistence, my clever retorts, my hours of patient explanation, my years of exhortation amount to little. If the scientific world has changed since I entered it, it is not because I managed to become an established psychologist within it. Rather, it is because a women's movement came along to change its character. It is true that as a member of that movement, I have acted to change the character of the scientific world. But without the movement, none of my actions would have brought about change. And now, as the strength of the women's movement ebbs, the old horrors are returning. This must not happen.

Science, knowledge, the search for fundamental understanding is part of our humanity. It is an endeavor that seems to give us some glimpse of what we might be and what we might do in a better world. To deny us the right to be scientists is to deny us our humanity. We cannot let that happen.

Notes

Somewhat different versions of this paper have appeared in *Federation Proceedings*, 35 (1976), 2226–31; and in *Working It Out*, S. Ruddick and P. Daniels, eds., (New York: Pantheon Books, 1977), pp. 242–250. I wish to thank Tobey Klass for her critical comments and her guidance through the recent literature. I also wish to thank Roger Burton for bringing the Campbell and Yarrow sudy to my attention.

1. Paul de Kruif, *Microbe Hunters* (New York: Brace & World, 1926).
2. I discovered later on that this in itself was unusual—by high school, if not before, girls are generally discouraged from showing an interest in science.
3. P. J. McGauley, Letter to the Editor, *Chem. Eng. News*, 48 (1970), pp. 8–9.
4. Erik H. Erikson, "Inner and Outer Space: Reflections on Womanhood," *Daedalus*, 93 (1964), pp. 585–606.
5. From a speech entitled "The Commitment Required of a Woman Entering a Scientific Profession in Present Day American Society," Massachusetts Institute of Technology, 1965.
6. J. Rheingold, *The Fear of Being a Woman* (New York: Grune & Stratton, 1964).
7. Naomi Weisstein, "Psychology Constructs the Female; or the Fantasy Life of the Male Psychologist," *J. Soc. Ed.*, 35 (1970), pp. 362–373.
8. J. Block, "Some Reasons for the Apparent Inconsistency of Personality," *Psychol. Bull.*, 70 (1968), 210–212; W. Mischel, *Personality and Assessment* (New York: Wiley, 1968); W. Mischel, "Toward a Cognitive Social Learning Reconceptualization of Personality," *Psychological Review*, 80 (1973), pp. 252–283.
9. K. B. Little and E. S. Schneidman "Congruences Among Interpretations of Psychological and Anamnestic Data," *Psychol. Monogr.*, 73 (1959), pp. 1–42. See also R. E. Tarter, D. I. Templer, and C. Hardy, "Reliability of the Psychiatric Diagnosis," *Diseases of the Nervous System*, 36 (1975), pp. 30–31.
10. E. Hooker, "Male Homosexuality in the Rorschach," *J. Projective Techniques*, 21 (1957), pp. 18–31.
11. Naomi Weisstein, "Psychology Constructs the Female . . . ."
12. K. B. Little and E. S. Schneidman, "Congruences among interpretations . . . ."
13. Dorothy Tennov, *Psychotherapy* (New York: Abelard-Schuman, 1975); M. L. Smith and G. V. Glass, "Meta-Analysis of Psychotherapy Outcome Studies," *American Psychologist*, 32 (1977), pp. 752–60; Sloan, Staples, Cristol, Yorkston and Whipple, "Short-term Analytically Oriented Psychotherapy Versus Behavior Therapy," *American Journal of Psychiatry*, 132:4 (1975), pp. 373–377.

14. See H. F. Harlow, "The Heterosexual Affectional System in Monkeys," *Am. Psychol.*, 17 (1962), pp. 1–9; L. Tiger, *Men in Groups* (New York: Random House, 1969); and L. Tiger, "Male Dominance? Yes. Alas. A Sexist Plot? No.," *New York Times Magazine Section*, 25 October 1970.

15. Naomi Weistein, "Psychology Constructs the Female . . . ." *See also* Hubbard's article in this collection.

16. R. Rosenthal, "On the Social Psychology of the Psychological Experiment: the Experimenter's Hypothesis as Unintended Determinant of Experimental Results," *Am. Sci.*, 51 (1963), 268–283; and R. Rosenthal, *Experimenter Effects in Behavioral Research* (New York: Appleton-Century-Crofts, 1966). A meticulous observation of behavior at a summer camp suggesting the same results is J. D. Campbell and M. R. Yarrow's "Perceptual and Behavioral Correlates of Social Effectiveness," *Sociometry*, 24 (1961), pp. 1–20.

17. H. F. Harlow, "The Heterosexual Affectional System . . . ."

18. John Money, "Sexual Dimorphism and Homosexual Gender Identity," *Psychol. Bull.*, 74 (1970), pp. 6, 425–440; and S. Schachter and J. E. Singer, "Cognitive, Social, and Physiological Determinants of Emotional State," *Psychol. Rev.*, 63 (1962), pp. 379–399.

19. John Money, "Sexual Dimorphism and Homosexual Gender . . . ."

20. S. Milgram, "Some Conditions of Obedience and Disobedience to Authority," *Human Relations*, 18 (1965), pp. 57–76; and S. Milgram, "Liberating Effects of Group Pressure," *J. Pers. Soc. Psychol.*, 1 (1965), pp. 127–134.

[Naomi Weisstein *is Professor of Psychology at SUNY at Buffalo. Her research is in vision, perception, cognition and brain theory, focusing on how images are represented in the visual system so as to lead to human visual understanding. She is a member of many professional organizations and has published numerous scientific papers and reviews. During the 1960's, she was active in CORE and formed women's caucuses in SDS and the New University Conference. In the early women's liberation movement, she was a founding member of Chicago West Side Group and the Chicago Women's Liberation Union. She has been active in guiding and defining an insurgent feminist culture, as organizer and pianist for the Chicago Women's Liberation Rock Band (1970–73), writer and critic, and currently, as feminist comedienne presenting themes ranging from rape, to the mores of the scientific profession, to attempts to deliver scientific data as stand-up monologue.*]

Figure XXVI. Dr. Ellen Swallow Richards, first woman to serve on M.I.T. teaching staff, with group of female students (1888). (M.I.T. Historical Collections)

Epilogue

> .... *Said the sages, "In the first place,*
> *The thing cannot be done!*
> *And, second, if it could be,*
> *It would not be any fun!*
> *And, third, and most conclusive,*
> *And admitting no reply,*
> You would have to change your nature!
> *We should like to see you try!"*
> *They chuckled then triumphantly,*
> *These lean and hairy shapes.*
> *For these things passed as arguments*
> *With the Anthropoidal Apes.*
> —From the poem, *Similar Cases*, by Charlotte Perkins Gillman

What can women expect from science? What can it do for us, and we for it?

The first thing it can do *for* us is to stop doing a number of unpleasant things *to* us. Or in the words of Sarah Grimke written in 1837 to her sister, Angelina:

> All I ask of our brethren is, that they will take their feet from off
> our necks and permit us to stand upright on that ground which
> God designed us to occupy.[1]

We offer the following to our brethren in science as a model for the expeditious removal of feet. Recently, it was brought to the attention of E. B. White that his *Elements of Style*, which has been a writers' bible for twenty years, was outrageously and gratuitously sexist. Among its hundreds of examples, illustrating every disservice we are likely to render the English language, not *one* presents a positive image of women. White, not one to waste words on superfluities, did not waste them here. "I'll fix it," he said according to a July 1977 news item, "I wasn't aware."

If every man who was ever confronted with the antifemale bias of his work—be he biology professor, grant reviewer, researcher, medical school admissions officer, textbook writer—were to adopt the above example as *method*, we could begin to redress past grievances and get on to the business of collectively building a better future. But that is unfortunately not likely to happen soon.

Quite the contrary, the last five or six years have witnessed the rebirth of biological determinism in the guise of the "new" science of sociobiology as well as in a flood of sex differences research, which purport to provide "scientific evidence" that women's and men's work roles and social positions in contemporary America are rooted in our innate, biological "propensities." (How do you scientifically prove a propensity?) For this purpose, we are served up *ad hoc* reconstructions of our evolutionary past as well as hypothetical genes, hormone effects, and differences in brain structure and functions, many of which are conjectural or have been observed only in that prototypic human, the white laboratory rat.

The theory in which much of this research is grounded, as well as many of the actual experiments, are so poorly thought out and conducted that the results would be rejected by reputable professional journals and book publishers if they did not provide so-called scientific legitimation of the *status quo* by "justifying" discriminary practices against women in employment, education, and other aspects of our public and private lives. Furthermore, each piece of research, however bad or encrusted in scientific jargon (and hence inaccessible without prior translation), immediately becomes front-page news in the daily papers and material for cover stories in *Time, Newsweek*, and other popular magazines.[2] This kind of "science" is, of course, enormously useful in sustaining the myth of equal opportunity, according to which everyone has the same chance to succeed, and differences in people's economic and social positions are interpreted to reflect innate biological differences between us.[3]

It is clear that progress in forcing a redefinition of what women are—or better yet, an end to the effort to define it for us—is going to be slow and will be met with vigorous attempts to prove again and again that our social disadvantages are grounded in biology. Indeed, "progressives" even offer us "compensatory" education for our (innate) leadership deficiencies while promising men remedial training for their (congenitally) defective "parenting."[4]

Sometimes it does not seem that we have moved very far beyond Darwin's Victorian credo, or the psychologist Bettelheim's restatement of it in 1965:

> We must start with the realization that, much as women want
> to be good scientists and/or engineers, they want first and
> foremost to be womanly companions of men and to be
> mothers.

We have not even won that first critical victory: to have our dissatisfaction with the current state of affairs be other than a source of amusement to those responsible for it. The routinely jocular dismissal of charges of misogyny brings to mind Kate Millett's remark that "sexism (unlike most other systems of oppression) is pleased with itself."

So, to answer the original questions, one of the most liberating things androcentric scientists could do for women is to stop telling us what we are, while a male-dominated society sets the limits to what we can be. Another is to let us control those aspects of science that most directly affect our lives. Yet here, too, a happy ending is not in sight. Women's health care is still delivered primarily by men. Midwifery is still illegal in most parts of the United States and, if the number of states that allow it is increasing, that is probably owing to the increasing expense of hospital-based medical care rather than to a commitment to women-controlled birthing. Indeed, many women's health activists worry that midwifery itself is being coopted into the male- and physician-dominated mainstream of the health care system, in which midwives will be allowed to operate so long as they do not offer a genuine alternative or challenge to physician-controlled birthing.

While, under the watchful eye of Title IX, there are now more women in medical and graduate schools than ever before, numbers tell only part of the story. Women are being admitted, but what will become of them after they graduate? It looks as though male doctors may cede the less lucrative and prestigious field of primary care to women (a kind of occupational "there goes the neighborhood") and attempt to retain their dominant position by concentrating in the more "elite" male bastions of surgery, medical research, and hospital administration. And the same pattern threatens in academic science, where, in the shrinking job market, more women are now permitted to enter as instructors, research fellows, or untenured assistant professors. But they are slow to rise to the tenured positions that bring greater freedom to choose what to study and more money with which to do it.

In and of themselves, increases in numbers will avail women little. For slowly we realize that present research trends often provide us with options we would rather not have and force us to make choices we would rather not make. (Do we want to get blood clots from the Pill or have our uteruses perforated by IUDs? Would we rather burn garbage, which wastes its nutrients while polluting the air, or dump it into the sea?) The increasing pace of science and its applications as technology ensure that the future will confront us with more such choices, not fewer.

Unless we participate in the processes that generate the options, when it comes to a final choice—among methods of birth control, childbirth, or abortion; among means of diagnosing, curing, preventing, or caring for illness; among methods of growing food, generating electric power, heating our houses—we will find ourselves always unhappy choosers between lesser and greater evils. And until we develop the expertise needed to create new kinds of options, we will probably not even be able to see beyond those evils, and will believe them to be unavoidable—which is, of course, the way they are usually portrayed.

Unfortunately, in the last three centuries science has become a monolith in which only certain ways of viewing nature and learning about it are acceptable. The technology that derives from it is equally monolithic; its main charge is: go forth and multiply profits. For this reason, we are usually presented with a very limited range of options, generated by "experts" and with essentially no public debate. It is not easy for feminists to provide alternatives. Women are still vastly underrepresented in the top ranks of science, where decisions are made about meanings, aims, strategies and tactics; we still cluster at the bottom of the workforce that makes science and technology "run." The few of us who enter the classrooms where we can learn to become scientists, doctors, and engineers rarely derive from our isolated and precarious situation the courage to question the system of presuppositions on which present-day science and technology are built. Indeed, one of the functions of the present-day educational system is to obscure the very existence of these questions.

Even when it has been our lot (as it often is for women) to have been sufficiently external to the traditional schools and disciplines that we are able to see beyond accepted views and question them, we are still in a double bind: if our questions are too out of the ordinary, they can be used to disqualify us as proper "material" for training; but if we suppress such questions sufficiently deep and long, we may stop thinking them and emerge from our educations as the monolith's true devotees. This is likely the reason why few women doctors have been much help to feminists working for health care alternatives. And it explains even more the orthodoxy of female scientists; for while "alternative" doctors have some measure of economic and intellectual independence as entrepreneurs, scientists work almost exclusively within the hierarchical, male-dominated academic-technical establishment where employment, research funds, ability to publish one's findings, and tenure and rank heavily depend upon conformity to the established view.

There is another problem women must face if we decide to effect change by becoming scientists: how to reconcile our lives with the traditional structure of the male-dominated professions. From the models of the old-world monastic scholar and the upper-class gentleman-scientist of independent means, has come the bourgeois scientist of today, whose success depends upon having assistants, mostly females (wife, secretary, technicians . . .), to nourish and nurture *him* in *his* single-minded search for truth.

Women are told that if we want to enter the highly competitive rat-race that is said to be the *only* way to do *real* science, we must adopt the life style and manner of the male researcher. What does that mean? That we must not form egalitarian relationships at home and at work, not marry or have children if we choose to? Obviously that is not enough. Prototypic male scientists not only reject family responsibilities (which is not to say that they do not have families; only that wives and children must know their place in the scientists' hierarchy of commitments and values); they also exploit women's labor, as unpaid wife and as the underpaid "support staff" that makes modern laboratories run.

The social structure of scientific work as well as its products reflect its male origins, and both will have to change before scientific work will be a genuine option for large numbers of women. E. O. Wilson may have been correct when he wrote in the *New York Times Magazine* that "Even with identical education and equal access to all professions, men are likely to play a disproportionate role in political life, business, and science." But this is not, as he would have us believe, because our evolutionary history has endowed women with domestic and nurturing genes and men with professional ones, but because conditions of work in the male-dominated professions are not easily integrated into the lives most women want or are able to live.

Another challenge awaiting feminist scientists is to redress the exploitative relationship of androcentric science and technology to nature. Can we come up with ways to think about nature and to probe its workings that do not degrade, kill and destroy it—ways that recognize that we, too, are its creatures? That is the chief methodological challenge that confronts the feminist revulsion against the Baconian warrior who has been riding astride nature like the bombadier-cowboy in Kubrick's *Doctor Strangelove* as he yodels off into space to destroy the world.

The man-nature antithesis was invented by men. Our job is to reinvent a relationship that will realize (in the literal sense of making real) the unity of people with nature, and will try to understand its workings from the inside— ever cognizant of the fact that our lives are inextricably bound up with it. Science is just one way of making sense of natural events; there are others. We must try to select from among them those that yield a sense that is consonant with our ideas of human dignity, free from sexual, racial, and economic oppressions. We must also insist that the technology we build must improve our lives as the people who must live with it define improvement, not as it is defined for them by the "experts."

In this book, we have tried to raise new questions and to move toward new answers; and we hope that the book will impel other women to look critically at what they accept as real and significant about our selves and the world.

If we want science and technology to serve us, we must change their present course. Both are human constructs that were produced under a particular set of historical conditions when men's domination of nature seemed a positive goal. The conditions have changed and we know now that the path we are traveling is more likely to destroy nature than to explain or improve it. Because of our different socialization and experiences, women have recognized more often than men that we are part of nature and that its fate is in human hands that have not cared for it well. We must now acquire the means and power to act on that knowledge.

October, 1981

Notes

1. Letter of July 17, 1837, reprinted in *Feminism: The Essential Historical Writings*, Miriam Schneir, ed. (New York: Knopf, Bantam Edition, 1972), p. 30.
2. *See* for example E. O. Wilson, "Human Decency is Animal," *New York Times Magazine* (October 12, 1975, pp. 38–50); "Why You Do What You Do—Sociobiology: A New Theory of Behavior," cover story in *Time* (August 1, 1977); "The Sexes: How They Differ—And Why," cover story in *Newsweek* (May 18, 1981); "The Sexes and the Brain," cover story in *Discover* (April 1981).
3. Some of the built-in biases in sociobiology and much contemporary sex differences research are discussed in the articles by Hubbard, Fried, Birke, and Lowe. For further detailed critiques and discussions of the social and political implications of this kind of socially laden research, see *Genes and Gender II: Pitfalls in Research on Sex and Gender*, Ruth Hubbard and Marian Lowe, eds. (New York: Gordian Press, 1979).
4. It is interesting that in present-day English, the verbs "to mother" and "to father" so completely reflect gender inequalities in our society, in which men contribute sperm and women provide care, that a third verb, "to parent," has come to be used to describe fathers caring for their children.

Mary Sue Henifin and
Joan Cindy Amatniek

Bibliography:
Women, Science, and Health

Contents

Figure XXVII. Dr. Ellen Swallow Richards conducting water pollution tests with an unidentified assistant (1870's). (M.I.T. Historical Collections)

Introduction

This bibliography is intended to help students, teachers, and other interested readers gain access to books and articles on science and health issues affecting women. It is an ongoing project begun by Mary Sue Henifin in 1976 and first published in 1979 in *Women Look at Biology Looking at Women*. The editors of the present volume decided that it would be useful to bring that bibliography up to date, as there has been a virtual explosion of literature on "women's health" and "sex differences" since the first bibliography was published. Much of the work on women's health has been stimulated by feminist perspectives while the burgeoning literature on sex differences in many instances is a reaction against such perspectives.

Joan Cindy Amatniek has done the majority of the work of revising the earlier bibliography. In the process, its length has increased by more than a third and we have rearranged the entries into what we hope is a more accessible format. Like the first bibliography, this one cannot be comprehensive; but we hope it will introduce you to this fast growing literature. We try to indicate the range in the available materials both historically and in terms of subject matter and viewpoint. Books or review articles with asterisks are those which seem particularly useful to us, but of course we have not read, or even seen, every reference.

The lot of the bibliographer is at times dull; the job of cataloging and arranging references remains after the excitement of discovery wanes. However, we have been interested to notice the ways in which political decisions are imbedded in the very categories we use to organize such a bibliography. For example, we include "Aging" and "Menopause" as two different categories, since we realized while working on the bibliography that it is reductionistic to assume that the health concerns of aging are necessarily related to menopause.

We look forward to further expansions and revisions of this bibliography as interest in women's role in biology, science, and health continues to grow. Hopefully, a book such as this one will embolden women to be more critical as we examine sexist theories about women's biology and health, as well as to develop theories and practices that acknowledge and encourage all of our capacities.

We would like to thank all those who have shared references and bibliographies with us. In particular we would like to acknowledge the important contributions and inspiration of Ruth Hubbard. Suggestions of references or format changes for future revisions should be sent to Professor Ruth Hubbard, Biological Laboratories, Harvard University, Cambridge, Mass. 02138.

I. What is a Woman?

EVOLUTION

Barash, David. *The Whispering Within: Evolution and the Origin of Human Nature.* New York: Harper and Row, 1978.
Blackwell, Antoinette Brown. "The alleged antagonism between growth and reproduction." *Popular Science Monthly* 29 (Sept. 1874): 606–609.
_____. "Letter [on sex and evolution]." *Popular Science Monthly* 31 (July 1876): 362–363.
_____. *The Sexes Throughout Nature.* New York: G. P. Putnam's Sons, 1875; reprinted Westport. Conn.: Hyperion Press, 1978. Excerpted in *The Feminist Papers.* Edited by Alice Rossi. New York: Bantam Books, 1973.
Brooks, W. K. "The condition of women from a zoological point of view." *Popular Science Monthly* 34 (June 1879): 145–155.
Bullough, V. L. *The Subordinate Sex.* Urbana, Ill: University of Illinois Press, 1973.
Campbell, B. *Sexual Selection and the Descent of Man, 1871–1971.* Chicago: Aldine-Atherton, 1972.
Conway, J. "Stereotypes of femininity in a theory of sexual evolution." *Victorian Studies* 14 (1970): 51.
Cutler, John Henry. *What About Women? An Examination of the Present Characteristics, Nature, Status, and Position of Women as They Have Evolved During this Century.* New York: I. Washburn, 1961.
Daly, Martin, and Margo Wilson. *Sex, Evolution, and Behavior: Adaptations for Reproduction.* North Scituate, Mass.: Duxbury Pr, 1978.
Darwin, Charles. *The Descent of Man and Selection in Relation to Sex.* New York: Appleton, 1872.
_____. *The Origin of Species.* 1st Ed. New York: Appleton, 1860.
Ehrlich, Carol. "Evolutionism and the place of women in the United States: 1885–1900." In *Women Cross-Culturally.* Edited by Ruby Rohrlich-Leavitt. Chicago: Aldine, 1975.
Ellis, Havelock. *Women and Marriage or Evolution in Sex: Illustrating the Changing Status of Women.* London: William Reeves, 1888.
Engels, Friedrich. *The Origin of the Family, Private Property, and the State.* Edited with an introduction by Eleanor Leacock. New York: International Publishers, 1972.
Fisher, Elizabeth. *Sexual Evolution and the Shaping of Society.* Garden City, N.Y.: Anchor Pr, 1980.
Gamble, Eliza Burt. *Evolution of Woman: an Inquiry into the Dogma of Her Inferiority to Man.* London, 1849.
Ghiselin, Michael T. *The Economy of Nature and the Evolution of Sex.* Berkeley, Calif.: University of California Press, 1974.
Glaser, Otto Charles. "The constitutional conservatism of women." *Popular Science Monthly* 64 (Sept. 1911): 299–302.

Goldberg, S. *The Inevitability of Patriarchy: Why the Biological Differences Between Men and Women Always Produce Male Domination.* New York: Morrow, 1973.

Hapgood, Fred. *Why Males Exist: An Inquiry into the Evolution of Sex.* New York: Morrow, 1979.

Hardaker, M. A. "Science and the woman question." *Popular Science Monthly* 35 (March 1882): 577–583.

Holliday, L. *The Violent Sex: Male Psychology and the Evolution of Consciousness.* New York: Bluestocking, 1978.

Hrdy, Sarah. *The Woman That Never Evolved.* Cambridge, Mass.: Harvard University Press, 1981.

Hubbard, Ruth. "Have Only Men Evolved?" In this volume.

Jacoby, Robin Miller. "Science and sex roles in the Victorian era." In *Biology as a Social Weapon.* Edited by the Ann Arbor Science for the People Collective. Minneapolis, Minn: Burgess, 1977.

Johnson, G. W. *The Evolution of Woman from Subjection to Comradeship.* London: Robert Holden, 1926.

Johnson, Robert. *Aggression in Man and Animals.* Philadelphia: Saunders, 1972.

Leacock. Eleanor. Review of *The Inevitability of Patriarchy*, by Steven Goldberg. *American Anthropologist* 76 (1974): 363–365.

Leakey, Richard E., and Roger Levin. *Origins.* New York: Dutton, 1977.

Leibowitz, Lila. "Desmond Morris is wrong about breasts, buttocks and body hair." *Psychology Today*, 9 March 1970.

Lewontin, R. C. *The Genetic Basis of Evolutionary Change.* New York: Columbia University Press, 1978.

_____. "Sociobology: another biological determinism." *International Journal of Health Services* 10 (1980): 347–364.

Maccoby, Eleanor E. "Sex in the social order." *Science* 182 (1973): 469–471.

Maudsley, Henry. "Sex in mind and in education." *Popular Science Monthly* 27 (June 1874): 1982–214.

Maynard-Smith, J. *Evolution of Sex.* Cambridge: At the University Press, 1978.

Morais, Nina. "The woman question." *Popular Science Monthly* 21 (May 1882): 70–78.

Morgan, Elaine. *The Descent of Woman.* New York: Stein and Day, 1972.

Nowak, Mariette. *Eve's Rib: A Revolutionary New View of the Female.* New York: St. Martin, 1980.

Reed, Evelyn. *Women's Evolution: From Matriarchal Clan to Patriarchal Family.* New York: Pathfinder Pubns, 1974.

Reyburn, Wallace. *The Inferior Sex.* New York: Pathfinder Pubns, 1974.

Rhodes, Philip. *Woman: A Biological Study of the Female Role in the Twentieth-Century Society.* London: Corgi, 1969.

*Shields, S. A. "Functionalism, Darwinism, and the psychology of women: A study in social myth." *American Psychologist*, July 1975, pp. 739–754.

Smith, A. Lapthorn. "Higher education of women and race suicide." *Popular Science Monthly* 58 (March 1905): 466–473.

Spencer, Herbert. "On the comparative psychology of the sexes." In his *The Study of Sociology*. New York: Appleton, 1875.

_____. *The Principles of Sociology*. Vol. 1, Part III: Westport, Conn.: Greenwood, 1975 (originally 1897).

Symons, Donald. *The Evolution of Human Sexuality*. New York: Oxford University Press, 1981.

_____. "Précis of the evolution of human sexuality." *The Behavioral and Brain Sciences* 3 (1980): 171–214.

*Tanner, Nancy. *On Becoming Human*. New York: Cambridge University Press, 1981.

Tanner, Nancy, and Adrienne Zihlman. "Women in evolution. Part I: Innovation and selection in human origins." *Signs* 1 (1976): 585–608.

Tiger, Lionel. *Men in Groups*. New York: Random House, 1969.

Van De Warke, Ely. "The genesis of woman." *Popular Science Monthly* 27 (July 1874): 269–276.

_____. "Sexual cerebration." *Popular Science Monthly* 28 (July 1875): 187–300.

Wallington, Emma. "The physical and intellectual capacities of woman equal to those of man." *Anthropologia* 1 (1874): 552–556.

Washburn, Sherwood L. "Tools and human evolution." *Scientific American* 203 (Sept. 1960): 63–77.

Weitz, Shirley. *Sex Roles: Biological, Psychological and Social Foundations*. Oxford: Oxford University Press, 1977.

Wertheim, W. F. *Evolution and Revolution: The Rising Waves of Emancipation*. London: Penguin Books, 1974.

White, Frances Emily. "Woman's place in nature." *Popular Science Monthly* 28 (Jan. 1875): 292–300.

SOCIAL BIOLOGY AND PRIMATE STUDIES

Abele, L. G., and S. Gilchrest. "Homosexual rape and sexual selection in ancanthocephalan worms." *Science* 197 (1977): 81–83.

Alper, J. et al. "The implications of sociobiology." *Science* 192 (1976): 424.

Ann Arbor Science for the People Collective: Sociobiology Study Group. "Sociobiology: A New Biological Determinism." In *Biology as a Social Weapon*. Edited by the Ann Arbor Science for the People Collective. Minneapolis, Minn.: Burgess, 1977.

Barash, David B. *Sociobiology and Behavior*. New York: Elsevier North Holland, 1977.

_____. "Sociobiology of rape in mallards: Responses of the mated male." *Science* 197 (1977): 788–789.

Breines, Wini, Margaret Cerullo, and Judith Stacey. "Social biology, family studies, and antifeminist backlash." *Feminist Studies* 4 (Feb. 1978): 43–68.

Broad, W. J. "Primate lust." *Science News* 113 (1978): 397.

Brothwell, Don. *Biosocial Man.* London: Eugenics Society, 1977.

*Caplan, A. L., ed. *The Sociobiology Debate: Readings on Ethical and Scientific Issues.* New York: Harper and Row, 1978.

Chagnon, Napoleon and William Irons, eds. *Evolutionary Biology and Human Social Behavior.* North Scituate, Mass.: Duxbury Pr (Wadsworth), 1979.

Chasin, Barbara. "Sociobiology: A sexist synthesis." *Science for the People* 9 (May-June 1977): 3.

Dawkins, Richard. *The Selfish Gene.* New York: Oxford University Press, 1976.

Doty, Robert L. "A cry for the liberation of the female rodent: Courtship and copulation in rodentia." *Psychological Bulletin* 81 (1974): 159-172.

Goodall, Jane. *In the Shadow of Man.* Boston: Houghton Mifflin, 1971.

Gould, S. J. "Biological potentiality versus biological determinism." *Natural History*, May 1976, p. 12.

_____. "On human nature" [book review]. *Human Nature* 1 (Oct. 1978): 20-33.

Gross, H. E. et al. "Considering 'a biosocial perspective on parenting.'" *Signs* 4.4 (Summer 1979): 695-717.

Hamilton, W. D. "Innate social aptitudes of man." In *Biosocial Anthropology.* Edited by R. Fox. London: Halstead Press, 1975.

Haraway, Donna. "The Biological Enterprise: Sex, Mind, and Profit from Human Engineering to Sociobiology." *Radical History Review* 20 (Spring-Summer 1979): 206-237.

*Herschberger, Ruth. *Adam's Rib.* New York: Har/Row, 1970 (originally 1948).

Hrdy, Sarah Blaffer. *The Langurs of Abu: Female and Male Strategies of Reproduction.* Cambridge, Mass.: Harvard University Press, 1977.

Hubbard, R. "From termite to human behavior" [book review of E. O. Wilson's *On Human Nature*]. *Psychology Today* 12 (Oct. 1978): 124-134.

Kass-Simon, G. "Female strategies: Adaptations and adaptive animal significance." In *Beyond Intellectual Sexism.* Edited by Joan L. Roberts. New York: McKay, 1976.

Kolata, G. B. "Primate behavior: Sex and the dominant male." *Science* 191 (Jan. 1976): 55-56.

Kurchison, Carl. "Social behavior in infrahuman primates." In *Handbook of Social Psychology*, 2nd ed., Vol. 3. Edited by Carl Kurchison. New York: Russell and Russell, 1968.

Lancaster, Jane B. "In praise of the achieving female monkey." In *The Female Experience.* Edited by Carol Tavris. Del Mar, Calif.: CRM, 1973, pp. 5-9.

*_____. *Primate Behavior and the Emergence of Human Culture.* New York: Holt, Rinehart, and Winston, 1975.

*Leibowitz, Lila. *Females, Males, Families: A Biosocial Approach.* North Scituate, Mass.: Duxbury Pr, 1978.

Leutenegger, W. "Scaling of sexual dimorphism in body size and breeding system in primates." *Nature* 272 (1978): 610–611.

Lewontin, R. C. "Biological determinism as a social weapon." In *Biology as a Social Weapon*. Edited by the Ann Arbor Science for the People Collective. Minneapolis, Minn.: Burgess, 1977.

_____. "The fallacy of biological determinism." *The Sciences*, March-April 1976, p. 6.

_____. "Slight of hand." [Review of *Genes, Mind, and Culture*] *The Sciences*, July/Aug. 1981, pp. 23–26.

Lowe, Marian, and Ruth Hubbard. "Sociobiology and biosociology: Can science prove the biological basis of sex differences in behavior?" In *Genes and Gender II: Pitfalls in Research on Sex and Gender*. Edited by R. Hubbard and M. Lowe. New York: Gordian Press, 1979.

Lumsden, Charles, and Edward O. Wilson. *Genes, Mind, and Culture: The Coevolutionary Process*. Cambridge, Mass.: Harvard University Press, 1981.

Miller, G. S. "The primate basis of human sexual behavior." *Quarterly Review of Biology* 6 (1931): 379–419.

Mitchell, G., and E. M. Brandt. "Paternal behavior in primates." In *Primate Socialization*. Edited by F. Poirier. New York: Random House, 1972.

*Montagu, Ashley, ed. *Sociobiology Examined*. New York: Oxford University Press, 1980.

Ralls, Katherine. "Mammals in which females are larger than males." *The Quarterly Review of Biology* 51 (June 1976): 245–276.

Reed, Evelyn. *An Answer to the Naked Ape and Other Books on Aggression*. New York: Pathfinder Press, 1971.

_____. *Sexism and Science*. New York: Pathfinder Pubns, 1978.

Rossi, Alice S. "A biosocial perspective on parenting." *Daedalus* 106 (Spring 1977): 1–31.

Ruse, Michael. *Is Science Sexist?* Dordrecht, The Netherlands: Reidel, 1981.

_____. *Sociobiology: Sense or Nonsense?* Hingham, Mass.: Kluwer Boston, 1979.

*Sahlins, Marshall. *The Use and Abuse of Biology: An Anthropological Critique of Sociobiology*. Ann Arbor: The University of Michigan Press, 1976.

Sherman, Paul W. "Nepotism and the evolution of alarm calls." *Science* 197 (1977): 1246–1253.

Tavris, Carol. "Male supremacy is on the way out: It was just a phase in the evolution of culture." Interview with Carol Tavris. *Psychology Today*, Jan. 1975.

Tiger, Lionel. "Male dominance? Yes. Alas. A sexist plot? No." *New York Times Magazine*, 25 October 1970.

Warshall, Peter. "*Sociobiology*" [book review]. *CoEvolution Quarterly* 9 (Spring 1976): 88–90.

White, Elliott, ed. *Sociobiology and Human Politics*. Lexington, Mass.: Lexington Bks, 1981.

Wickler, Wolfgang. *The Sexual Code: The Social Behavior of Animals and Men*. Garden City, N.Y.: Anchor Pr (Doubleday), 1973.

Williams, George C. *Sex and Evolution*. Princeton: Princeton University Press, 1975.

Williams, Charlotte Neely. "The limitations of the male/female activity distinction among primates: An extension of Judith K. Brown's 'A note on the division of labor by sex.'" *American Anthropologist* 73 (1971): 805–806.

Wilson, E. O. *On Human Nature*. Cambridge, Mass.: Harvard University Press, 1978.

_____. *Sociobiology: The New Synthesis*. Cambridge, Mass.: Belknap Press of Harvard University Press, 1975.

Yerkes, R. M. "Social behavior of chimpanzees: Dominance between mates in relation to sexual status." *Journal of Comparative Psychology* 30 (Aug. 1940): 174–186.

_____. "Social dominance and sexual status in the Chimpanzee." *Quarterly Review of Biology* 14 (1939): 115–136.

Zihlman, A., and N. Tanner. "Gathering and the hominid adaptation." In *Female Hierarchies*. New York: Harry Frank Guggenheim Foundation Third International Symposium, April 1974.

Zuckerman, Solly. "The social life of primates." *The Realist* 6 (1929): 72–88.

SEX DIFFERENCES: HORMONAL, GENETIC, PHYSIOLOGICAL, AND PSYCHOLOGICAL

Al-Issa, Ihsan. *The Psychopathology of Women*. Englewood Cliffs, N.J.: Prentice-Hall, 1980.

Annals of the New York Academy of Sciences [Language, Sex and Gender, special issue]. 327 (June 25, 1979).

Arnold, Franz Xaver. *Woman and Man: Their Nature and Mission*. Translated by Rosaleen Brennan. New York: Herder and Herder, 1963.

Austin, Helen S., Allison Parelman, and Anne Fisher. *Sex Roles: A Research Bibliography*. Washington, D.C.: U.S. Government Printing Office, 1978.

Bart, Pauline B. "Biological determinism and sexism: Is it all in the ovaries?" In *Biology as a Social Weapon*. Edited by the Ann Arbor Science for the People Collective, Minneapolis, Minn.: Burgess, 1977.

Belotti, E. G. *What are Little Girls Made Of*. New York: Schocken, 1976.

Benbow, Camilla, and Julian Stanley. "Sex differences in mathematical ability: Fact or artifact?" *Science* 210 (12 Dec. 1980): 1262–1264.

Bernard, Jesse. *Sex Differences: An Overview*. In *Psychology: A Programmed Modular Approach*. Module No. 6. Homewood, Ill.: Learning Syst., 1975.

Bleier, Ruth H. "Brain, Body and Behavior." In *Beyond Intellectual Sexism*. Edited by Joan L. Roberts. New York: McKay, 1976.

_____. "Myths of the biological inferiority of women: An exploration of the sociology of biology research." *University of Michigan Papers in Women's Studies* 2 (1976):39–63.

Burnham, Dorothy. "Biology and gender: False theories about women and blacks." *Freedom Ways* 17 (1977): 8–13.

Burns, Robert B. "Male and female perceptions of their own and the other sex." *The British Journal of Social and Clinical Psychology* 16.3 (Sept. 1977): 213–220.

Burstyn, Joan N. "Brain and intellect: Science, applied to a social issue: 1860–1875." XII Congrés International D'Histoire des Sciences. Paris 1968. *Actes*, Tome IX.

_____. "Education and sex: The medical case against higher education for women in England, 1870–1900." *Proceedings American Philosophical Society* 117 (1973): 79.

Chafetz, J. S. *Masculine/Feminine or Human: An Overview of Sex Roles.* Itasca, Ill.: Peacock Pubs, 1974.

Chetwynd, J., and O. Harnett. *The Sex Role System: Psychological and Sociological Perspectives.* London: Routledge, Kegan Paul, 1978.

Clarke, Edward H. *Sex in Education.* Boston: James R. Osgood and Company, 1874.

Delauney, G. "Equality and inequality in sex." *Popular Science Monthly* 20 (Dec. 1881): 184–192.

De Leeuw, Hendrick. *Women: The Dominant Sex.* New York: A. S. Barnes, Thomas Yoseloff, 1957.

Deux, Kay. *The Behavior of Women and Men.* Monterey, Calif.: Brooks/ Cole, 1977.

deWied, David, and Pieter A. van Keep. *Hormones and the Brain.* Baltimore: University Park, 1980.

Distant, W. L. "On the mental differences between the sexes." *Journal of the Anthropological Institute* 4 (1875): 78–85.

Dörner, G., and M. Kawakami. *Hormones and Brain Development.* New York: Elsevier, 1978.

Duberman, L., ed. *Gender and Sex in Society.* New York: Praeger, 1975.

Dyer, Ken. "Female athletes are catching up: sex differential in track athletics and swimming performances are declining steadily." *New Scientist*, 22 September 1977, pp. 722–723.

Eichler, Margrit. *The Double Standard.* New York: St. Martin Press, 1979.

Eisenstein, Hester, and Alice Jardine. *The Future of Difference.* Boston: G. K. Hall, 1980.

Ellis, Havelock. *Man and Woman: A Study of Human Secondary Sexual Characters.* London: Walter Scott, 1894.

Fairweather, Hugh. "Sex differences in cognition." *Cognition* 4 (1976): 231–280.

Fee, Elizabeth. "Nineteenth century craniology: The study of the female skull." *Bulletin of the History of Medicine* 53 (1979): 415–433.

Filene, P. G. *Him Her Self: Sex Roles in Modern America*. New York: New American Library, 1975.

Fried, Barbara. "Boys Will Be Boys Will Be Boys: The Language of Sex and Gender." In this volume.

Friedman, H. S. "The scientific error of sexist research: a case study." *Sex Roles* 6 (October 1980): 747–750.

Friedman, R. C., R. M. Richart, and R. L. Van de Wille, eds. *Sex Differences in Behavior*. A Conference Sponsored by the International Institute for the Study of Human Reproduction. College of Physicians and Surgeons of Columbia University. New York: Wiley, 1974.

Frieze, I. H. et al. *Women and Sex Roles: A Social Psychological Perspective*. New York: Norton, 1978.

George, W. L. *Intelligence of Woman*. Boston: Little, Brown, 1916.

Giele, J. Z. *Women and the Future: Changing Sex Roles in Modern America*. New York: Free Pr, 1978.

Gomberg, Edith S., and Violet Franks. *Gender and Disordered Behavior: Sex Differences in Psychopathology*. New York: Brunner/Mazel, 1979.

Gooch, Stan. "Right brain, left brain." *New Scientist* 87 (1980): 790–792.

Gould, Stephen. *The Mismeasure of Man*. New York: Norton, 1981.

_____. "Women's brains." *Natural History* 87 (1978): 44–50.

Goy, R. W. "Early hormonal influences on the development of sexual and sex-related behavior." In *The Neurosciences: Second Study Program*. Edited by Francis O. Schmitt. New York: Rockefeller University Press, 1970.

Goy, Robert W., and Bruce S. McEwen. *Sexual Differentiation of the Brain*. Cambridge: MIT Press, 1980.

Hall, Diana Long. "Biology, sex hormones and sexism in the 1920's." In *Women and Philosophy*. Edited by C. C. Gould and M. W. Wartofsky. New York: Putnam, 1976.

Heckerman, Carol Landau. *The Evolving Female: Women in Psychosocial Context*. New York: Human Sci Pr, 1980.

Henley, Nancy. *Body Politics: Power, Sex, and Nonverbal Communication*. Englewood Cliffs, N.J.: Prentice Hall, 1977.

Hochschild, Arlie Russel. "A review of sex role research." *American Journal of Sociology* 78 (Jan. 1973): 1011–1029.

*Hubbard, Ruth, and Marian Lowe, eds. *Genes and Gender II: Pitfalls in Research on Sex and Gender*. New York: Gordian, 1979.

Hutt, Corinne. *Males and Females: Sex Differentiation—Genetic, Hormonal, Psychological*. Middlesex, England: Penguin, 1972.

International Political Science Association, *Research Committee on Sex Roles and Politics Newsletter*. London: Available from: Margheirta Rendel, University of London, Institute of Education, 55 Gordon Square, London WCIH ONU England.

Israel, S. Leon. "The essence of womanhood: Biological inequality." *Obstetrics and Gynecology* 29 (May 1967): 750–756.

*Kaplan, Alexandra G., and Joan P. Bean, eds. *Beyond Sex-Role Stereotypes: Readings Toward a Psychology of Androgyny*. Boston: Little, Brown, 1976.

Kopp, Claire B., ed. *Becoming Female: Perspectives on Development*. New York: Plenum Pub, 1979.

Laws, Judith Long. *The Second X: Sex Role and Social Role*. New York: Elsevier, 1979.

*Lee, Patrick C., and Robert S Stewart. *Sex Differences: Cultural and Developmental Dimensions*. New York: Urizen Bks, 1976.

Lips, Hilary M. and Colwill, Nina Lee. *The Psychology of Sex Differences*. Englewood Cliffs, N.J.: Prentice–Hall Spectrum Book, 1978.

Lowe, Marian. "Social Bodies: The Interaction of Culture and Women's Biology." In this volume.

Luria, Zella and Ruse, Mitchel P. *Psychology of Human Sexuality*. New York: Wiley, 1979.

Maccoby, Eleanor E., ed. *The Development of Sex Differences*. Stanford, Calif.: Stanford University Press, 1966.

Maccoby, Eleanor E., and Carol Nagy Jacklin. *The Psychology of Sex Differences*, 2 vols. Stanford, Calif.: Stanford University Press, 1974.

McEwen, Bruce S. "Interactions between hormones and nerve tissues." *Scientific American* 235 (July 1976): 48–58.

McGlone, Jeannette. "Sex differences in human brain asymmetry: a critical survey." *The Behavioral and Brain Sciences* 3 (1980): 215–263.

Miller, Alan V. *The Genetic Imperative: Fact and Fantasy in Sociobiology*. Toronto: Pink Triangle Press, 1979. Available from: Canadian Gay Archives, Box 639, Station A, Toronto, Ontario M5W K2, Canada.

Money, J., ed. *Sex Research, New Developments*. New York: Holt, Rinehart, and Winston, 1965.

_____, and P. Tucher. *Sexual Signatures: On Being a Man or A Woman*. Boston: Little, Brown, 1975.

Money, John, and Anke A. Ehrhardt. *Man and Woman, Boy and Girl: Differentiation and Dimorphism of Gender Identity, from Conception to Maturity*. Baltimore: The Johns Hopkins University Press, 1972.

Naftolin, Frederick, and Eleanore Butz, eds. *Sexual Dimorphism*. [Special Issue of *Science* Vol. 211, March 1981].

Oakley, Ann. *Sex, Gender, and Society*. New York: Harper Colophon Books, 1972.

Parsons, Jacquelynne E. *The Psychology of Sex Differences and Sex Roles*. Washington, D.C.: Hemisphere Publishing Corporation, 1980.

Pfaff, D. et al. "Neurophysiological analysis of mating behavior response of hormone sensitive reflexes." In *Progress in Physiological Psychology*, vol. 5. Edited by E. Stellar and J. M. Sprague. New York: Academic Press, 1973.

Psychology of Women Quarterly. New York: Human Sci Pr.

Ramey, Estelle P. "Men's Cycles (They Have Them Too, You Know)." In

Beyond Sex-Role Stereotypes. Edited by A. Kaplan and J. Bean. Boston: Little, Brown, 1976.

_____. "Sex hormones and executive ability." In *Women and Success*. Edited by Ruth B. Kundsin. New York: Morrow, 1974, pp. 248–256.

Reinisch, June. "Prenatal exposure to synthetic progestins increases potential for aggression in humans." *Science* 211 (13 March 1981): 1171–1173.

Rohrbaugh, Joanna Bunker. *Women: Psychology's Puzzle*. New York: Basic, 1979..

Romanes, George J. "Mental differences in men and women." *Popular Science Monthly* 31 (July 1887): 372–382.

Rosenberg, Miriam. "The biological basis for sex role stereotypes." *Contemporary Psychoanalysis* 9 (1973): 374–391.

Sayers, Janet. "Biological determinism, psychology, and the division of labour by sex." *International Journal of Women's Studies* 3.3 (May/June 1980): 241–257.

Schafer, Alice and Mary W. Gray. "Sex and Mathematics." *Science* 211 (16 Jan. 1981): front page editorial.

Schaffer, Kay F. *Sex-Role Issues in Mental Health*. Reading, Mass.: Addison Wesley, 1980.

Seward, John, and Georgene Seward. *Sex Differences: Mental and Temperamental*. Lexington, Mass.: Lexington Bks, 1980.

Sex Roles: A Journal of Research. New York: Plenum Pub.

Sherman, Julia. *On the Psychology of Woman: A Survey of Empirical Studies*. Springfield, Ill.: C. C. Thomas, 1975.

_____. *Sex-Related Cognitive Differences*. Springfield, Ill.: C. C. Thomas, 1978.

Spence, Janet and Robert Helmreich. *Masculinity and Femininity: Their Psychological Dimensions, Correlates, and Antecedents*. Austin, Texas: University of Texas Press, 1978.

Star, Susan Leigh. "Sex Differences and the Dichotomization of the Brain: Methods, Limits, and Problems in Research on Consciousness." In *Genes and Gender II: Pitfalls in Research on Sex and Gender*. Edited by R. Hubbard and M. Lowe. New York: Gordian Press, 1979.

*Tavris, Carol, and Carole Offir. *The Longest War: Sex Differences in Perspective*. New York: Harcourt, Brace and Jovanovich, 1977.

Teitelbaum, M. *Sex Differences: Social and Biological Perspectives*. New York: Anchor Pr, 1976.

Thompson, Helen Bradford. *The Mental Traits of Sex*. Chicago: University of Chicago Press, 1903.

Tiger, L. "The possible biological origins of sexual discrimination." *Impact of Science on Society* 20 (1970): 29–44.

*Tobach, Ethel, and Betty Rosoff, eds. *Genes and Gender I: First in a Series on Hereditarianism and Women*. New York: Gordian, 1978.

_____. *Genes and Gender III: Third in a Series on Hereditarianism and Women*. New York: Gordian, 1980.

Tobach, Ethel, ed. *Genes and Gender IV.* New York: Gordian, in press.

*Unger, Rhoda K. *Female and Male: Psychological Perspectives.* New York: Harper and Row, 1979.

Unger, R. K. and F. L. Denmark, eds. *Woman: Dependent or Independent Variable.* New York: Psychological Dimensions, 1975.

Van Den Berghe, Pierre L. *Another Perspective on Man. . . . and Woman and Child: Age and Sex Differentiation.* Belmont, Calif.: Wadsworth, 1973.

Walum, Laurel. *The Dynamics of Sex and Gender: A Sociological Perspective.* Chicago: Rand McNally, 1977.

Weitz, Shirley. *Sex Roles: Biological, Psychological and Social Foundations.* New York: Oxford University Press, 1977.

Weitzman, Lenore J. *Sex Role Socialization.* Palo Alto, Calif.: Mayfield Publishing Co., 1979.

Whalen, Richard E. "Sexual differentiation: Models, methods, and mechanisms." In *Sex Differences in Behavior.* Edited by R. C. Friedman, R. M. Richart, and R. L. Van de Wiele. New York: Wiley, 1974.

Whitbeck, Caroline. "Theories of sex differences." In *Women and Philosophy.* Edited by C. C. Gould and M. W. Wartofsky. New York: Putnam, 1976.

Williams, Juanita H. *Psychology of Women: Behavior in a Biosocial Context.* New York: Norton, 1977.

Williams, Juanita H., ed. *Psychology of Women: Selected Readings.* New York: Norton, 1979.

Wittig, M. A., and Peterson, A. C. *Sex-related Differences in Cognitive Functioning: Developmental Issues.* Los Angeles: Academic Press, 1979.

WOMEN AND MEN: CROSS-CULTURAL PERSPECTIVES

Andreski, Iris, ed. *Old Wives' Tales: Life Stories from Ibibioland.* New York: Schocken, 1970.

Ardener, Shirley, ed. *Perceiving Women.* New York: Halstead, 1975.

Bachofen, Johann J. *Myth, Religion and Mother Right.* Translated by Ralph Manheim. Princeton: Princeton University Press, 1967 (originally 1861).

Beidelman, T. O. *The Kaguru, a Matrilineal People of East Africa.* New York: Holt, Rinehart and Winston, 1971.

Bettelheim, Bruno. *Symbolic Wounds: Puberty Rites and the Envious Male.* New York: Collier, 1962.

Blurton-Jones, N. G., and M. J. Konner. "Sex differences in behavior of London and Bushmen children." In *Comparative Ecology and Behavior of Primates.* Edited by R. P. Michael and J. H. Cook. New York: Academic Press, 1973.

Bricker, Victoria Reifler. "Sex roles in cross-cultural perspective." *American Ethnologist* 2 (Nov. 1975): 4.

Briffault, Robert. *The Mothers.* New York: Grosset and Dunlap, 1963 (originally 1927).

Brown, Judith K. "Cross-cultural perspectives on the female life cycle." In *Handbook of Cross Cultural Human Development.* Edited by Robert

Munroe, Ruth Munroe, and Beatrice Whiting. New York: Garland Pub, in press.

Caplan, Patricia, and Janet Bujra, eds. *Women United, Women Divided: Cross-Cultural Perspectives on Female Solidarity*. London: Tavistock Publications, 1978.

Caulfield, Mina. "Universal sex oppression? A critique from Marxist anthropology." *Catalyst* (Summer 1977): 10–11.

Clignet, Remi. *Many Wives, Many Powers: Authority and Power in Polygynous Families*. Evanston, Ill.: Northwestern University Press, 1970.

Cornelison, Ann. *Woman of the Shadows*. Boston: Little, Brown, 1976.

Curtin, Katie. *Women in China*. New York: Pathfinder Pubns, 1975.

Davis, Elizabeth Gould. *The First Sex*. Baltimore: Penguin, 1971.

Diner, Helen. *Mothers and Amazons*. New York: The Julia Press, 1965.

Draper, P. "Kung women: Contrasts in sexual egalitarianism in the foraging and sedentary contexts." In *Toward an Anthropology of Women*. Edited by R. Reiter. New York: Monthly Rev, 1975.

Endholm, Felicity, Olivia Harris, and Kate Young. "Conceptualizing women." *Critique of Anthropology* 3 (1977): 9–10.

*Etienne, Mona, and Eleanor Leacock, eds. *Women and Colonization: Anthropological Perspectives*. New York: Praeger, 1980.

Evans-Pritchard, E. E. *The Position of Women in Primitive Societies and Other Essays in Social Anthropology*. New York: Free Pr, 1965.

Farber, Seymour M., and H. L. Roger, eds. *Man and Civilization: The Potential of Women: A Symposium*. New York: Free Pr, 1963.

Farmer, Claire R., ed. *Women and Folklore*. Austin. Tex.: University of Texas Press, 1975.

Fee, Elizabeth. "The sexual politics of Victorian social anthropology." In *Clio's Consicousness Raised: New Perspectives on the History of Women*. Edited by M. Hartman and L. Banner. New York: Harper and Row, 1974.

Food and Agricultural Organization of the United Nations. *The Missing Half: Woman 1975*. New York: United Nations, 1975.

Ford, Clellan S., and Frank A. Beach. *Patterns of Sexual Behavior*. New York: Harper and Bros, 1951.

*Friedl, Ernestine. *Women and Men: An Anthropologist's View*. New York: Holt, Rinehart, Winston, 1975.

Gale, Fay, ed. *Women's Role in Aboriginal Society*. Canberra: Australian Institute of Aboriginal Studies, 1970.

Giele, Janet Zoflinger, and Audrey Chapman Smock. *Women: Roles and Status in Eight Countries*. Somerset, N.J.: Wiley-Interscience, 1977.

Goffman, Erving. *Gender Advertisements*. New York: Harper and Row, 1979.

Goody, Jack, ed. *The Character of Kinship*. London: Cambridge University Press, 1973.

Hall, Nor. *The Moon and the Virgin: Reflections on the Archetypal Feminine*. New York: Harper and Row, 1980.

Hammond, Dorothy, and Alta Jablow. *Women: Their Economic Role in Traditional Societies*. Reading, Mass.: Addison-Wesley, 1973.

_____. *Women: Their Familial Roles in Traditional Societies*. Menlo Park, Calif.: Cummings, 1975.

Harris, Marvin. *Cannibals and Kings: The Origins of Cultures*. New York: Random House, 1977.

_____. *Culture, People, Nature*. 2nd edition. New York: Crowell, 1975.

Hoch-Smith, Judith, and Anita Spring, eds. *Women in Ritual and Symbolic Roles*. London: Plenum Pub, 1978.

Hogbin, Herbert Ian. *The Island of Menstruating Men*. Scranton, Pa.: Chandler Publishing Co., 1970.

Huffer, Virginia. *The Sweetness of the Fig: Aboriginal Women in Transition*. Seattle: University of Washington Press, 1981.

Iglitzin, Lynn B., and Ruth Ross, eds. *Women in the World: A Comparative Study*. Santa Barbara, Calif.: American Bibliographical Center-Clio Press, 1976.

*Jacobs, Sue-Ellen. "Women in perspective: a guide for cross-cultural studies" [a bibliography]. Urbana Ill.: University of Illinois Press, 1976.

Joyce, Thomas, ed. *Women of All Nations: A Record of their Characteristics, Habits, Manners, Customs and Influence*. New York: Funk and Wagnalls, 1912.

Kaberry, Phyllis Mary. *Aboriginal Women: Sacred and Profane*. London: Routledge and Sons, 1970.

_____. *Women of the Grassfields*. London: H. M. Stationery Office, 1952.

Kessler, Evelyn S. *Women: An Anthropological View*. New York: Holt, Rinehart and Winston, 1976.

Kessler, S. T., and W. McKenna. *Gender, An Ethnomethodological Approach*. New York: Wiley, 1978.

Kitzinger, Sheila. *Women as Mothers: How They See Themselves in Different Cultures*. New York: Random House, 1980.

Johnson, Barclay D. "Durkheim on women." In *Woman in a Man-Made World*. 2nd Edition. Edited by Nona Y. Glazer and Helen Waehrer. Chicago: Rand McNally, 1977.

Kolata, Gina B. "Kung hunter-gatherers: Feminism, diet, and birth control." *Science* 185 (1974): 932–934.

*Lamphere, Louise. Review essay on anthropology. *Signs* 2 (Spring 1977): 612–627.

Landes, Ruth. *The City of Women: Negro Women Cult Leaders of Babia, Brazil*. New York: 1947.

_____. *The Ojibwa Women*. New York: Norton, 1971 (originally 1938).

Lapidus, Gail Warshofsky. *Women in Soviet Society: Equality, Development, and Social Change*. Berkeley: University of California Press, 1978.

Leacock, Eleanor Burke. "Class, Commodity and the Status of Women." In *Women Cross-Culturally: Change and Challenge*. Edited by Ruby Rohrlich-Leavitt. Chicago: Aldine, 1975.

* _____. *Myths of Male Dominance: Collected Articles*. New York: Monthly Rev, 1981.

_____. Review of Louise Spindler, *Menomini Women and Cultural Change.* *American Anthropologist* 65 (1963): 4.

_____. "Women in egalitarian societies." In *Becoming Visible: Women in European History.* Edited by Bridenthal and Koonz. New York: Houghton Mifflin, 1977.

Lee, Richard B., and Irvin DeVore. *Kalahari Hunter-Gatherers: Studies of the !Kung and their Neighbors.* Cambridge, Mass.: Harvard University Press, 1976.

MacCormack, Carol P. *Biological Events and Cultural Control.* Chicago: University of Chicago Press, 1977.

Malinowski, Bronislaw. *Sex and Depression in Savage Society.* New York: Harcourt, Brace and Co., 1927.

_____. *The Sexual Life of Savages in North-Western Melanesia.* New York: Halcyon, 1941.

*Martin, M. Kay, and Barbara Voorhies. *Female of the Species.* New York: Columbia University Press, 1975.

Mason, Otis T. "Environment in relation to sex in human culture." *Popular Science Monthly* 55 (Feb. 1902): 336–345.

_____. *Women's Share in Primitive Culture.* New York: Gordon Press, 1976 (originally 1895).

*Matthiasson, C. J., ed. *Many Sisters: Women in Cross-Cultural Perspective.* New York: Free Pr, 1974.

*Mead, Margaret. *Male and Female: A Study of the Sexes in a Changing World.* New York: Dell, 1949.

*_____. *Sex and Temperment in Three Primitive Societies.* New York: Dell, 1935.

Mead, Margaret, and Rhoda Metraux. *Aspects of the Present.* New York: Morrow, 1980.

Minturn, Leigh, and William Lambert. *Mothers of Six Cultures.* New York: Wiley, 1964.

Montagu, Ashley. *The Natural Superiority of Women.* New York: Collier, 1968.

Murphy, Yolanda, and Robert F. Murphy. *Women of the Forest.* New York: Columbia University Press, 1974.

*Ortner, Sherry and Harriet Whitehead, eds. *The Cultural Construction of Gender and Sexuality.* New York: Cambridge University Press, 1981.

Parker, Seymour, and Hilda Parker. "The myth of male superiority: rise and demise." *American Anthropologist* 81(2): 289–309.

Paulme, Denise, ed. *Women of Tropical Africa.* Berkeley and Los Angeles: University of California Press, 1963.

Quinn, Naomi. "Anthropological studies on women's status." *Annual Review of Anthropology* 6(1977): 181–225.

Raphael, Dana, ed. *Being Female: Reproduction, Power, and Change.* Chicago: Aldine, 1973.

Rapp, Rayna. "Anthropology." *Signs* 4.3 (Spring 1979): 497–513.

*Reiter, R., ed. *Toward an Anthropology of Women*. New York: Monthly Rev, 1975.

Rogers, Susan Carol. "Woman's place: a critical review of anthropological theory." *Comparative Studies in Society and History* 20.1 (1978).

Rohrlich-Leavitt, Ruby. *Anthropological Approaches to Women's Status*. New York: Harper and Row, 1975.

*Rohrlich-Leavitt, Ruby, ed. *Women Cross-Culturally: Change and Challenge*. Chicago: Aldine, 1975.

Ronhaar, J. H. *Woman in Primitive Motherright Societies*. Holland: J. B. Wolters, 1931.

Rosaldo, M. Z. "The use and abuse of anthropology: reflections on feminism and cross-cultural understanding." *Signs* 5.3 (Spring 1980): 389–417.

*Rosaldo, M. Z., and L. Lamphere, eds. *Women, Culture, and Society*. Stanford, Calif.: Stanford University Press, 1974.

*Sacks, Karen. *Sisters & Wives: The Past and Future of Sexual Equality*. Greenwood, Conn.: Greenwood, 1979.

*Sanday, Peggy. *Female Power and Male Dominance: On the Origins of Sexual Inequality*. New York: Cambridge University Press, 1981.

Schlegel, Alice. *Male Dominance and Female Autonomy: Domestic Authority in Matrilineal Societies*. New Haven: HRAF Press, 1972.

Schlegel, Alice, ed. *Sexual Stratification: A Cross-Cultural View*. New York: Columbia University Press, 1977.

Schneider, David M., and Kathleen Gough, eds. *Matrilineal Kinship*. Berkeley and Los Angeles: University of California Press, 1961.

Schreiner, Olive. *Women and Labour*. London: Virago, 1978 (originally 1911).

*Shostak, Marjorie. *Nisa: The Life and Words of a !Kung Woman*. Cambridge, Mass.: Harvard University Press, 1981.

*Stack, C. B. et al. Review Essay on Anthropology. *Signs* 1 (Autumn 1975): 147–160.

*Steady, Filomina C. *Black Women Cross-Culturally*. Cambridge, Mass.: Schenkman, 1981.

Talbot, D. A. *Women's Mysteries of a Primitive People: The Ibibios of South Africa*. New York: Cassel and Co., 1915.

Thomas, Elizabeth Marshall. *The Gentle People*. New York: Knopf, 1959.

Tiffany, Sharon W., ed. *Women and Society: An Anthropological Reader*. St. Albans, Vt.: Eden Women, 1979.

Tilly, Louise. "The social sciences and the study of women: a review article." *Comparative Studies in Society and History* 20 (January 1978): 163–173.

VanBaal, J. *Reciprocity and the Position of Women*. Amsterdam, Holland: VanGorcum, 1975.

VanLeeuwen, M. S. "A cross-cultural examination of psychological differentiation in males and females." *International Journal of Psychology* 13 (1978): 87–122.

The Western Canadian Journal of Anthropology [Cross-Sex Relations: Native Peoples, special issue] 6.3 (1976).

Wolf, Margaret, and Roxanne Witke, eds. *Women in Chinese Society*. Stanford: Stanford University Press, 1975.

LESBIANISM

Abbott, Sidney, and Barbara Love. *Sappho Was a Right-on Woman: A Liberated View of Lesbianism*. New York: Stein and Day, 1978.

Bell, A. P., and M. S. Weinberg. *Homosexualities: A Study of Diversity Among Men and Women*. London: Mitchell Beazley, 1978.

Bell, Alan. Martin Weinberg, and Sue Hammersmith. *Sexual Preference: Its Development in Men and Women*. Indianapolis: Indiana University Press, 1981.

Birke, Lynda. "From Sin to Sickness: Hormonal Theories of Lesbianism." In this volume.

Bullough, Vern L. *Homosexuality: A History*. New York: Meridian, 1979.

Cedar and Nelly, eds. *A Woman's Touch: An Anthology of Lesbian Eroticism and Sensuality for Women Only*. Eugene, Oreg.: Womanshare Books, 1979.

Cruikshank, Margaret, ed. *The Lesbian Path*. Ottawa, Ill.: Caroline Hse, 1980.

Frontiers [Lesbian History, special issue] 4.3 (Fall 1979).

Jay, Karla, and Allen Young. *Lavender Culture*. New York: Harcourt, Brace and Jovanovich, 1978.

*Katz, Jonathan. *Gay American History: Lesbians and Gay Men in the U.S.A.* New York: Avon, 1976.

Klaich, Dolores. *Woman + Woman: Attitudes toward Lesbianism*. New York: Morrow, 1974.

*Martin, Del, and Phyllis Lyon. *Lesbian/Woman*. New York: Bantam, 1972.

Masters, William H., and Virginia E. Johnson. *Homosexuality in Perspective*. Boston: Little, Brown, 1979.

Ponse, Barbara. *Identities in the Lesbian World: The Social Construction of Self*. Westport, Conn.: Greenwood, 1978.

Rosen, David H. *Lesbianism: A Study of Female Homosexuality*. Springfield, Ill.: C. C. Thomas, 1974.

Sisley, Emily L., and Bertha Harris. *The Joy of Lesbian Sex*. New York: Crown, 1977.

Stanley, Julia P., and Susan J. Wolfe, eds. *The Coming Out Stories*. Watertown, Mass.: Persephone, 1980.

*Vida, Ginny, ed. *Our Right to Love: A Lesbian Resource Book*. Englewood Cliffs, N.J.: Prentice-Hall, 1978.

Weeks, Jeffrey. *Coming Out: Homosexual Politics in Britain from the Nineteenth Century to the Present*. London: Quartet Books; New York: Horizon, 1977.

Wolff, Charlotte. *Love between Women*. New York: Harper and Row, 1971.

Wysor, Betty. *The Lesbian Myth*. New York: Random House, 1974.

SEXUALITIES

Adams, C., and R. Laurikietis. *The Gender Trap: Book 2—Sex and Marriage.* London: Virago Books, 1976.

Archives of Sexual Behavior: An Interdisciplinary Research Journal. New York: Plenum Pub Co.

Barbach, Lonnie Garfield. *For Yourself: The Fulfillment of Female Sexuality.* New York: A Signet Book, New American Library, 1975.

_____. "Group treatment for pre-orgasmic women." *Journal of Sex and Marital Therapy* (1975): 2.

Barbach, Lonnie, and Toni Ayers. "Group process for women with orgasmic difficulties." *Personnel and Guidance Journal* 54 (March 1976): 389-391.

Beach, Frank A., ed., *Human Sexuality in Four Perspectives.* Baltimore, Md.: Johns Hopkins University Press, 1977.

Belliveau, Fred, and Lin Richter. *Understanding Human Sexual Inadequacy.* New York: Bantam, 1970.

Benedek, Therese. *Psychosexual Function in Women.* New York: Ronald Press, 1952.

_____. "Sexual functions in women and their disturbance." *American Handbook of Psychiatry*, vol. 1. Edited by S. Arietta. New York: Basic, 1959.

Berghe, Pierre van den. *Age and Sex in Human Society: A Biosocial Perspective.* Belmont, Calif.: Wadsworth, 1973.

Bergler, E., and W. S. Korger. *Kinsey's Myth of Female Sexuality.* New York: Grune and Stratton, 1954.

Blank, J., and H. L. Cottrell. *I Am My Lover.* Burlingame, Calif.: Down There Pr, 1978.

Bonaparte, Marie. *Female Sexuality.* New York: Intl Univs Pr, 1956.

Bullough, Vern L., and Bonnie. *Sin, Sickness, and Sanity. A History of Sexual Attitudes.* New York: New American Library, 1977.

Carey, Emily A. "Women: sexuality, psychology, and psychotherapy." Boston: Womanspace, a Feminist Therapy Collective available from: 636 Beacon St., Boston, MA 02215.

Chassequet-Smirgel, J. *Female Sexuality: New Psychoanalytic Views.* Ann Arbor: University of Michigan Press, 1970.

Christenson, C. V., and J. H. Gagnon. "Sexual behavior in a group of older women." *Journal of Gerontology* 20 (1965): 351-356.

Cott, Nancy F. "Passionlessness: an interpretation of Victorian sexual ideology." *Signs* 4 (1978): 219-236.

Country Women [issue on sexuality]. Issue No. 15 (April 1975). Available from Box 51. Albion, California 95410.

Demeter, Kass. *Women's Sexuality: Myth and Reality.* E. Palo Alto, Calif.: UP Press, 1977.

Dickinson, R. L., and H. H. Pierson. "The average sex life of American women." *Journal of the American Medical Association* 85 (1925): 1113-1117.

Dodson, Betty. *Liberating Masturbation*. Available from Box 1933. New York, N. Y. 10001 (1974).

Dworkin, Andrea. *Woman Hating: A Radical Look at Sexuality*. New York: Dutton, 1976.

Elkan, E. "Evolution of female orgastic ability—a biologic survey." *International Journal of Sexology* 2 (1948): 84-93.

Ellis, Albert. "Is the vaginal orgasm a myth." In *Sex, Society and the Individual*. Edited by A. P. Pillay and A. Ellis. Bombay: International Journal of Sexology Press, 1953.

Ellis, Havelock. *The Erotic Rights of Women*. London: British Society for the Study of Sex Psychology, 1918.

_____. *Studies in the Psychology of Sex*. New York: Random House, 1936.

Feinbloom, Deborah Heller. *Transvestites and Transsexuals*. New York: Dell, 1976.

Fisher, S. *The Female Orgasm: Psychology, Physiology, Fantasy*. New York: Basic, 1973.

_____. *Understanding the Female Orgasm*. New York: Bantam, 1973.

Foucault, Michel. *The History of Sexuality. Volume I: An Introduction*. New York: Vintage Books, 1980.

Gill, Derek. *Illegitimacy, Sexuality and the Status of Women*. Oxford: Basil Blackwell, 1977.

Haller, John S., and Robin M. Haller. *The Physician and Sexuality in Victorian America*. New York: Norton, 1974.

*Hammer, Signe, ed. *Women: Body and Culture—Essays on the Sexuality of Women in a Changing Society*. New York: Perennial Library, Harper and Row, 1974.

Harrison, Fraser. *The Dark Angel: Aspects of Victorian Sexuality*. Azusa, Calif.: Sheldon Press, 1977.

Heiman, J., L. LoPiccolo, and J. LoPiccolo. *Becoming Orgasmic: A Sexual Growth Program for Women*. New York: Prentice-Hall, 1976.

Herculine Barbin: Being the Recently Discovered Memoirs of a Nineteenth-Century French Hermaphrodite. Introduced by Michel Foucault. New York: Pantheon, 1980.

Heresies. Sex Issue. Vol. 3 No. 4 (1981) Issue 12.

*Hite, Shere. *The Hite Report, A Nationwide Study of Female Sexuality*. New York: Dell, 1976.

* _____. *The Hite Report on Male Sexuality*. New York: Knopf, 1981.

_____. *Sexual Honesty*. New York: Warner Bks, 1974.

Horney, K. "The denial of the vagina: A contribution to the problem of genital anxieties specific to women." *International Journal of Psychoanalysis* 14 (1933): 55-70.

Human Sexuality. New York: Haworth Pr.

Journal of Sex Research. New York: Society for the Scientific Study of Sex.

Hutchinson, J. B. *Biological Determinants of Sexual Behavior*. New York: Wiley, 1978.

Kelly, G. L. *Sexual Feeling in Woman*. Augusta, Georgia: Elkay Press, 1930.

*Kinsey, A. C. et al. *Sexual Behavior in the Human Female*. Philadelphia: Saunders, 1953.

*_____. *Sexual Behavior in the Human Male*. Philadelphia: Saunders, 1948.

Kline-Graber, Georgia, and Benjamin Graber. *Woman's Orgasm*. New York: Popular Lib, 1976.

Koedt, Anne. "The myth of the vaginal orgasm." In *Liberation Now*. Edited by Babcor and Belkin. New York: Dell, 1971, pp. 311–320.

Kronhausen, Phyllis, and Eberhard Kronhausen. *The Sexually Responsive Woman*. New York: Ballantine, 1965.

Lemay, Helen Rodnits. "Some thirteenth and fourteenth century lectures on female sexuality." *International Journal of Women's Studies* 1 (July/August 1978): 391–400.

LoPiccolo, J., and M. A. Lobitz. "The role of masturbation in the treatment of sexual dysfunction." *Archives of Sexual Behavior* 2 (1972): 163–171.

Lowry, Thomas and Thea, eds. *The Clitoris*. St. Louis: Green, 1976.

Maslow, A. H., H. Rand, and S. Newman. "Some parallels between sexual and dominance behavior of infrahuman primates and the fantasies of patients in psychotherapy." *Journal of Nervous and Mental Disease* 313 (1960): 202–212.

Masters, William H., and Virginia Johnson. *Human Sexual Inadequacy*. Boston: Little, Brown, 1970.

*_____. *Human Sexual Response*. Boston: Little, Brown, 1966.

Maxwell, R. J. "Quiz: Female sexuality in primitive cultures." *Medical Aspects of Human Sexuality* 1 (Jan. 1973).

Parsons, Elsie Clews. *The Old Fashioned Woman: Primitive Fancies About Sex*. New York: Putnam, 1913.

Pierson, E. C., and W. V. D'Antonia. *Female and Male: Dimensions of Human Sexuality*. Philadelphia: Lippincott, 1974.

Ploss, Herman Heinrich. *Woman in the Sexual Relation: An Anthropological and Historical Survey*. New York: Medical Press of New York, 1964.

Radical History Review [Sexuality in History, special issue] 20 (Spring/Summer 1979).

Raymond, Janice G. *The Transsexual Empire: The Making of the She-Male*. Boston: Beacon Pr, 1979.

Robbins, Jhan, and Jane Robbins. *An Analysis of Human Sexual Inadequacy*. New York: Signet, 1970.

Robinson, Paul. *The Modernization of Sex: Havelock Ellis, Alfred Kinsey, William Masters and Virginia Johnson*. New York: Harper and Row, 1976.

Robinson, W. J. *Woman: Her Sex and Love Life*. New York: Eugenics Publishing Co., 1929.

Rossi, William P. *The Sex of the Foot and Shoe*. Boston: Routledge, 1977.

Rowbotham, Sheila, and Jeffrey Weeks. *Socialism and the New Life: The Personal and Sexual Politics of Edward Carpenter and Havelock Ellis*.

London: Pluto Press, 1977.

Rush, Anne Kent. *Getting Clear: Body Work for Women.* New York: Random House, 1973.

*Seaman, Barbara. *Free and Female: The New Sexual Role of Women.* Greenwich, Conn.: Fawcett Crest Book, 1972.

Sears, Hal D. *The Sex Radicals: Free Love in High Victorian America.* Lawrence, Kans.: Regents Press of Kansas, 1977.

Seidenberg, Robert. "Is sex without sexism possible?" *Sexual Behavior* 46 (1972).

Sherfey, Mary Jane. "Female sexuality and psychoanalytic theory." In *Woman in a Man-Made World.* 2nd Edition. Edited by Nona Glazer and Helen Waehrer. Chicago: Rand McNally, 1972.

*_____. *The Nature and Evolution of Female Sexuality.* New York: Random House. 1966.

Signs. [Sex and Sexuality Issue 5] (1980): 569-856; 6 (1980): 1-186.

*Singer, June. *Androgyny: Toward a New Theory of Sexuality.* Garden City. N.Y.: Anchor Pr, 1977.

Smart, C., and B. Smart. *Women, Sexuality, and Social Control.* London: Routledge and Kegan Paul, 1978.

Soble, Alan, ed. *Philosophy of Sex: Contemporary Readings.* Totowa, N.J.: Littlefield, Adams, 1981.

Stekel, Wilhelm. *Frigidity in Women.* 2 volumes. London: Boni and Liveright, 1926.

Teeters, Kass. *Female Sexuality.* Women Inc., San Jose, California 95132.

Tiefer, Leonore. "The context and consequences of contemporary sex research: A feminist perspective." In *Sex and Behavior.* Edited by Thomas E. McGille et al. New York: Plenum Pub, 1978.

Tyson, Joanne, and Richard. "Sex and the black woman: They are now seeking advice." *Ebony*, August 1977, p. 103.

Trudgill, Eric. *Madonnas and Magdalens: The Origins and Development of Victorian Sexual Attitudes.* New York: Homes and Meier, 1976.

von Krafft-Ebing, Richard. *Psychopathia Sexualis.* New York: Putnam, 1965.

Wallace, F. *Masturbation: A Woman's Handbook.* Bloomfield, N.J.: R. J. Williams, 1975.

Wear, Jennifer, and King Holmes. *How to Have Intercourse without Getting Screwed.* Seattle: Madiona Pub., 1978.

Weinberg, Martin S., and Colin J. Williams. "Sexual embourgeoisement? Social class and sexual activity, 1938-1970." *American Sociological Review* 45 (February 1980): 33-48.

Wolff, Charlotte. *Bisexuality: A Study.* London: Quartet Bks, 1979.

FUTURES

Birkby, Phyllis, and Leslie Weisman. "Patritecture and feminist fantasies." *Liberation Magazine* (1976): 48-52.

Bryant, Dorothy. *The Kin of Ata are Waiting for You*. San Francisco: Moon Books, Random House, 1976.

Gearhardt, Sally. *The Wanderground*. Watertown, Mass.: Persephone Press, 1979.

Gilman, Charlotte P. *Herland: A Lost Feminist Utopian Novel*. New York: Pantheon, 1979.

LeGuin, Ursula. *The Dispossessed*. New York: Avon, 1974.

_____. *The Left Hand of Darkness*. New York: Ace Bks, 1969.

_____. "The space crone." *The CoEvolution Quarterly* (Summer 1976): 108–110.

Madsen, Catherine. "Commodore Bork and the compost." *Women: A Journal of Liberation* 5 (1977): 32–35.

Piercy, Marge. *Woman on the Edge of Time*. New York: Knopf, 1976.

GENERAL

*The Brighton Women and Science Group. *Alice through the Microscope: The Power of Science Over Women's Lives*. London: Virago, 1980.

Clark, Lorenne M. G., and L. Lange, eds. *The Sexism of Social and Political Theory: Women and Reproduction from Plato to Nietzsche*. Toronto: University of Toronto Press, 1979.

*Griffin, Susan. *Woman and Nature: The Roaring Inside Her*. New York: Harper and Row, 1978.

Kern, S. *Anatomy and Destiny: Cultural History of the Human Body*. Indianapolis, Ind.: Bobbs-Merrill, 1976.

Leiss, William. *The Domination of Nature*. Boston, Mass.: Beacon Pr, 1974.

*Merchant, Carolyn. *The Death of Nature: Women, Ecology, and the Scientific Revolution*. New York: Harper and Row, 1981.

Popular Science Monthly [1870–1900, Debate on the Biological and Evolutionary Potentials of Women].

Trenfeld, Karen. "On the role of biology in feminist ideology." *Hecate III* 2 (July, 1977): 41–56.

Vetterling-Braggin, Mary, Frederick A. Elliston, and Jane English, eds. *Feminism and Philosophy*. Totowa, N.J.: Littlefield, Adams, 1977.

Vetterling-Braggin, Mary, ed. *Sexist Language: A Modern Philosophical Analysis*. Totowa, N.J.: Littlefield, Adams, 1981.

Weisstein, Naomi, Virginia Blaisdell, and Jesse Lemisch. *The Godfathers: Freudians, Marxists, the Scientific and Political Protection Societies*. New Haven, Conn.: Bella Donna Publishing, 1975.

II. Our Lives and Our Health

ADDICTIONS

Allen, Chaney. *I'm Black and I'm Sober: A Minister's Daughter Tells Her Story about Fighting the Disease of Alcoholism—and Winning*.

Minneapolis: CompCare, 1978.

Belfer, M. D. et al. "Alcoholism in women." *Archives of General Psychiatry* 25 (1971): 540–544.

Camberwell Council on Alcoholism. *Women and Alcohol.* New York: Tavistock, 1980.

Christenson, Susan, and Gayle Inlenfeld with assistance from Janice Kinsolving. *Lesbians, Gay Men and their Alcohol and Other Drug Use: Resources.* Madison, Wisc.: Wisconsin Clearinghouse for Alcohol and Other Drug Information.

Corrigan, Eileen M. *Alcoholic Women in Treatment.* New York: Oxford University Press, 1979.

Cuskey, Walter and Richard Wathey. *Female Addicts.* Lexington, Mass.: Lexington Bks, 1981.

Dowsling, Janet and Anne MacLennan. *The Chemically Dependent Woman. Rx: Recognition, Referral, Rehabilitation.* Toronto: Addiction Research Foundation, Toronto, 1978.

Focus on Women: Journal of Addictions and Health. George F. Stickley Co., 210 W. Washington Sq., Philadelphia, PA 19106.

Frontiers [Equal Opportunity Addiction: Women, Alcohol, and Drugs, special edition] 4.2 (Summer 1979).

Figure XXVIII.
(Photo: A. Henifin)

Hall, Nancy Lee. *A True Story of a Drunken Mother*. Houston, Tex.: Daughters, 1974.

Kirkpatrick, Jean. *Turnabout: Help for a New Life*. Garden City, N.Y.: Doubleday, 1978.

Lindbeck, V. L. "The woman alcoholic—a review of the literature." *International Journal of Addiction* 7 (1972): 567–580.

Nellis, Muriel. *The Female Fix*. Boston: Houghton Mifflin, 1980.

Nero, Jack. *Drink Like a Lady, Cry Like a Man*. Minneapolis: CompCare, 1977.

Sandmaier, Marian. *The Invisible Alcoholics: Women and Alcohol Abuse in America*. New York: McGraw-Hill, 1980.

Tyler, Joanna and Marian Thompson. "Patterns of drug abuse among women." *International Journal of the Addictions* 15 (1980): 309–321.

Women Alcohol Abusers: Subject Area Bibliography. Rockville, Md.: National Clearinghouse for Alcohol Information, 1979.

Women and Drugs: An Annotated Bibliography. Rockville, Md.: National Clearinghouse for Drug Abuse Information, 1975.

Youcha, Geraldine. *A Dangerous Pleasure*. New York: Dutton, 1978.

AGING

Bibliography on Older Women. Los Angeles. Available from: the Publications Office, Andrus Gerontology Center, University of Southern California, Los Angeles, CA: 90007.

Block, Marilyn R. et al. *Uncharted Territory: Issues and Concerns of Women Over 40*. College Park, Mich.: University of Michigan.

Block, Marilyn, Janice Davidson and Jean Gramps. *Women Over Forty: Visions and Realities*. New York: Springer Pub, 1980.

Bluh, Bonnie. *The "Old" Speak Out*. New York: Horizon, 1979.

Broomstick: A Periodical Supporting Options for Women over Forty. Available from: The Women's Building, 3543 18th St., San Francisco, CA 94110.

*Butler, Robert N. *Why Survive? Being Old in America*. New York: Harper and Row, 1975.

de Beauvoir, Simone. *The Coming of Age*. New York: Warner Bks, 1973.
_____. *Old Age*. Harmondsworth, England: Penguin Books, 1970.

Fuchs, Estelle. *The Second Season: Love and Sex—Women in the Middle Years*, New York: Doubleday, 1978.

Gray, Madeline. *The Changing Years: The Menopause Without Fear*. New York: Doubleday & Co., 1981.

Helsing, Knud, Moyses Szklo and George Comstock. "Factors associated with mortality after widowhood." *American Journal of Public Health* 71 (Aug. 1981): 802–807.

Jacobs, H. *Life after Youth—Female, Forty—What Next?* Boston: Beacon Pr, 1979.

Kuhn, Maggie. *Maggie Kuhn on Aging: A Dialogue.* Edited by Dieter Hessel. Philadelphia: Westminster Press, 1977.

Lake, Alice. *Our Own Years: What Women over 35 Should Know about Themselves.* New York: A Woman's Day/Random House Book, 1979.

Luce, Gay Gaer. *Your Second Life.* New York: Delta, 1980.

National Action Forum for Older Women Newsletter. Available from: School of Allied Health Professions, Health Sciences Center, SUNY, Stony Brook, NY 11794.

Nudel, Adele. *For the Woman Over 50.* New York: Avon, 1978.

The Older Women's Rights Committee. *Age Is Becoming: An Annotated Bibliography on Women and Aging.* Available from: Interface Bibliographers, 3018 Hillegass Ave., Berkeley, CA 94705.

Prime Time [By and for Older Women]. Available from 420 W. 46 St., New York, NY 10036.

Rubin, Lillian B. *Women of a Certain Age: The Midlife Search for Self.* New York: Harper and Row, 1979.

Rulude, Louise. *Women and Aging: A Report on the Rest of Our Lives.* Ottawa, Canada: Advisory Council on the Status of Women, 1978.

Shields, Laurie. *Displaced Homemakers.* New York: McGraw-Hill, 1981.

Slovik, David et al. "Deficient production of 1,25-dihydroxyvitamin D in elderly osteoporotic patients." *New England Journal of Medicine* 305 (August 13, 1981): 372–374.

Strugnell, Cecile. *Adjustment to Widowhood and Some Related Problems: A Selective and Annotated Bibliography.* New York: Health Sciences Publishing Corp. 1974.

Troll, Lillian, Joan Israel, and Kenneth Israel. *Looking Ahead: A Woman's Guide to the Problems and Joys of Growing Older.* Englewood Cliffs, N.J.: Prentice-Hall, 1977.

Vining, Elizabeth. *Being Seventy.* New York: Viking Pr, 1978.

BATTERED WOMEN

Dobash, E. Emerson and Russell P. Dobash. *Violence Against Wives.* New York: Free Pr, 1979.

_____. "Wives: The 'appropriate' victims of marital violence." *Victimology* 2.3–4 (1977–1978): 426–442.

D'Ogley, Vincent, ed. *Domestic Violence: Issues and Dynamics.* Toronto, Canada: Ontario Institute for Studies in Education, 1978.

Fleming, Jennifer. *Stopping Wife Abuse.* Garden City, N.Y.: Anchor Pr, 1979.

Langley, Roger, and Richard Levy. *Wife Beating: The Silent Crisis.* New York: Pocket Bks, 1977, Institute for Studies in Education, 1978.

Pethick, Jane. "A Bibliography on Battered Wives." In *Domestic Violence: Issues and Dynamics.* Edited by Vincent D'Ogley. Toronto, Canada: Ontario Institute for Studies in Education, 1978.

Roberts, Albert R. *Sheltering Battered Women: A National Study and Guide to Services*. New York: Springer Pub, 1980.

Roy, Maria. *Battered Women: A Psychosocial Study of Domestic Violence*. New York: Van Nostrand Reinhold, 1977.

Stahly, Geraldine Butts. "A review of select literature of spousal violence." *Victimology* 2.3–4 (1978–1979): 591–607.

Strauss, M., R. Gelles, and S. Steinmetz. *Behind Closed Doors: Violence in the American Family*. Garden City, N.Y.: Doubleday, 1980.

Sutton, Jo. "The growth of the British movement for battered women." *Victimology* 2.3–4 (1977–1978): 576–584.

BIRTH CONTROL, ABORTION, AND STERILIZATION

Abortion Law Reform Association Newsletter/A Woman's Right to Choose Campaign. Available from: ALRA, 88A Islington High St., London N1 8 EG England.

Alexander, Daryl. "A Montgomery tragedy: the Relf family refused to be the nameless victims of involuntary sterilization." *Essence* (September 1973): 44.

Arnstein, Helene S. *What Every Woman Needs to Know About Abortion*. New York: Scribner, 1973.

Association for the Study of Abortion. *Bibliography Reprint List*. Available from: 120 W. 57 St., New York, N.Y. 10019.

Banks, Joseph Ambrose, and Banks, Olive. *Feminism and Family Planning in Victorian England*. New York: Schocken Books, 1964.

Barr, S. J. *A Woman's Choice* [abortion]. New York: Rawson Assoc, 1977.

Beral, Valerie. "Cardiovascular-disease mortality trends and oral-contraceptive use in young women." *The Lancet*, 13 Nov. 1976, pp. 1047–1051.

Billings, Evelyn, and Ann Westmore. *The Billings Method: Controlling Fertility Without Drugs or Devices*. New York: Random House, 1981.

Birdsall, Nancy. "Women and population studies." *Signs* 1 (Spring 1976): 699–712.

Calderone, M. S., ed. *Manual of Family Planning and Contraceptive Practice*. Baltimore: Williams and Wilkins, 1970.

Caress, Barbara. "Sterilization guidelines." *Health Pac Bulletin* 65 (July/ August 1975).

Carter, Luther J. "New feminism: potent force in birth control policy." *Science* 167 (1970): 1234–1236.

Cates, Willard et al. "The effect of delay and method choice on the risk of abortion morbidity." *Family Planning Perspectives* 9 (Nov./Dec. 1977): 266–273.

Cohen, M., T. Nagle, and T. Scanlon. *The Rights and Wrongs of Abortion*. Princeton, N.J.: Princeton University Press, 1974.

*Committee for Abortion Rights and against Sterilization Abuse. *Women under Attack: Abortion, Sterilization Abuse, and Reproductive Freedom.* New York: CARASA, 1979.

Cone, Jim. "Forced sterilization and the poor." *Synapse* (Spring 1975).

"Contraceptive hormones and human welfare." *Comment* 10.2 (October 1977). Available from Office of Women, American Council on Education, One Dupont Circle, Washington, D.C. 20036.

Demography. Washington, D.C.: Population Association of America.

Denes, Magda. *In Necessity and Sorrow: Life and Death in an Abortion Hospital.* Harmondsworth, England: Penguin Books, 1976.

Devereux, George. *A Study of Abortion in Primitive Societies.* New York: Julian Press, 1955.

Djerassi, Carl. *The Politics of Contraception.* New York: Norton, 1979.

Draper, Elizabeth. *Birth Control in the Modern World.* Baltimore: Penguin, 1965.

Dreifus, Claudia. "Sterilizing the poor." *The Progressive* 39 (Dec. 1975): 12.

Dumund, Don E. "The limitation of human population: A natural history." *Science* 187 (1974): 713-721.

Edelman, E., G. S. Berger, and L. Keith. *Intrauterine Devices and their Complications.* Boston: G. K. Hall, 1979.

Eliot, Johan N. "Fertility control and coercion." *Family Planning Perspectives* 5 (Summer 1973): 3.

English, Deirdre. "The war against choice: inside the antiabortion movement." *Mother Jones* 6.2 (February/March 1981): 16-26.

Erlien, Marla et al. *More Than a Choice: Women Talk about Abortion.* Somerville, Mass.: New England Free Press, 1979.

Family Planning Perspectives. New York: Alan Guttmacher Institute.

Francke, L. B. *The Ambivalence of Abortion.* New York: Random House, 1978.

Frankfort, Ellen. *Rosie: The Investigation of a Wrongful Death.* New York: Dial, 1978.

Freeden, Michael. "Eugenics and progressive thought: a study in ideological affinity." *Historical Journal* 22 (September 1979): 645-715.

Frisch, Rose E. "Population, food intake, and fertility." *Science* 199 (Jan. 1978): 22-33.

Gage, Suzann. *When Birth Control Fails. . . .* Hollywood, Calif.: Speculum Press/Self-Health Circle, 1979.

Gallagher, Janet. "Abortion rights: critical issue for women's freedom." *WIN* (March 8, 1979).

Garfink, C., and H. Pizer. *The New Birth Control Program: The Safe and Sure Method of Natural Contraception.* New York: A Bolder Book, 1977.

Gebhard, Paul H. *Pregnancy, Birth, and Abortion.* New York: Wiley, 1958.

Gillette, Paul. *Vasectomy.* New York: Paperback Library, 1972.

Glass, Robert H., and Nathan G. Kase. *Women's Choice: A Guide to Contraception, Fertility, Abortion, and Menopause.* New York: Basic, 1970.

Gordon, Linda. "The politics of birth control, 1920–1940: The impact of professionals." *International Journal of Health Services* 5 (1975): 253–278.

*_____. *Woman's Body Woman's Right: A Social History of Birth Control in America*. New York: Grossman Publishers, 1976.

Greep, Roy O., and Marjorie Koblinsky, eds. *Frontiers in Reproduction and Fertility Control*. Cambridge, Mass.: 1977.

Guay, Terrie. *Creation of Life . . . Your Choice*. Ashland, Oreg.: Emergence, 1979.

Guren, D., and N. Gillette. *The Ovulation Method: Cycles of Fertility—A Natural Birth Control Method that Outdates Rhythm*. Ovulation Method Teachers Association, 760 Aldrich Road, Bellingham, Washington 98255 (1978).

Health Pac Bulletin #62. "Sterilization of women: The facts." (Jan/Feb 1975).

Hennekens, Charles, and Brian MacMahon. "Oral contraceptives and myocardial infarction." *New England Journal of Medicine* 296 (May 1977): 1166–1167.

*Himes, Norman E. *A Medical History of Contraception*. New York: Schocken Books, 1970.

_____. "Note on the early history of contraception in America." *New England Journal of Medicine* 205 (1931).

Hodgson, Jane. *Abortion and Sterilization: Medical and Social Aspects*. New York: Academic Press, 1981.

*Holmes, Helen B., Betty B. Hoskins, and Michael Gross, eds. *Birth Control and Controlling Birth: Women Centered Perspectives*. Clifton, N.J.: Humana, 1979.

Jaffe, Frederick S., Barbara L. Lindheim, and Phillip R. Lee. *Abortion Politics: Private Morality and Public Policy*. New York: McGraw-Hill, 1981.

Jick, Hersehl et al. "Vaginal spermicides and congenital disorders." *Journal of the American Medical Association* 245 (April 1981): 1329–1332.

Jordan, Brigitte. *Birth in Four Cultures: A Cross Cultural Investigation of Childbirth in Yucatan, Holland, Sweden and the United States*. St. Albans, Vt.: Eden Press, Women's Publications, 1978.

Kaye, Archiprete et al. *The Abortion Business: A Report on Free-Standing Abortion Clinics. 1975*. Women's Research Action Project, Box 119, Porter Square Station, Cambridge, Mass., 02140.

Keith, Louis G., ed. *Focus on Women. Vol. 1: The Safety of Fertility Control*. New York: Springer Pub, 1980.

Kennedy, David M. *Birth Control in America: The Career of Margaret Sanger*. New Haven: Yale University Press, 1970.

Knight, Patricia. "Women and abortion in Victorian and Edwardian England." *History Workshop* 4 (Autumn, 1977): 57–69.

Knodel, John. "Breast feeding and population growth." *Science* 198 (1977): 1111–1115.

Kupinsky, Stanley, ed. *The Fertility of Working Women: A Synthesis of International Research.* New York: Praeger, 1977.

Lacey, Louise. *Lunaception.* New York: Coward, McCann and Geoghegan, 1975.

Lader, L. *Abortion.* Boston: Beacon Pr, 1966.

Lane, Mary. "Successful use of the diaphragm and jelly by a young population: Report of a clinical study." *Family Planning Perspectives* 8 (March/April 1976): 81-86.

Lang, Frances. *The Sickle Cell and the Pill.* Unpublished, n.d. Available from: Women's Community Health Center, Inc., 639 Massachusetts Ave., Cambridge, MA 02139.

Lanson, Lucienne. *Woman to Woman.* New York: Random House, 1981.

Ledbetter, Rosanna. *A History of the Malthusian League, 1877-1927.* Columbus, Ohio: Ohio State University Press, 1976.

Leudon, Henri. *Human Fertility: The Basic Components.* Translated from French by Judith Helzner. Chicago: University of Chicago Press, 1977.

Levine, Carol. "Depo-provera and contraceptive risk: a case study of values in conflict—third world pressures for FDA approval." *The Hastings Center Report* 9.4 (August 1979): 8-12.

Loebel, S. *Conception, Contraception: A New Look.* New York: McGraw-Hill, 1974.

Luker, Kristin. *Taking Chances: Abortion and the Decision Not to Contracept.* Berkeley: University of California Press, 1975.

Mass, Bonnie. *Population Target: The Political Economy of Population Control in Latin America.* Ontario, Canada: Charters Publishing Co., 1976.

McCoy, Terry L. et al. *The Dynamics of Population Policy in Latin America.* Cambridge, Mass.: Bulinger, 1974.

McLaren, Angus. *Birth Control in Nineteenth Century England.* London: Croon Helms, 1978.

Minkin, Stephen. "Depo-provera: a critical analysis." *Women and Health.* 5.2 (Summer 1980): 49-69.

Mohr, J. C. *Abortion in America: The Origins and Evolution of National Policy.* New York: Oxford University Press, 1978.

*National Women's Health Network. *Abortion: Resource Guide 8.* Washington, D.C.: National Women's Health Network, 1980.

* _____. *Birth Control: Resource Guide 5.* Washington, D.C.: National Women's Health Network, 1980.

* _____. *Sterilization: Resource Guide 9.* Washington, D.C.: National Women's Health Network, 1980.

Newland, Kathleen. *Women and Population Growth: Choice Beyond Childbearing.* Washington, D.C.: Worldwatch Institute, 1977.

Newman, Sidney H., and Zanvel E. Klein, eds. *Behavioral-Social Aspects of Contraceptive Sterilization.* Lexington, Mass.: Lexington Bks, 1978.

Noiziger, M. *A Cooperative Method of Birth Control,* 2nd ed., rev. Summertown, Tenn.: Book Pub Co., 1976.

Norsigian, Judy. "Redirecting contraceptive research." *Science for the People* (Jan./Feb., 1979): 27–30.

Osofky, Howard J., and Joy D., eds. *The Abortion Experience: Psychological and Medical Impact.* Hagerstown, Md.: Harper and Row, 1973.

Outhwaite, R. B. "Population change, family structure and the good of counting." *Historical Journal* 22 (March 1979): 229–237.

Perkins, R. L. *Abortion: Pro and Con.* Cambridge, Mass.: Schenkman, 1974.

Piers, M. W. *Infanticide, Past and Present.* New York: Norton, 1978.

Potts, Malcolm, Peter Diggory, and John Pell. *Abortion.* London: Cambridge University Press, 1977.

Psychology of Women Quarterly [The Motherhood Mandate, special issue] 4.1 (1979).

*Reed, James. *From Private Vice to Public Virtue: The Birth Control Movement and American Society Since 1830.* New York: Basic, 1978.

Reedy, Juanita. "Diary of a Prison Birth." *Majority Report* (May 31, 1975).

Report Prepared for the American Friends Service Committee. *Who Shall Live? Man's Control Over Birth and Death.* New York: Hill and Wang, 1970.

Reproductive Rights Newsletter. Available from: Reproductive Rights National Network, 41 Union Square West, New York, 10003.

Ris, H. W. "The essential emancipation: The control of reproduction." In *Beyond Intellectual Sexism.* Edited by Joan L. Roberts. New York: McKay, 1976.

Roberts, Helen. *Women, Health and Reproduction.* London: Routledge and Kegan Paul, 1981.

_____. "Women, social class and IUD use." *Women's Studies International Quarterly* 2.1 (1979): 49–56.

Rodriguez-Trias, Helen. "Sterilization Abuse." In this volume.

Rosen, Harold. *Abortion in America.* Boston: Beacon Press, 1967.

Royal College of General Practitioners' Oral Contraceptive Users Study. "Mortality among oral-contraceptive users." *The Lancet*, 8 Oct. 1977, pp. 728–733.

Sanger, Margaret. *What Every Girl Should Know.* New York: Bel Vedere, republished 1980.

Sauer, R. "Infanticide and abortion in nineteenth century Britain." *Population Studies* 32 (March 1978): 81–94.

Schulder, Diane, and Florynce Kennedy. *Abortion Rap: Testimony by Women Who Have Suffered the Consequences of Restrictive Abortion Laws.* New York: McGraw-Hill, 1971.

*Seaman, Barbara. *The Doctors' Case against the Pill.* Garden City, N.Y.: Doubleday, 1980.

Shapiro, Howard I. *The Birth Control Book.* New York: St. Martin, 1977.

Short, R. V. "The evolution of human reproduction." *Proceedings of the Royal Society, London Series* B 195 (December 1976): 3–24.

Skowronski, M. *Abortion and Alternatives.* Millbrae, Calif.: Les Femmes Publishing, 1977.

Slone, Dennis et al. "Risk of myocardial infarction in relation to current and discontinued use of oral contraceptives." *New England Journal of Medicine* 305 (Aug. 20, 1981): 420–424.

"Spermicides–simplicity and safety are major assets," *Barrier Method Series* H.5 (September 1979). Available from: Population Information Program, The Johns Hopkins University, 624 North Broadway, Baltimore, MD 21205.

Stadel, Bruce. "Oral contraceptives and cardiovascular disease." [Review Article] *New England Journal of Medicine* 305 (Sept. 10, 1981): 612–618.

Stokes, Bruce. "Men and Family Planning." *World Watch Paper* 41 (December 1980). Available from: Worldwatch Institute, 1776 Massachusetts Ave., N.W., Washington, D.C. 20036.

Studies in Family Planning. Available from: The Population Council, 245 Park Avenue, New York, NY 10017.

Tucker, Tarvez. *Birth Control.* New Canaan, Conn.: Tobey Publishing Co., 1975.

van der Tak, Jean. *Abortion, Fertility and Changing Legislation: An International Review.* Lexington, Mass.: Lexington Bks, 1974.

Weissman, Steve. "Why the population bomb is a Rockefeller baby." *Ramparts* 8 (May, 1970): 42–47.

Wharton, Lawrence Richardson. *The Ovarian Hormones: Safety of the Pill, Babies after Fifty.* Springfield, Ill.: C. C. Thomas, 1967.

Women Workers at Two Abortion Clinics. *Getting Stronger: Women Workers Organize the Abortion Clinics.* Cambridge, Mass.: Red Sun Pr, 1978.

Wylie, Evan McLeod. *All About Voluntary Sterilization.* New York: Berkeley Medallion Books, 1977.

Zimmerman, Mary K. *Passage through Abortion: The Personal and Social Reality of Women's Experience.* New York: Praeger, 1977.

CANCER AND SURGERY

American Journal of Nursing [Breast Cancer, special issue] 77.9 (September 1977).

Bichler, Joyce. *DES Daughter.* New York: Avon, 1981.

Brand, P., and P. A. VanKeep. *Breast Cancer: Psychosocial Aspects of Early Detection and Treatment.* Baltimore, Md.: University Park, 1978.

Campion, R. *The Invisible Worm* [alternatives to breast surgery]. New York: Avon, 1975.

Centerwall, Brandon. "Premenopausal hysterectomy and cardiovascular disease." *American Journal of Obstetrics and Gynecology* 139 (Jan. 1981): 58–61.

Cocke, William. *Breast Reconstruction Following Mastectomy.* Boston: Little, Brown, 1977.

*Cope, Oliver. *The Breast: A Health Guide for Women of All Ages.* Boston: Houghton Mifflin, 1978.

Coweles, J. *Informed Consent* [breast cancer]. New York: Coward, McCann and Geoghegan, 1976.

*Crile, George, Jr. *What Women Should Know About the Breast Cancer Controversy*. New York: Pocket Bks, 1973.

Eagen, Andrea. "Breast cancer, facts a woman needs to know." *Healthright* 2 (Summer, 1976).

Faulder, Carolyn. *Breast Cancer*. London: Pan Books, 1979.

Folta, Anne-Marie, and Jennifer Kelsey. "The annual pap test: a dubious policy success," *The Milbank Memorial Fund Quarterly* 56.4 (1978).

Greenfield, N. S. *First Do No Harm: A Dying Woman's Battle Against the Physicians and the Drug Companies Who Misled Her About the Pill*. New York: Sun River 1976.

Guistini, F. G., and F. J.Keefer. *Understanding Hysterectomy: A Woman's Guide*. New York: Walker & Co., 1979.

Jameson, DeeDee, and Roberta Schwalb. *Every Woman's Guide to Hysterectomy: Taking Charge of Your Own Body*. Englewood Cliffs, N.J.: Prentice-Hall, 1978.

*Kushner, Rose. *Why Me: What Every Woman Should Know About Breast Cancer to Save Her Life*. New York: A Signet Book. New American Library, 1977.

*Lorde, Audre. *The Cancer Journals*. Argyle, N.Y.: Spinsters Ink, 1980.

McCauley, Carole Spearin. *Surviving Breast Cancer*. New York: Dutton, 1979.

Mansfield, Carl M. *Early Breast Cancer: Its History and Results of Treatment*. Basel, Switzerland: S. Karger, 1976.

Milan, Albert. *Breast Self Examination*. New York: Workman Pub, 1980.

Morgan, S. B. *Hysterectomy*. Los Angeles: Feminist History Research Project and Our Bodies Ourselves, 1978.

*National Women's Health Network. *Breast Cancer: Resource Guide 1*. Washington, D.C.: National Women's Health Network, 1980.

* ____. *DES (Diethylstilbesterol): Resource Guide 6*. Washington, D.C.: National Women's Health Network, 1980.

* ____. *Hysterectomy: Resource Guide 2*. Washington, D.C.: National Women's Health Network, 1980.

Newman, J. H. *What Every Woman Should Know About Breast Cancer*. Canoga Park, Calif.: Major Bks, 1976.

Nugent, N. *Hysterectomy*. Garden City, N.Y.: Doubleday, 1972.

*Seaman, Barbara, and Gideon Seaman. *Women and the Crisis in Sex Hormones*. New York: Rawson Assoc, 1977.

Seaman, S. S. *Always A Woman: What Every Woman Should Know About Breast Surgery*. Larchmont, N.Y.: Argonaut Bks, 1965.

Senie, Ruby et al. "Breast self-examination and medical examination related to breast cancer stage." *American Journal of Public Health* 71 (June 1981): 583–590.

Shapiro, Lucy. *Never Say Die: A Doctor and Patient Talk About Breast Cancer*. New York: Prentice-Hall, 1981.

Smith, Donald C. et al. "Association of exogenous estrogen and endometrial carcinoma." *New England Journal of Medicine* 293 (Dec. 1975): 1164-1167.

Stoll, Basil ed. *Risk Factors in Breast Cancer.* Chicago: Year Bk Med, 1976.

Strax, P. *Early Detection: Breast Cancer is Curable.* New York: New American Library, 1974.

Veronesi, Umberto et al. "Comparing radical mastectomy with quadrant-ectomy, axillary dissection, and radiotherapy in patients with small cancers of the breast." *New England Journal of Medicine* 305 (July 2, 1981): 6-11.

Weiss, Kay. "Cancer and estrogens—a review." *Women and Health* 1 (1976): 3-4.

_____. "Vaginal cancer: An iatrogenic disease?" *International Journal of Health Services* 5 (1975): 235-252.

Zalon, J., with J. L. Block. *I Am Whole Again: A Case for Reconstruction After Breast Surgery.* New York: Random House, 1978.

Ziel, Harry K., and William D. Finkle. "Increased risk of endometrial carcinoma among users of conjugated estrogens." *New England Journal of Medicine* 293 (Dec. 1975): 1167-1170.

DISABILITIES

Campling, Jo, ed. *Images of Ourselves: Women with Disabilities Talking.* Boston: Routledge and Kegan Paul, 1981.

Task Force on Concerns of Physically Disabled Women. *Toward Intimacy.* New York: Human Sci Pr, 1978.

Task Force on Concerns of Physically Disabled Women. *Within Reach.* New York: Human Sci Pr, 1978.

Women's Educational Equity Communications Network. *Resource Roundup: Disabled Women and Equal Opportunity, a Bibliography.* San Francisco: Women's Educational Equity Communications Network, 1979.

EATING

Bruch, Hilde. *The Golden Cage: The Enigma of Anorexia Nervosa.* Cambridge, Mass.: Harvard University Press, 1978.

Druss, Vicki, and Mary Sue Henifin. "Why Are So Many Anorexics Women?" In *Women Look at Biology Looking at Women.* Edited by R. Hubbard, M. S. Henifin, and B. Fried. Boston, Mass.: G. K. Hall; Cambridge, Mass.: Schenkman, 1979.

Food and Nutrition Group, Boston Science for the People. *Feed, Need, Greed. Food, Resources & Population. A High School Curriculum.* Cambridge, Mass.: Science for the People.

Josephs, Rebecca. *Early Disorder* [anorexia nervosa, a novel]. New York: Farrar, Strauss and Giroux, 1980.

Kaplan, Jane R., ed. *A Woman's Conflict: The Special Relationship Between Women and Food.* Englewood Cliffs, N.J.: Prentice-Hall, 1979.

Lawrence, Marilyn. "Anorexia nervosa—the control paradox." *Women's Studies International Quarterly* 2.1 (1979): 93–101.

Levenkron, Steven. *The Best Little Girl in the World* [Anorexia Nervosa, a novel]. New York: Warner Bks, 1979.

Liu, Aimée. *Solitaire: The Compelling Story of a Young Woman Growing up in America and Her Triumph over Anorexia.* New York: Harper and Row, 1979.

Millman, M. *Such a Pretty Face: Being Fat in America.* New York: Norton, 1980.

Orbach, Susie. *Fat Is a Feminist Issue.* New York: Pattington Press, 1978.

Palazzoli, Mara S. *Self-starvation: From Individual to Family Therapy in the Treatment of Anorexia Nervosa.* New York: Aronson, 1978.

Vigersky, Robert A., ed. *Anorexia Nervosa: a Monograph of the National Institute of Child Health and Human Development.* New York: Raven, 1977.

Winick, Myron, ed. *Nutritional Disorders of American Women. 5th in Series. Current Concepts in Nutrition. Proceedings of a Symposium.* New York: 1975.

EXERCISE

American Alliance for Health, Physical Education and Recreation. *Women's Athletics: Coping with Controversy.* Washington, D.C.: AAHPER, 1974.

Balazs, Eva. *In Quest of Excellence: A Psycho-social Study of Female Olympic Champions.* Warwick, N.J.: Hoctor Products for Education, 1975.

Barilleaux, D., and J. Murray. *Inside Weight Training for Women.* New York: Contemporary, 1978.

*Blum, Arlene. *Anapurna: A Woman's Place.* San Francisco: Sierra, 1980.

Boslooper, Thomas and Marcia Hayes. *The Femininity Game* [Women & Sports]. New York: Stein and Day, 1974.

Butt, D. S. *Psychology of Sport.* New York: Van Nos Reinhold, 1976. [See Chapter 4: Sex Roles in Sport.]

Carr, D. et al. "Physical conditioning facilitates the exercise-induced secretion of beta-endorphin and beta-lipotropin in women. *New England Journal of Medicine* 305 (1981): 560–563.

Costanza, Betty. *Women's Track and Field.* New York: Hawthorne Bks, 1978.

Douglas, J., and M. Miller. "Record breaking women." *Science News* 112 (1977): 172–174.

Dowling, Claudia. "The Tomboy's Who's Who." *WomenSport* 4 (1977): 33–40.

Drinkwater, Barbara et al. "Menstrual changes in athletes." *Sportsmedicine* 9 (Nov. 1981): 99–112.

Dyer, Ken. "Female athletes are catching up." *New Scientist* (22 Sept. 1977): 722–723.

Garrick, J. G., and R. K. Regman. "Girls' sports injuries in high school' athletics." *Journal of the American Medical Association* 239 (1978): 2245–2248.

Gerber, E. W. et al. *The American Woman in Sport.* Menlo Park, Calif.: Addison Wesley, 1974.

Haycock, C. E., and Gillette, J. "Susceptibility of women athletes to injury: Myth vs. reality." *Journal of the American Medical Association* 236 (1976): 163–165.

Jacobs, K. F. *Girlsports.* New York: Bantam, 1978.

Kaplan, Janice. *Women and Sports.* New York: Avon, 1979.

Klafs, Carl, and M. Joan Lyon. *The Female Athlete: A Coach's Guide to Conditioning & Training.* St. Louis, Mo.: Mosby, 1978.

LaBastille, Anne. *Women and Wilderness.* San Francisco: Sierra, 1980.

Lance, Kathleen. *Getting Strong: A Woman's Guide to Realizing Her Physical Potential.* Indianapolis, Ind.: Bobbs-Merrill Co., 1978.

McWhirter, Norris. *Guinness Book of Women's Sports Records.* New York: Bantam, 1979.

Marshall, John L. *The Sports Doctor's Fitness Books for Women.* New York: Delacorte, 1981.

Moffet, Martha. *Great Women Athletes.* New York: Platt & Munk, 1974.

Myers, Anita M., and Hilary M. Lips. "Participation in competitive amateur sports as a function of psychological androgyny." *Sex Roles* 41 (1978).

Oglesby, C. A. *Women and Sport: From Myth to Reality.* Philadelphia: Lea and Febiger, 1978.

Phillips, Louis, and Karen Markoe. *Women in Sports.* New York: Harcourt Brace Jovanovich, 1979.

Runner's World Magazine. *The Female Runner.* Mountain View, Calif.: World Pubns, 1974.

Rush, Cathy. *Women's Basketball.* New York: Hawthorne Bks, 1976.

Seghers, Carroll. *The Peak Experience: Hiking and Climbing for Women.* New York: Indianapolis Pr, 1979.

*Twin, Stephanie. *Out of the Bleachers: Writings on Women and Sport.* Old Westbury, N.Y.: Feminist Pr, 1979.

Ullyot, Joan. *Women's Running.* Mountain View, Calif.: World View Pubns, 1976.

Wood, P. S. "Sex differences in sports: Is anatomy destiny?" *New York Times Magazine.* Section 6 (May 18, 1980).

Zaharieva, E. "Olympic participation by women: Effects on pregnancy and childbirth." *Journal of the American Medical Association* 221 (1972): 992–995.

LESBIAN HEALTH

Brooks, Virginia R. *Minority Stress and Lesbian Women*. Lexington, Mass.: Lexington Bks, 1981.

Chico Feminist Women's Health Center. *Health Care for Lesbians*. Chico Feminist Women's Health Center, 330 Flume St., Chico, California 95926.

Christenson, Susan, Gayle Ihlenfeld, and Janice Kinsolving. *Lesbians, Gay Men and their Alcohol and Other Drug Use: Resources*. Madison, Wis.: Wisconsin Clearinghouse for Alcohol and Other Drug Information, 1980.

Fenwick, R. D. *The Advocate Guide to Gay Health*. New York: Dutton, 1978.

Good, R. S. "The gynaecologist and the lesbian." *Clinical Obstetrics & Gynaecology* 19 (1976): 473–482.

Hornstein, Frances. *Lesbian Health Care*. OFWHC. 2930 McClure Street, Oakland, California 94609.

Kenyon, F. Edwin. "Homosexuality in gynaecological practice." *Clinics in Obstetrics and Gynaecology* 7 (August 1980): 363–386.

National Gay Health Coalition. *The National Gay Health Directory*. Available from: the Coalition, P. O. Box 677, Old Chelsea Station, New York, NY 10011.

O'Donnell, Mary et al. *Lesbian Health Matters!* Santa Cruz, Calif.: Santa Cruz Women's Health Collective, 1979.

Radicalesbians Health Collective. "Lesbians and the health care system." In *Out of the Closets: Voices of Gay Liberation*. Edited by Karla Jay and Allen Young. Moonachie, N.J.: Pyramid, Pubns, 1974.

*Santa Cruz Women's Health Center. *Annotated Bibliography on Lesbian Health Issues*. Santa Cruz Women's Health Center, 250 Locust Street, Santa Cruz, California 95060.

MENOPAUSE

Anderson, Barrie et al., ed. *The Menopause Book*. New York: Hawthorne Bks, 1977.

Beard, Robert J., ed. *The Menopause—a Guide to Current Research and Practice*. Baltimore, Md.: University Park, 1976.

Clay, Vidal S. *Women: Menopause and Middle Age*. Pittsburgh, Pa.: Know Inc., 1977.

Cooper, W. *Don't Change: A Biological Revolution for Women* [menopause]. New York: Stein and Day, 1975.

Evans, Barbara. *Change of Life*. Pomona, Calif.: Hunter House, 1980.

Flint, Marsha. "The menopause—reward or punishment?" *Psychosomatics* 16 (1975): 4.

Grossman, Marlyn, and Pauline Bart. "Taking the men out of menopause." In this volume.

McKinlay, S. M. "Annotated bibliography: selected studies of the menopause." *Journal of Biosocial Sciences* 5 (1973): 533–555.

Nachtigall, L., with I. Hellman. *The Nachtigall Report* [menopause and estrogen therapy]. New York: Putnam, 1977.

*National Women's Health Network. *Menopause: Resource Guide 3*. Washington, D.C.: National Women's Health Network, 1980.

Newman, J. H. *Our Own Harms: The Startling Facts Behind Menopause and Estrogen Therapy*. Newport Beach, Calif.: Quail Street Publishing Co., 1974.

Page, J. *The Other Awkward Age: Menopause*. Berkeley, Calif.: Ten Speed Press, 1977.

*Reitz, Rosetta. *Menopause: A Positive Approach*. Radnor, Pa.: Chilton Book Co., 1977.

Rose, Louise. *The Menopause Book*. New York: Hawthorn, 1977.

*Seaman, Barbara, and Gideon Seaman. *Women and the Crisis in Sex Hormones*. New York: Rawson Assoc, 1977.

*Weideger, Paula. *Menstruation and Menopause: The Physiology and Psychology, The Myth and the Reality*. New York: Knopf, 1976.

Womanspirit Magazine 1 (WinterSolstice, 1974): 2 [Box 263 Wolf Creek, Oregon, 97497].

MENSTRUATION

Birke, Lynda, with Sandy Best. "Changing Minds: Women, Biology, and the Menstrual Cycle." In this volume.

Budoff, Penny W. *No More Menstural Cramps and Other Good News*. New York: Putnam, 1980.

Can, Alice J., Effie A. Graham, and Carol P. Beecher, eds. *The Menstrual Cycle: Volume 1/A Synthesis of Interdisciplinary Research*. New York: Springer Pub, 1980.

Culpepper, Emily. "Exploring Menstrual Attitudes." In *Women Look at Biology Looking at Women*. Edited by R. Hubbard, M. S. Henifin and B. Fried. Cambridge, Mass.: Schenkman Publishing Co., 1979.

_____. *Period Piece* [10 minute. 16 mm color film]. Available from 64R Sacramento St., Cambridge MA 02138.

Dalton, Katharine. *The Menstrual Cycle*. New York: Warner Paperback Library, 1972.

_____. *Once a Month*. Pomona, Calif.: Hunter Hse, 1979,

_____. *The Premenstrual Syndrome and Progesterone Theory*. New York: William Heinemann Medical Books, 1972.

Davis, Jeffrey et al. "Toxic-shock syndrome." *New England Journal of Medicine* (Dec. 18, 1980): 1429–1435.

*Delany, Janice, Mary Jane Lupton, and Emily Toth. *The Curse: A Cultural History of Menstruation.* New York: A Mentor Book, New American Library, 1976.

Duke, Alexander. "The use of the sponge pessary during menstruation." *The Medical Press* 58 (Nov. 1894).

Edgar, J. Clifton. "Bathing during the menstural period." *American Journal of Obstetrics* 1 (1904).

Ernster, Virginia L. "American menstrual expressions." *Sex Roles* 1 (March 1975): 3–13.

*Friedman, Nancy. *Everything You Must Know About Tampons.* New York: Berkeley Books, 1981.

Friedrich, Eduard and Siegesmund, Kenneth. "Tampon-associated vaginal ulcerations." *Obstetrics & Gynecology* 55 (February 1980): 149–156.

Frisch, Rose E. "Demographic implications of the biological determinants of female fecundity." *Social Biology* 22 (1975): 17–22.

———. "Fatness, puberty, and fertility." *Natural History* 89.10 (October 1980): 16–27.

_____. "Pubertal adipose tissue: is it necessary for normal sexual maturation? Evidence from the rat and the human female." *Federation Proceedings* 39.7 (May 15, 1980).

Frisch, Rose E., Jacob A. Canick, and Dan Tulchinsky. "Comments: human fatty marrow aromatizes androgen to estrogen." *Journal of Clinical Endocrinology and Metabolism* 51.2: 394–396.

Frisch, Rose E., and Janet W. McArthur. "Menstrual cycles: fatness as a determinant of minimum weight for height necessary for their maintenance and onset. *Science* 185 (1974): 949–951.

Frisch, Rose E., and R. Revelle. "Height and weight at menarche and a hypothesis of menarche." *Archives of Diseases of Childhood* 46 (1971): 695–701.

Frisch, Rose E., Grace Wyshak, and Larry Vincent. "Delayed menarche and amenorrhea in ballet dancers." *New England Journal of Medicine* 303 (July 3, 1980): 17–19.

Golub, Sharon. "The magnitude of premenstrual anxiety and depression." *Psychosomatic Medicine* 38 (1976); 1.

Gruba, Glen H., and Michael Rohrbaugh. "MMPI correlates of menstrual distress." *Psychosomatic Medicine* (1975).

Komnenich, Pauline et al., eds. *The Menstrual Cycle: Volume 2/Research and Implications for Women's Health.* New York: Springer Pub, 1980.

Lein, Allen. *The Cycling Female: Her Menstrual Rhythm.* San Francisco: Freeman, 1979.

Logan, Deana Dorman. "The menarche experience in twenty-three foreign countries." *Adolescence* 15 (1980): 247–256.

Maddox, Hilary. *Menstruation.* New Canaan, Conn.: Tobey Publishing Co., 1975.

Merriman, Georgia. "Do women require mental and bodily rest during menstruation." *Columbus Medical Journal* 13 (1894).

*Parlee, Mary Brown. "Stereotypic beliefs about menstruation: A methodological note on the Moos questionnaire and some new data." *Psychosomatic Medicine* 36 (1974): 229–240.

Rome, Esther R. and Emily E. Culpepper. "Menstruation." West Somerville, Mass.: 1977. Available from: Boston Women's Health Book Collective, Inc., Dept. B, Box 192, West Somerville, MA 02144.

Shands, Kathryn et al. "Toxic-shock syndrome in menstruating women." *New England Journal of Medicine* (December 18, 1980): 1436–1442.

Walsh, R. N. et al. "The menstrual cycle, personality and academic performance." *Archives of General Psychiatry* 38 (1981) 219–221.

*Weideger, Paula. *Menstruation and Menopause: The Physiology and Psychology, The Myth and the Reality.* New York: Knopf, 1976.

Wheat, Valerie. "The red rains: A period piece." *Crysalis* (Spring 1977).

Womanspirit Magazine 1 (WinterSolstice, 1974): 2 [Box 263 Wolf Creek. Oregon, 97497].

Women Health International. "Forty years later: Are tampons really safe?" Available from: TSS Packet, NWHN, 224 Seventh St., SE, Washington, D.C. 20003.

OCCUPATIONAL SAFETY AND HEALTH

Art Hazards Newsletter. New York: Art Hazards Project, 5 Beekman St., 10038.

Baer, Judith A. *The Chains of Protection: The Judicial Response to Women's Labor Legislation.* Westport, Conn.: Greenwood, 1978.

Baetjer, Anna M. *Women in Industry: Their Health and Efficiency.* Philadelphia: Saunders, 1946.

Bergman, Tobi. *Health Protection for Operators of VDT's/CRT's.* New York: New York Committee on Occupational Safety and Health, 1980.

*Bingham, Eula, ed. *Proceedings: Conference on Women and the Workplace.* Washington, D.C.: Society for Occupational and Environmental Hazards, 1977.

British Society for Social Responsibility. *Office Workers' Survival Handbook: A Guide to Fighting Health Hazards in the Office.* London: British Society for Social Responsibility, 1981.

Chavkin, Wendy and Laurie Welch. *Occupational Hazards to Reproduction: An Annotated Bibliography.* 1980. Available from: Program in Occupational Health, Department of Social Medicine and the Residency Program in Social Medicine. Montefiore Hospital and Medical Center, 111 E. 210th St., Bronx, NY 10467.

Coalition for the Medical Rights of Women. *Radiation on the Job: A Manual for Health Workers on Ionizing Radiation.* San Francisco: Coalition for the Medical Rights of Women, 1981.

*Coalition for the Reproductive Rights of Workers. *Reproductive Hazards in the Workplace: A Resource Guide.* Washington, D.C.: 1980. Available from: CRROW, 1917 Eye St., N.W. Washington, D.C. 20006.

Feminist Studies [Workers, Reproductive Hazards, and the Politics of Protection, special issue] 5.2 (Summer 1979).

Finneran, Hugh M. "Title VII and restrictions on employment of fertile women." *Labor Law Journal* 31 (April 1980): 223-31.

George, Anne. "Occupational health hazards to women: a synoptic view." *Labour Research Bulletin* 6. 5 (May 1978): 13-27.

Gloss, David. "Ergonomics and the working woman." *Occupational Safety and Health* (May/June 1979): 20-23.

Haynes, S. G. and Feinleib, M. "Women, work and coronary heart disease: Prospective findings from the Framingham heart study." *American Journal of Public Health* 70 (1980): 133-141.

*Hricko, Andrea, and Melanie Brunt. *Working for Your Life: A Woman's Guide to Job Health Hazards*. Berkeley, Calif.: LOHP, University of California, 1976.

Hunt, Vilma. *Health of Women at Work: A Bibliography*. Chicago: Program on Women, Northwestern University, 1977.

*_____. *Work and the Health of Women*. Boca Raton, Fla.: CRC Press, 1979.

Kessler-Haris, Alice. *Women Have Always Worked: A Historical Overview*. Old Westbury, N.Y.: Feminist Pr, 1981.

Kinzer, Nora Scott. *Stress and the American Woman*. Garden City, N.Y.: Doubleday, 1979.

Kuplinsky, Stanley, ed. *The Fertility of Working Women: a Synthesis of International Research*. New York: Praeger, 1977.

Linnes, Patricia M. "Update: new rights for pregnant employees." *Personnel Journal* 58 (January 1979): 33-37.

McKiever, Margaret I. *The Health of Women Who Work*. Washington, D.C.: U.S. Department of Health, Education, and Welfare, Public Health Service, Government Printing Office, 1965.

MacKinnon, Catharine A. *Sexual Harassment of Working Women*. New Haven, Conn.: Yale University Press, 1979.

McLaren, Angus. "Women's work and regulation of family size." *History Workshop* 4 (1977): 70-81.

Makower, Joel. *Office Hazards*. Washington, D.C.: Tilden Press, 1981.

Preventive Medicine [Forum on Women's Occupational Health, special issue] 7 (September 1978).

Rashke, Richard. *The Killing of Karen Silkwood: The Story Behind the Kerr McGee Plutonium Case*. Boston: Houghton Mifflin, 1981.

Stellman, Jeanne. "The effects of toxic agents on reproduction." *Occupational Health and Safety* (April 1977): 36-43.

*_____. *Women's Work, Women's Health: Myths and Realities*. New York: Pantheon, 1977.

Stellman, Jeanne, and Mary Sue Henifin. "No fertile women need apply: Employment discrimination and reproductive hazards in the workplace." In this volume.

Stellman, Steve, and Stellman, Jeanne. "Women's occupations, smoking, and cancer and other diseases." *Ca- A Cancer Journal for Clinicians* 31 (January/February 1981): 29–45.

Waldron, Ingrid. "Employment and women's health: an analysis of causal relationships." *International Journal of Health Services* 10.3 (1980).

Walsh, Diana, and Richard Egdahl, eds. *Women, Work and Health: Challenges to Corporate Policy.* New York: Springer-Verlag, 1980.

Wardle, Miriam Gayle, and David Samuel Gloss. "Women's capacities to perform strenuous work." *Women and Health* 5.2 (Summer 1980): 5–15.

Women's Occupational Health Resource Center News. Available from: WOHRC, Columbia University School of Public Health, 60 Haven Avenue, New York, NY 10032.

Working Women. *Race Against Time: Automation of the Office.* Cleveland, Ohio: Working Women, 1980.

PREGNANCY, CHILDBIRTH, AND MOTHERHOOD

Adler, Norman T., ed. *Neuroendocrinology of Reproduction: Physiology and Behavior.* New York: Plenum Pub, 1981.

American Journal of Obstetrics and Gynecology. St. Louis, Mo.: C. V. Mosby Co.

*Arms, Suzanne. *Immaculate Deception: A New Look at Women and Childbirth in America.* San Francisco: San Francisco Book Co./Houghton Mifflin, 1975.

Ashdown-Sharp, P. *A Guide to Pregnancy and Parenthood for Women on Their Own.* New York: Vintage Books, 1977.

Association of Radical Midwives Newsletter. Available from: Pepper MacKeith, 8 Mount Hooton Terrace. Forest Rd E., Nottingham, UK.

Aveling, James H. *English Midwives: Their History and Prospects.* London: Hugh K. Elliott, 1967.

Baldwin, Rahima. *Special Delivery.* Millbrae, Calif.: Les Femmes, 1979.

Ballou, Judith. *The Psychology of Pregnancy.* Lexington, Mass.: Lexington Bks, 1978.

Bean, C. A. *Labor and Delivery.* Garden City, N.Y.: Doubleday, 1977.

———. *Methods of Childbirth.* Garden City, N.Y.: Dolphin Books, 1974.

*Bernard, Jessie. *The Future of Motherhood.* New York: Penguin, 1974.

Bing, Elizabeth. *The Adventure of Birth.* New York: Ace Books, 1970.

———. *Six Practical Lessons for an Easier Childbirth: The Lamaze Method.* New York: Bantam, 1967, 1977.

Bing, Elizabeth, and Libby Colman. *Having a Baby after 30.* New York: Bantam, 1980.

———. *Making Love During Pregnancy.* New York: Bantam, 1977.

Blais, Madeleine, *They Say You Can't Have a Baby: the Dilemma of Infertility.* New York: Norton, 1979.

Borg, Susan and Judith Lasker. *When Pregnancy Fails: Families Coping with Stillbirth and Infant Death*. Boston: Beacon Pr, 1981.

Boston's Children's Medical Center. *Pregnancy, Birth, and the Newborn Baby*. New York: Delacorte, 1972.

*Boston Women's Health Book Collective. *Ourselves and our Children: A Book by and for Parents*. New York: Random House, 1978.

Brack, Datha Clapper. "Displaced–the Midwife by the Male Physician." In this volume.

_____. "Social forces, feminism and breastfeeding." *Nursing Outlook* 23 (Sept., 1975): 9.

Bradley, Robert. *Husband Coached Childbirth*. New York: Harper and Row, 1981.

Brennan, Barbara, and Joan Rattner Heilman. *The Complete Book of Midwifery*. New York: Dutton, 1976.

Brewer, Gail Sforza. *The Pregnancy after 30 Workbook*. Emmaus, Pa.: Rodale Press, 1978.

Brewer, Gail Sforza, with T. Brewer. *What Every Pregnant Woman Should Know*. New York: Penguin, 1979.

Brewster, D. P. *You Can Breastfeed Your Baby . . . Even in Special Situations*. Emmaus, Pa.: Rodale Press, 1979.

Brook, Danae. *Naturebirth*. New York: Pantheon Books, 1976.

Callahan, Daniel et al. "In vitro fertilization: four commentaries." *The Hastings Center Report* 8.5 (October 1978): 7–14.

Caplan, Ronald, and William I. Sweeney. *Advances in Obsteterics and Gynecology*. Baltimore, Md.: Williams and Wilkins, 1978.

Cartwright, Ann. *The Dignity of Labour: A Study of Childbearing and Induction*. Andover, Hants., England: Tavistock, 1979.

Chard, Tim, and Martin Richards, eds. *Benefits and Hazards of the New Obstetrics*. Philadelphia: Lippincott, 1977.

Chayen, Sonya, ed. "An assessment of the hazards of amniocentesis: Report to the Medical Research Council by their working party on amniocentesis." *British Journal of Obstetrics and Gynecology* 85 (1978), supplement 2.4.

Clyne, Douglas G. *A Concise Textbook for Midwives*. London: Faber and Faber, 1975.

Chesler, Phyllis. *With Child: A Diary of Motherhood*. New York: Berkeley Bk, 1979.

Clark, Ann L. *Culture and Childbearing*. Philadelphia: F. A. Davis Co., 1981.

Cole, I. C. *What Only a Mother Can Tell You about Having a Baby*. Garden City, N.Y.: Doubleday, 1980.

Colman, Arthur, and Libby Colman. *Pregnancy: The Psychological Experience*. New York: Bantam, 1977.

Cooke, Robert. *Improving on Nature: The Brave New World of Genetic Engineering*. New York: MacMillan, 1977.

Corea, Gena. "The Caesarean Epidemic." *Mother Jones* (July 1980).

_____. "Ethical issues in human reproductive technology: analysis by

women." Somerville: 1979. Available from: Boston Women's Health Book
Collective, Box 192, Somerville, MA 02144.

Courtney, Terry, Jill Wolhandler, and Laura Punnett. "Ethical issues in
reproductive technology: An analysis by women—almost." *Off Our Backs*
9 (December 1979).

Cutter, Irving S., and Henry R. Viets. *A Short History of Midwifery*.
Philadelphia: Saunders, 1964.

Davies, Margaret Llewelyn, ed. *Maternity: Letters form Working Women*.
New York: Norton, 1978.

Davin, Ann. "Imperialism and motherhood." *History Workshop—A Journal
of Socialist Historians* 5 (1978).

Demeter, Anna. *Legal Kidnapping. A Mother's Account of What Happens to a
Family When the Father Kidnaps Two Children*. Boston: Beacon, 1977.

Devitt, Neal. "The transition from home to hospital birth in the United
States, 1930-1960." *Birth and the Family Journal* 1 (Summer 1977):
47–58.

Dick-Read, Grantly. *Childbirth Without Fear: The Principles and Practices
of Natural Childbirth*. New York: Dell, 1962 (originally 1944).

Dilfer, Carol Stahman. *Your Baby, Your Body: Fitness During Pregnancy*.
New York: Crown, 1977.

Donegan, Jane B. *Women and Men Midwives: Medicine, Morality and
Misogyny in Early America*. Westport, Conn.: Greenwood, 1978.

Donovan, B. *The Caesarean Experience*. Boston: Beacon Pr, 1977.

Dowie, Mark and Carolyn Marshall. "The Bendectin cover-up." *Mother
Jones* (November 1980).

Dowrick, Stephanie and Sibyl Grundber. *Why Children?* New York: Harcourt
Brace Jovanovich, 1980.

Eiger, Marvin S., and Sally Wendkos Olds. *The Complete Book of Breast-
feeding*. New York: Bantam, 1972.

Elkins, Valmai Howe. *The Rights of the Pregnant Parent*. New York: Two
Continents, 1976.

Emnions, Arthur Brewster, and James Lincoln Huntington, "A review of the
midwife situation." *Boston Medical and Surgical Journal* 164 (1911).

Ewy, Donna, and Rodger Ewy. *Preparation for Childbirth: A Lamaze Guide*.
Garden City, N.Y.: Doubleday, 1975.

Fabe, Marilyn, and Norma Wikler. *Up against the Clock*. New York: Warner
Bks, 1980.

Feldman, S. *Choices in Childbirth*. New York: Grosset and Dunlap, 1978.

Feminist Studies [Toward a Feminist Theory of Motherhood, special issue]
4.2 (June 1978).

Fenton, Judith Alsofrom, Aaron Lifchez, and Clarkson Potter. *The Fertility
Handbook*. New York: Potter, 1980.

Forbes, Thomas Rogers. *The Midwife and the Witch*. New Haven: Yale
University Press, 1966.

_____. "Midwifery and witchcraft." *Journal of the History of Medicine*
17 (1962).

_____. "The regulation of English midwives in the sixteenth and seventeenth centuries." *Medical History* 8 (1964).

Fraiberg, Selma. *Every Child's Birthright: In Defense of Motherhood.* New York: Basic, 1977.

Francoeur, Robert. *Utopian Motherhood—New Trends in Human Reproduction.* Garden City, N. Y.: Doubleday, 1970.

Galana, Laurel. "Radical reproduction: X without Y." In *The Lesbian Reader.* Edited by Gina and Laurel. Oakland. California: Amazon Pr, 1975.

Gallagher, Eugene B. *Infants, Mothers, and Doctors.* Lexington, Mass.: Lexington Bks, 1978.

Gaskin, I. M. *Spiritual Midwifery*, rev. ed. Summertown, Tenn.: Book Pub Co., 1978.

Gilgoff, A. *Home Birth.* New York: Coward, McCann and Geoghegan, 1978.

Greep, R. O., M. A. Koblinsky, and R. S. Jaffe, eds. *Reproduction and Human Welfare: A Challenge to Research.* Cambridge, Mass.: MIT Press, 1976.

Gregory, Samuel. *Man Midwifery Exposed and Corrected.* Boston, 1848.

Gubernick, David and Peter Klopper, eds. *Parental Care in Mammals.* New York: Plenum Pub, 1981.

Hansen, Caryl. *Your Choice: A Young Woman's Guide to Making Decisions About Unmarried Pregnancy.* New York: Avon, 1980.

Heffner, Elaine. *Mothering.* Garden City, N.Y.: Doubleday, 1980.

Heinonen, Olli P., Dennis Slone, and Samuel Shapiro. *Birth Defects and Drugs in Pregnancy.* Publishing Sciences Group, 1977.

Hausknecht, R., and J. H. Heilman. *Having a Caesarean Baby.* New York: Dutton, 1978.

*Holmes, Helen, Betty Hoskins and Michael Gross, eds. *The Custom-Made Child? Women-Centered Perspectives.* Clifton, N.J.: Humana, 1981.

Home Oriented Birth Experience [H.O.M.E.]. *A Comprehensive Guide to Home Birth.* Available form 511 New York Ave., Takoma Park, Washington, D.C., 1976.

Howard, Janes, and Dodi Schultz. *We Want to Have a Baby: The Couple's Complete Guide to Overcoming Infertility.* New York: Dutton, 1979.

Jex-Blake, Sophia. "Women as practitioners of midwifery." *Lancet* 2 (1870).

Kippley, Sheila. *Breast Feeding and Natural Child Spacing: The Ecology of Natural Mothering.* Middlesex, England: Penguin Books, 1974.

*Kitzinger, Sheila. *Birth at Home.* Oxford England: Oxford University Press, 1979.

* _____. *The Complete Book of Pregnancy and Childbirth.* New York: Knopf, 1981.

_____. *Education and Counseling for Childbirth.* New York: Schocken, 1979.

_____. *The Experience of Breastfeeding.* Harmondsworth, Middlesex, England: Penguin Books, 1980.

_____. *The Experience of Childbirth.* Middlesex, England: Penguin Books, 1977.

_____. *Giving Birth: The Parents' Emotions in Childbirth*. New York: Schocken, 1977.

_____. *Women as Mothers: How They See Themselves in Different Cultures*. New York: Random House, 1980.

Kippley, Sheila, and J. A. Davis. *The Place of Birth*. Oxford: University Press, 1978.

Kolata, Gina Bari. "Mass screening for neural tube defects." *The Hastings Center Report* 10.6 (December 1980): 8-10.

Kramer, R. *Giving Birth: Childbearing in America Today*. Chicago: Contemp Bks, 1978.

Kupinsky, Stanley, ed. *The Fertility of Working Women: A Synthesis of International Research*. New York: Praeger, 1977.

The Lactation Review [Mothers in Poverty, Breast-Feeding and the Maternal Struggle for Infant Survival, special issue] 2 (1977).

Lamaze, Fernand. *Painless Childbirth: The Lamaze Method*. New York: Pocket Bks, 1955.

Lazare, Jane. *The Mother Knot*. New York: Dell Publishing Co., 1976.

Leboyer, Frederick. *Birth Without Violence*. New York: Knopf, 1975.

Lewi, Maurice J. "What shall be done with the professional midwife." *Transactions of the Medical Society of the State of New York* (1902).

Lichtendorf, Susan S., and Phyllis L. Gillis. *The New Pregnancy: The Active Woman's Guide to Work, Legal Rights, Health Care, Travel, Sports, Dress, Sex, and Emotional Well-Being*. New York: Random House, 1979.

Linnes, Patricia M. "Update: New rights for pregnant employees." *Personnel Journal* 58 (January 1979): 33-37.

Long, Raven, ed. *Birth Book*. Felton, Calif.: Genesis Press, 1972.

Lubin, Bernard. Sprague H. Gardener, and Aleda Roth. "Mood and somatic symptoms during pregnancy." *Psychosomatic Medicine* 37 (1975).

Marieskind, Helen I. *An Evaluation of Caesarean Section in the United States: Final Report Submitted to: Department of Health, Education and Welfare Office of the Assistant Secretary for Planning and Evaluation/Health*. Washington, D.C.: Department of Health, Education, and Welfare, 1979.

Marshall, Joan F., Susan Morris, and Steven Polgar. "Culture and natality: a preliminary classified bibliography." *Current Anthropology* 13 (April 1972): 268-277.

Maternal Health News. Available from: Maternal Health Society, Box 46563, Station G, Vancouver VGR 4G8.

McBride, Angela Barron. *The Growth and Development of Mothers*. New York: Perennial Library, Harper and Row, 1973.

McCauley, Carole Spearin. *Pregnancy After 35*. New York: Dutton, 1976.

McFarlane, Aida. *The Psychology of Childbirth*. Cambridge, Mass.: Harvard University Press, 1977.

McTaggart, Lynn. *The Baby Brokers: The Marketing of White Babies in America*. New York: Dial, 1980.

Marilus, Esther. *Natural Childbirth the Swiss Way*. Englewood Cliffs, N.J.: Prentice-Hall, 1979.

Medvin, Jeanne O'Brien. *Prenatal Yoga and Natural Childbirth*. Albion, Calif.: Freestone Pub Co., 1974.

Mehl, Lewis E. "Options in maternity care." *Women and Health* 2 (Sept./Oct., 1977): 29–42.

Mengert, William F. "The origin of the male midwife." *Annals of Medical History* 4 (1932).

Menning, Barbara. *Infertility: A Guide for the Childless Couple*. Englewood Cliffs, N.J.: Prentice-Hall, 1977.

Milinaire, Caterine. *Birth*. New York: Harmony Bks, 1974.

Mitchell, Kathleen and Marty Nason. *Caesarean Birth*. San Francisco; Harbor, 1981.

Moore, Kristin A. et al. *Teenage Motherhood: Social and Economic Consequences*. Washington, D.C.: 1979.

Movland, Egbert. *Alice and the Stork: Or the Rise in the Status of the Midwife as Exemplified in the Life of Alice Gregory, 1867–1944*. London: Hodder and Stoughton, 1951.

*National Women's Health Network. *Maternal Health and Childbirth: Resource Guide 4*. Washington, D.C.: National Women's Health Network, 1980.

Noble, Elizabeth. *Essential Exercises for the Childbearing Years*. Boston: Houghton Mifflin, 1976.

Oakley, Ann. "A Case of Maternity: Paradigms of Women as Maternity Cases." *Signs* 4.4 (1979): 607–631.

_____. *Becoming a Mother*. New York: Schocken, 1980.

* _____. *Women Confined: Towards a Sociology of Childbirth*. New York: Schocken, 1980.

Pierce, Ruth. *Single and Pregnant*. Boston: Beacon Pr, 1970.

Pizer, Hank, and Christine Palinsk. *Coping with Miscarriage*. New York: Dial, 1980.

"Prenatal diagnosis for sex choice." *The Hastings Center Report*. 10.4 (February 1980).

Radcliff, Walter. *Milestones in Midwifery*. Bristol, England: John Wright, 1967.

Rapaport, Rhona et al., eds. *Fathers, Mothers, and Society: Toward New Alliances*. New York: Basic, 1977.

*Rich, Adrienne. *Of Woman Born: Motherhood as Experience and Institution*. New York: Norton, 1976.

Rindfuss, Ronald R. et al. "Convenience and the occurrence of births. Induction of labor in the United States and Canada." *International Journal of Health Services* 9.3 (1979): 439–460.

Rose, Hilary, and Jalna Hanmer. "Women's liberation and the technological fix." *Ideology of/in the Natural Sciences*. Edited by Hilary Rose and S. Rose. Cambridge, Mass.: Schenkman, 1980.

Rosengren, W. R. "Social sources of pregnancy as illness or normality." *Social Forces* 39 (March 1961): 260–267.

Sablosky, A. H. "The power of the forceps: A comparative analysis of the midwife—historically and today." *Women and Health* 1 (Jan./Feb., 1976): 10–13.

Schultz, Terri. *Women Can Wait: The Pleasures of Motherhood after Thirty.* New York: Doubleday, 1979.

Shainess, Natalie. "Psychological problems associated with motherhood." In *American Handbook of Psychiatry,* vol. III. Edited by Silvano Arieti. New York: Basic, 1966.

Shandler, Nina, and Michael Shandler. *Yoga for Pregnancy and Birth: A Guide for Expectant Parents.* New York: Schocken, 1979.

Shaw, N. S. *Forced Labor: Maternity Care in the United States.* New York: Pergamon, 1974.

Sinai, N. and O. Anderson. *EMIC (Emergency Maternity and Infant Care).* New York: Arno, 1974.

Smith, David W. *Mothering Your Unborn Child.* Philadelphia, Pa.: Saunders, 1979.

Sousa, Marion. *Childbirth at Home.* New York: Bantam, 1976.

Steinfels, Margaret O'Brien. "New childbirth technology: A clash of values and report from a Hastings Center Conference." *The Hastings Center Report* 8.1 (February 1978).

Stengel, J. J. *Fertility and Conception: An Essential Guide for Childless Couples.* New York: New American Library, 1980.

Stewart, D., and L. Stewart, eds. *Safe Alternatives in Childbirth.* Chapel Hill, N.C.: NAPSAC, 1976.

Tanzer, Debora. *Why Natural Childbirth.* New York: Schocken, 1972.

Tucker, Tarvez, with Elizabeth Bing. *Prepared Childbirth.* New Canaan, Conn.: Tobey Publishing Co., 1975.

Ward, C., and F. Ward. *The Home Birth Book.* Garden City, N.Y.: Dolphin Bks, 1977.

*Wertz, R. W., and D. C. Wertz. *Lying-In: A History of Childbirth in America.* New York: Free Pr, 1977.

Willoughby, Percival. *Observations in Midwifery.* Edited by Henry Blenkinsop. Yorkshire, England: S. R. Publishers, 1863.

Wilson, Christine Coleman, and Wendy Roe Hovey. *Cesarean Childbirth: A Handbook for Parents.* Garden City, N.Y.: Doubleday, 1980.

Woolfolk, W., and J. Woolfolk. *The Great American Birth Rite: Babies as Big Business.* New York: Dial, 1975.

Wright, Erma. *The New Childbirth.* New York: Pocket Bks, 1976.

PSYCHOLOGY

APA Task Force on Sex Bias and Sex-Role Stereotyping in Psychotherapeutic Practice. "Report of the Task Force." *American Psychologist* (Dec. 1975): 1169–1175.

Aslin, Alice L. "Feminist and community mental health center psychotherapists: Expectations of mental health for women." *Sex Roles* 3 (Dec.
1977).

Bardwick, Judith M. *Feminine Personality and Conflict.* Monterey, Calif.:
Brooks-Cole, 1970.

_____. *Psychology of Women: A Study of Bio-Cultural Conflicts.* New York:
Harper and Row, 1971.

Bardwick, Judith M., ed. *Readings on the Psychology of Women.* New York:
Harper and Row, 1972.

Bart, Pauline. "Depression in middle-aged women." In *Women in Sexist
Society.* Edited by Vivian Gornick and Barbara Moran. New York: Basic,
1971.

Bem, S. L. "Sex-role adaptability: one consequence of psychological
androgyny." *Journal of Personality and Social Psychology* 31 (1975):
634–643.

Berk, Juliene. *The Down Comforter: How to Beat Depression and Pull
Yourself out of the Blues.* New York: St. Martin, 1980.

Bonaparte, Marie. "Passivity, masochism and femininity." *International Journal
of Psychoanalysis* 16 (1935): 235–333.

Bosma, Barbara J. "Attitudes of women therapists toward women clients, or
a comparative study of feminist therapy." *Smith College Studies in Social
Work* 46 (Nov. 1975): 1.

*Broverman, Inge K. et al. "Sex role stereotypes and clinical judgements of
mental health." *Journal of Consulting and Clinical Psychology* 35 (Feb.
1970): 1–7.

Brown, George W., and Tirrel Harris. *Social Origins of Depression: A Study of
Psychiatric Disorder in Women.* London: Tavistock, 1978.

Brown, Judith. "Feminism and its implications for therapy." *Radical
Therapist* 1 (1970): 5–6.

Brown, Phil, ed. *Radical Psychology.* New York: Colophon Books, Harper
and Row, 1973.

Castillejo, Claremont. *Knowing Woman: A Feminine Psychology.* New York:
Colophon Books, Harper and Row, 1973.

*Changing Directions in Treatment of Women: A Mental Health Bibliog-
raphy.* Washington, D.C.: National Institute of Mental Health, U.S. GPO,
1980.

Chapman, Joseph Dudley. *The Feminine Mind and Body: The Psychosexual
and Psychosomatic Reactions of Woman.* New York: Philosophical
Library, 1967.

Chesler, Phyllis. *About Men: A Psycho-Sexual Meditation.* New York: Simon
and Schuster, 1978.

*_____. *Women and Madness.* New York: Doubleday, 1972.

*Chodorow, N. *The Reproduction of Mothering: Psychoanalysis and the
Sociology of Gender.* Berkeley: University of California Press, 1978.

Clancey, K., and W. Gove. "Sex differences in mental illness." *American
Journal of Sociology* 80 (1974):204–216.

Cole, J. D., B. F. Pennington, and H. H. Buckley. "Effects of situational stress and sex roles on the attribution of psychological disorder." *Journal of Consulting and Clinical Psychology* 4 (1974): 559–568.

Cox, Sue, ed. *Female Psychology: The Emerging Self*. Chicago: Science Research Associates, 1976.

Cromwell, Phyllis E. *Women and Mental Health: Selected Annotated References, 1970–1973*. Washington, D.C.: N.I.M.H., 1974.

DeRosis, *Women and Anxiety*. New York: Delta, 1979.

Deutsch, Helene. *The Psychology of Women*, vols. I and II. New York: Grune and Stratton, 1944.

*Dinnerstein, Dorothy. *The Mermaid and the Minotaur*. New York: Harper/ Colophon Books, 1976.

Dohrenwend, B. P., and B. S. Dohrenwend. "Sex differences and psychiatric disorders." *American Journal of Sociology* (May 1976): 1447–1454.

Donelson, E., and J. E. Gullahorn. *Women: A Psychological Perspective*. New York: Wiley, 1977.

Franks, V., and W. Burtle. *Women in Therapy: New Psychotherapies for a Changing Society*. New York: Brunner/Mazel Publishers, 1974.

Freidman, Susan Stanford. *A Woman's Guide to Therapy*. Englewood Cliffs, N.J.: Prentice-Hall, 1979.

Freud, Sigmund. "Female sexuality." In *Collected Papers*, vol. 5. London: Hogarth, 1950.

_____. "Femininity." In *New Introductory Lectures on Psychoanalysis*. Edited by J. Strachey. New York: Norton, 1965 (originally 1933), pp. 112–135.

_____. "The psychology of women: biology as destiny." In *Women in a Man-Made World*. Edited by Nona Glazer-Malbin and Helen Youngelson Wachrer. Chicago: Rand McNally, 1972, pp. 58–61.

_____. "Some psychical consequences of the anatomical distinction between the sexes." In *Women and Analysis*. Edited by Jean Strauss. New York: Dell, 1974.

_____. *Three Essays on the Theory of Sexuality*. Translated and edited by James Strachey. New York: Basic, 1963.

Freud, Sigmund, and J. Breuer. "Studies in hysteria." In *The Complete Freud*, vol. II. London: Hogarth Press, 1955.

Friedan, Betty. "The sexual solipsism of Sigmund Freud." In *The Feminine Mystique*. New York: Dell, 1963.

Gede, Eva. "Women and psychiatry." *Atlantis* 4.2 (1979): 81–87.

Gove, W. "Adult sex roles and mental illness." *American Journal of Sociology* 78 (Jan. 1973): 812–835.

_____. "The relationship between sex roles, marital status and mental illness." *Social Forces* 51 (1972): 34–44.

Greek, Frances E. "A serendipitous finding: Sex roles and schizophrenia." *Journal of Abnormal and Social Psychology* 69 (1964): 392–400.

Gullahorn, Jeanne. *Psychology and Women: In Transition*. Washington, D.C.: V. H. Winston & Sons, 1979.

Gump, Janice Porter. "Sex role attitudes and psychological well-being." *Journal of Social Issues* 29 (1973): 79–92.

H. D. [Hilda Doolittle] *Tribute to Freud.* New York: McGraw-Hill, 1973 (originally 1944).

Halas, Celia, and Matteson Robert. *I've Done so Well—Why Do I Feel so Bad?* New York: Macmillan, 1978.

Hansson, Laura. *The Psychology of Woman.* London: G. Richards, 1899.

Harding, Mary Esther. *The Way of All Women: A Psychological Interpretation.* New York: Harper/Colophon Books, 1970.

_____. *Woman's Mysteries: Ancient and Modern.* New York: Putnam, 1971.

Hays, H. R. *The Dangerous Sex: The Myth of Feminine Evil.* New York: Putnam, 1964.

Hinkle, Beatrice. "On the arbitrary use of the terms 'masculine' and 'feminine.'" *Psychoanalytic Review* 7 (1920): 15–30.

Horner, Matina. "The motive to avoid success." *Psychology Today* 36 (1969).

_____. "Toward an understanding of achievement related conflicts in women." *Journal of Social Issues* 28 (1972): 157–176.

*Horney, Karen. *Feminine Psychology.* New York: Norton, 1967.

Hyde, J. S., and B. G. Rosenberg. *Half the Human Experience: The Psychology of Women.* Lexington, Mass.: D. C. Health, 1976.

Jung, Carl G. "Psychological aspects of the mother archetype." In *Four Archetypes.* Princeton: Bollingen, 1973.

*Kaplan, Alexandra, and Joan P. Bean, eds. *Beyond Sex-Role Stereotypes: Readings Toward a Psychology of Androgyny.* Boston: Little, Brown, 1976.

Kinzer, Nora Scott. *Stress and the American Woman.* New York: Ballantine, 1978.

Lennane, K. Jean, and R. J. Lennane. "Alleged psychogenic disorders in women—a possible manifestation of sexual prejudice." *New England Journal of Medicine* 288 (1973): 288–292.

Levine, Saul, Louise Karin, and Eleanor Lee Levine. "Sexism and psychiatry." *American Journal of Orthopsychiatry* 44 (April 1974).

Levy, R. Psychosomatic symptoms and women's protest: Two types of reaction to structural strain in the family." *Journal of Health and Social Behavior* 17 (1976): 121–133.

Manalis, Sylvia A. "The psychoanalytic concept of feminine passivity: A comparative study of psychoanalytic and feminist views." *Comprehensive Psychiatry* 17 (1976).

Mander, Anica Vesel, and Anne Kent Rush. *Feminism as Therapy.* New York: Random House, 1974.

Miller, Jean Baker, ed. *Psychoanalysis and Women.* New York: Penguin, 1973.

_____. *Toward a New Psychology of Women.* Boston: Beacon Pr, 1976.

Millett, Kate. "Freud and the influence of psychoanalytic thought." In *Sexual Politics.* New York: Doubleday, 1970, pp. 176–233.

Mitchell, Juliet. *Psychoanalysis and Feminism.* New York: Pantheon, 1974.

Nelson, Marie Coleman, and Jean Ikenberry, eds. *Psychosexual Imperatives: Their Roles in Identity Formation.* New York: Human Sci Pr, 1978.

Neumann, Erich. *Amor and Psyche: The Psychic Development of the Feminine.* Princeton: Princeton University Press, 1956.

Patrick, T. T. W. "The psychology of woman." *Popular Science Monthly* 47 (1895): 209-224.

Psychology of Women Quarterly. New York: Human Sci Pr.

Radloff, Lenore. "Sex differences in depression." *Sex Roles* 1 (Sept. 1975): 249-265.

Rieff, Philip. *Freud: The Mind of the Moralist.* New York: Doubleday, 1959. [See Chapter V: Sexuality and Domination.]

Robson, Elizabeth and Gwenyth Edwards. *Getting Help: A Woman's Guide to Therapy.* New York: Dutton, 1980.

Rohrbough, Joanna. *Women: Psychology's Puzzle.* New York: Basic, 1979.

Scarf, Maggie. *Unfinished Business: Psychology of Depression.* Garden City, N.Y.: Doubleday, 1980.

Schaffer, Kay F. *Sex Role Issues in Mental Health.* Menlo Park, Calif.: Addison Wesley, 1980.

*Seiden, Anne M. "Overview: Research on the psychology of women. I. Gender differences and sexual and reproductive life." *American Journal of Psychiatry* 133 (1976): 995-1007.

*_____. "Overview: Research on the psychology of women. II. Women in families, work, and psychotherapy." *American Journal of Psychiatry* 133 (1976): 1111-1123.

Sherman, Julia A. *On the Psychology of Women: A Survey of Empirical Studies.* Springfield, Ill.: Charles C. Thomas, 1971.

Smith, Dorothy F., and Sara J. David, eds. *Women Look at Psychiatry.* Vancouver, B.C.: Press Gang Publishers, 1975.

Smith-Rosenberg, Carroll. "The hysterical woman: Sex roles and role conflict in nineteenth-century America." *Social Research* 39 (1972).

Stein, Robert. "Phallos and feminine psychology." In *Incest and Human Love: The Betrayal of the Soul in Psychotherapy.* Edited by Robert Stein. Baltimore: Penguin, 1973.

Strauss, Jean, ed. *Women and Analysis: Dialogues on Psychoanalytic Views of Femininity.* New York: Dell, 1974.

Sturdivant, Susan. *Focus on Women. Volume 2. Therapy with Women: A Feminist Philosophy of Treatment.* New York: Springer Pub, 1980.

Thompson, Clara. "Cultural pressures in the psychology of women." *Psychiatry* 5 (1942): 331-339.

Vieth, Ilza. *Hysteria: The History of a Disease.* Chicago: University of Chicago Press, 1965.

Walstedt, Joyce Jennings. *The Psychology of Women: A Partially Annotated Bibliography.* Pittsburgh: KNOW, Inc., 1972.

*Weisstein, Naomi. *Kinder, Kuche, Kirche as Scientific Law: Psychology Constructs the Female.* Boston: New England Free Press, 1968.

*_____. "Psychology constructs the female, or the fantasy life of the male psychologist (with some attention to the fantasies of his friends, the male biologist and the male anthropologist)." *Social Education* 35 (1971): 362–373.

Wesley, Carol. "The women's movement and psychotherapy." *Social Work* 20 (March 1975): 120–125.

Williams, Elizabeth Friar. *Notes of a Feminist Therapist.* New York: Dell, 1976.

Williams, Juanita H. *Psychology of Women.* New York: Norton, 1977.

Women and Therapy Collective. *Off the Couch: A Woman's Guide to Psychotherapy.* Cambridge, Mass.: Goddard Cambridge Graduate Program in Social Change, 1976.

Zeldow, Peter. "Clinical judgment: A search for sex differences." *Psychological Reports* 37 (1975): 1135–1142.

_____. "Psychological androgyny and attitudes towards feminism. *Journal of Consulting and Clinical Psychology* 44 (Feb. 1976): 1.

RAPE AND INCEST

Adams, Caven, and Jennifer Fay. *No More Secrets: Protecting Your Child From Sexual Assault.* San Luis Obispo, Calif.: Impact Pubs Cal, 1981.

Allen, Charlotte. *Daddy's Girl.* New York: Wyndham Bks, 1980.

American Journal of Orthopsychiatry [Sexual Abuse, section on] 49.4 (October 1979): 634–704.

Amir, Menachem. *Patterns in Forceable Rape.* Chicago: University of Chicago Press, 1971.

Armstrong, L. *Kiss Daddy Goodnight: Speak Out on Incest.* New York: Hawthorn, 1978.

Bakan, David et al. *Child Abuse: A Bibliography.* Toronto: Canadian Council on Children and Youth, 1975. Available from: Canadian Council on Children and Youth, 323 Chapel St., Ottowa K1N 7Z2.

Barnes, D. L. *Rape: A Bibliography 1965–1975.* Troy, N.Y.: Whiteson Publishers, 1977.

Barry, Kathleen. *Female Sexual Slavery.* New York: Avon, 1979.

Bode, Janet. *Fighting Back: How to Cope With the Medical, Emotional and Legal Consequences of Rape.* New York: Macmillan, 1978.

Brady, Katherine. *Father's Days: A True Story of Incest.* New York: Seaview, 1979.

*Brownmiller, Susan. *Against Our Will: Men, Women, and Rape.* New York: Bantam, 1975.

Burgess, Ann Wolbert et al. *Sexual Assault of Children and Adolescents.* Lexington, Mass.: Lexington Bks, 1978.

Burgess, A., and L. Holmstrom. *Rape: Victims of Crisis.* Bowie, Md.: Brady (Prentice-Hall), 1974.

Butler, S. *Conspiracy of Silence: The Trauma of Incest.* San Francisco: New Glide, 1978.

Chappell, D., R. Geis, and G. Geis, eds. *Forcible Rape: The Crime, the Victim, and the Offender.* New York: Columbia University Press, 1977.

Chetin, Helen. *Frances Ann Speaks Out: My Father Raped Me.* Stanford, Calif.: New Seed Press, 1977.

Clark, L., and D. Lewis. *Rape: The Price of Coercive Sexuality.* Toronto, Canada: The Women's Press, 1977.

Connell, Noreen, and Cassandra Wilson. *Rape: The First Source Book for Women.* New York: A Plume Book, New American Library, 1974.

Davis, Angela. "The dialectics of rape." *Ms.* (June 1974): 74.

Dworken, Andrea: *Pornography: Men Possessing Women.* New York: Perigee, 1981.

Edwards, Alison. *Rape, Racism, and the White Women's Movement: An Answer to Susan Brownmiller. Second Edition, with a New Afterword.* Chicago, Ill.: Soujourner Truth Organization, 1979.

Faust, Beatrice. *Women, Sex and Pornography.* New York: Macmillan, 1980.

Feminist Alliance against Rape and the National Communications Network. *Aegis: Magazine on Organizing to Stop Violence against Women.* Available from: FAAR, P. O. Box 21033, Washington, D.C. 20009.

Finkelhor, David. *Sexually Victimized Children.* New York: Free Pr, 1979.

Forward, Susan and Craig Buck. *Betrayal of Innocence: Incest and its Devastation.* New York: Penguin, 1979.

Friedman, Deb. "Rape, racism—and reality." *FAAR and NCN Newsletter* (July/August 1978): 17–26. [Now called *AEGIS: Magazine on Ending Violence against Women.* Refer to Feminist Alliance against Rape.]

Gager, Nancy, and Kathleen Schurr. *Sexual Assault: Confronting Rape in America.* New York: Grosset and Dunlap, 1976.

Geiser, Robert. *Hidden Victims: The Sexual Abuse of Children.* Boston: Beacon Pr, 1979.

Giovannoni, Jeanne and Becerra, Rosina. *Defining Child Abuse.* New York: Free Pr, 1979.

Griffin, Susan. *Pornography and Silence: Culture's Revenge Against Nature.* New York: Harper and Row, 1981.

Griffin, Susan. *Rape: The Power of Consciousness.* New York: Harper and Row, 1979.

Hanmer, Jalna. "Violence and the social control of women." In *Power and the State.* Edited by Gary Littlejohn et al. London: Croon Helm.

*Herman, Judith. *Father-Daughter Incest.* Cambridge, Mass.: Harvard University Press, 1981.

Hilberman, E. *The Rape Victim.* New York: Basic Books, 1976.

Horos, C. V. *Rape.* New Canaan, Conn.: Tobey Publishing Co., 1974.

Jackson, Stevi. "The social context of rape: sexual scripts and motivation." *Women's Studies International Quarterly* 1.1 (1978): 27–38.

Katz, Sedelle and Maryann Mazur. *Understanding the Rape Victim: A Synthesis of Research Findings.* New York: Wiley, 1979.

Kemmer, Elizabeth. *Rape and Rape Related Issues: An Annotated Bibliography.* New York: Garland Pub, 1977.

McCahill, Thomas, Linda Meyer, and Arthur Fischman. *The Aftermath of Rape.* Lexington, Mass.: Lexington Books, 1979.

Malamuth, N. M., M. Heim, and S. Feshbach. "Sexual responsiveness of college students to rape depictions: inhibitory and disinhibitory effects." *Journal of Personality and Social Psychology* (in press).

_____. "Testing hypotheses regarding rape: exposure to sexual violence, sex differences and the 'normality' of rapists." *Journal of Research in Personality* (in press).

Medea, Andra, and Kathleen Thompson. *Against Rape.* New York: Farrar, Straus, and Giroux, 1974.

National Institute for Mental Health. *Rape and Older Women, a Guide to Prevention and Protection.* GPO Stock #017–024–00849–4. Washington, D.C.: U. S. GPO.

*Rush, Florench. *The Best Kept Secret: Sexual Abuse of Children.* Englewood Cliffs, N.J.: Prentice-Hall, 1980.

Russell, Diane E. H. *The Politics of Rape: The Victim's Perspective.* New York: Stein and Day, 1976.

Russell, D. E. H., and N. Van de Ven. *The Proceedings of the International Tribunal on Crimes Against Women.* Millbrae, Calif.: Les Femmes, 1976.

St. Louis Feminist Research Project. *The Rape Bibliography: A Collection of Abstracts.* St. Louis, Mo.: Edy Netter, 1976.

Smith, Diana and Veronica Woollacolt. *Breaking Hold: A Booklet about Rape Prevention.* Vancouver, B. C.: Press Gang Publishers, 1977.

Toronto Rape Crisis Center. *An Annotated Bibliography on Rape.* Toronto: n. p., 1977. Available from: TRCC, P. O. Box 6596, Station A, Toronto, Ontario M5W 1X4.

Tscherhat, Sanford Linda, and Ann Fetter. *In Defense of Ourselves: A Rape Prevention Handbook for Women.* Garden City, N.Y.: Doubleday, 1979.

Wheat, Patte. *By Sanction of the Victim.* New Milford, Conn.: Timely Bks, 1978.

WOMEN IN THE THIRD WORLD

Birdsall, N. *An Introduction to the Social Science Literature on "Woman's Place" and Fertility in the Developing World. Annotated Bibliography* 2.1. Washington, D. C.: Interdisciplinary Communications Programs, Smithsonian Institution, 1975.

_____. "Women and population studies." *Signs* 1 (Spring 1976): 699–712.

Boserup, Ester. *Women's Role in Economic Development.* London: Allen and Unwin, 1970.

Davin, Anna. "Imperialism and motherhood." *History Workshop—A Journal of Socialist Historians* 5 (1978).

"Development begins with women." *UNICEF News.* 104 (1980).

Dreifus, Claudia. "Sterilizing the poor." *The Progressive* 39 (December 1975).

Ehrenreich, Barbara et al. "The charge: Gynocide, the accused: the U.S. government." *Mother Jones* (November 1979).

Eliot, Johan N. "Fertility control and coercion." *Family Planning Perspectives* 5 (Summer 1973).

*Etienne, Mona, and Eleanor Leacock, ed. *Women and Colonization: Anthropological Perspectives*. New York: Praeger, 1980.

Food and Nutrition Group, Boston Science for the People. *Feed, Need, Greed. Food, Resources & Population. A High School Curriculum*. Cambridge, Mass.: Science for the People, 1980.

Hosken, Franziska P. "The epidemology of female genital mutilations." *Tropical Doctor*. (July 1978): 150–156.

_____. *The Hosken Report: Genital/Sexual Mutilation of Females*. WIN News, 1979.

Levine, Carol. "Depo-provera and contraceptive risk: a case study of values in conflict—third world pressures for FDA approval." *The Hastings Center Report* 9.4 (August 1979): 8–12.

Lindsay, Beverly, ed. *Comparative Perspectives of Third World Women: The Impact of Sex, Race, and Class*. New York: Praeger, 1980.

Mass, Bonnie. *Population Target: The Political Economy of Population Control in Latin America*. Ontario, Canada: Charters Publishing Co., 1976.

McCoy, Terry L. et al. *The Dynamics of Population Policy in Latin America*. Cambridge, Mass.: Bulinger, 1974.

Rodriguez-Trias, Helen. "Sterilization abuse." In this volume.

Saulniers, Suzanne Smith, and Cathy A. Rakowski, eds. *Women in the Development Process: A Select Bibliography of Women in Sub-Saharan Africa and Latin America*. Austin, Tex.: University of Texas Press, 1977.

"Some ideas from women technicians in small countries." *Impact of Science on Society* 30.1 (January-March 1980).

Tinker, Irene et al., eds. *Women and World Development*. New York: Praeger, 1976.

Weissman, Steve. "Why the population bomb is a Rockefeller baby." *Ramparts* 8 (May, 1970): 42–47.

Williams, Kathleen N. *Health and Development: An Annotated, Indexed Bibliography*. Baltimore, Md.: Department of International Health, 1972.

"Women and appropriate technology." *International Women's Tribune Center Newsletter* 7 (July 1978, revised 1979). Available from IWTC, 305 E. 46th St., New York, NY 10017.

WOMEN OF COLOR

Alexander, Daryl. "A Montgomery tragedy: The Relf Family refused to be the nameless victims of involuntary sterilization." *Essence* (September 1973).

Ardern, S., ed. *Defining Females: The Nature of Women in Society*. London: Croom Helm, 1978.

The Black Scholar. Available from: Black World Foundation, Box 908, Sausalito, CA 94965.

Black Women's Community Development Foundation. *Mental and Physical Health Problems of Black Women.* Washington, D.C.: 1975.

Boston Women's Health Book Collective. *Nuestros Cuerpos, Nuestras Vidas.* 1979. Available from: Boston Women's Health Book Collective, Box 192, West Somerville, MA 02144.

Combahee River Collective. "Eleven black women—why did they die?" [pamphlet] 1979. Available from: Combahee River Collective, c/o AASC, P. O. Box 1, Cambridge, MA 02139.

Committee on Black Women's Concerns. "Psychological perspectives on black women. A selected bibliography of recent citations." Division of Psychology of Women (Division 35), American Psychological Association, August 1977.

"Crimes in the clinic: a report on Boston City Hospital." *The Second Wave* 2.3 (1973): 17-20.

Cutright, P., and E. Snorter. "Effects of health on the completed fertility of nonwhite and white United States women born between 1867 and 1935." *Journal of Social History* 13 (Winter 1979): 191-218.

*Davis, Angela. *Women, Race and Class.* New York: Random House, 1981.

Edwards, Alison. *Rape, Racism, and the White Women's Movement: An Answer to Susan Brownmiller.* Second Edition, with a New Afterword. Chicago, Ill.: Sojourner Truth Organization, 1979.

Friedman, Deb. "Rape, racism—and reality." FAAR and NCN Newsletter (July/August 1978): 17-26. [Now called *AEGIS: Magazines on Ending Violence against Women.*] Available from FAAR, P. O. Box 21033, Washington, D. C. 20009.

Green, Rayna. "Review essay: Native American women," *Signs* 6.2 (Winter 1980): 248-267.

Leffall, LaSalle. "Breast cancer in black women." *Ca-A Cancer Journal for Clinicians* 31 (July/August 1981): 208-211.

Medicine, Beatrice. "The role of women in Native American societies." [bibliography] *The Indian Historian* 8.3 (1975): 51-53.

Melville, Margarita, ed. *Twice a Minority: Mexican American Women.* St. Louis: C. V. Mosby Co., 1981.

*Moraga, Cherríe and Gloria Anzaldúa, eds. *This Bridge Called My Back: Writings by Radical Women of Color.* Watertown, Mass.: Persephone Press, 1981.

Poor Black Women. A Statement by the Black Unity Party, Peekskill, NY. Boston: New England Free Press, n.d.

Rodriguez-Trias, Helen. "Sterilization abuse." In this volume.

Slater, Jack. "Suicide: a growing menace to Black Women." *Ebony* (September 1973): 152.

Smith, Beverly. "Black Women's Health: Notes for a course." In this volume.

"Sterilization of women: The facts." *Health/Pac Bulletin* 62 (January/ February 1975).

Swartz, Donald P. "The Harlem Hospital Center experience." In *The Abortion Experience: Psychological and Medical Impact.* Edited by Howard J. Osofky and Joy D. Osofky. Hagerstown, Md.: Harper and Row, 1973.

Tyson, Joanne, and Richard. "Sex and the black woman: they are now seeking advice." *Ebony* (August 1977): 103.

Wood, Rosemary. "Health problems facing Native American women." Paper presented at the Invitational Conference of Native American Women, National Institute of Education, Albuquerque, 1978.

GENERAL

Ayalah, Daphna, and Isaac Weinstock. *Breasts: Women Speak About Their Breasts and Their Lives.* New York: Summit Bks, 1979.

Bermosk, Loretta. *Women's Health and Human Wholeness.* New York: Appleton-Century-Crofts, 1981.

Biermann, J., and B. Toohey. *The Women's Holistic Headache Relief Book.* New York: St. Martin, 1979.

*Boston Women's Health Book Collective. *Our Bodies, Ourselves: A Book By and For Women.* 2nd Edition. New York: Simon and Schuster, 1976.

Boston Women's Health Book Collective and ISIS (Geneva). *International Women and Health Resource Guide.* 1980.

Brooks, Beth, and Jackie Niemand. *A Woman's Guide to Treating and Preventing Bladder Infections.* Edited by Liz King. Iowa City, Iowa: Emma Goldman Clinic for Women, 1979.

Cooke, Cynthia, and Susan Dworkin. *The Ms. Guide to Women's Health.* Garden City, N.Y.: Doubleday, 1979.

Cowan, Belita. *Women's Health Care: Resources, Writings, Bibliographies.* Ann Arbor, Mich.: Anshen Publishing, 1977.

Dennerstein, Lorraine et al. *Gynaecology, Sex and Psyche.* Melbourne, Australia: Melbourne University Press, International Scholarly Book Services, 1978.

Derbyshire, Caroline. *The New Woman's Guide to Health and Medicine.* New York: Appleton-Century-Crofts, 1981.

deRosis, Helen. *Women and Anxiety.* New York: Delacorte, 1979.

The Diagram Group. *Woman's Body: An Owner's Manual.* New York: Bantam, 1977.

Dreifus, Claudia. *Seizing Our Bodies: The Politics of Women's Health.* New York: Vintage Bks, 1978.

Fairfield, L. "Health of professional women." *Medical Woman's Journal* 34 (1927).

The Federation of Feminist Women's Health Centers. *How to Stay Out of the Gynecologist's Office.* Culver City, Calif.: Peace Press, 1981.

*____. *A New View of a Woman's Body*. New York: Simon and Schuster, 1981.

The Female Patient. New York: P. W. Publications, 34 (1927).

*Frankfort, Ellen. *Vaginal Politics*. New York: Quadrangle, 1972.

Freeman, Jo., ed. *Women: A Feminist Perspective*. Palo Alto, Calif.: Mayfield Pub, 1975.

Futoran, Jack M., and May Annexton. *Your Body: A Reference Book for Women*. New York: Ballantine, 1976.

Grissum, Marlene and Carol Spengler. *Womanpower and Health Care*. Boston: Little, Brown, 1976.

Haire, Doris. "How the FDA determines the 'safety' of drugs—just how safe is 'safe'?" Available from: National Women's Health Network, 224 Seventh St., SE, Washington, D.C. 20003.

Health/Pac. *Women and Health*. New York: Health/Pac, 1980.

Hilliard, Marion. *Women and Fatigue: A Woman Doctor's Answer*. Garden City, N.Y.: Doubleday, 1960.

Horos, C. V. *Vaginal Health*. New Canaan, Conn.: Tobey Publishing Co., 1975.

Hull, Bonnie. *In Our Hands: A Woman's Health Manual*. Melbourne: Hyland Hse, 1980.

International Journal of Health Services. "Women and Health: Special Issue." 5 (1975): 167-346.

ISIS International Bulletin [Women and Health] 2.8 (Summer 1978).

Jackson, June. *Everywoman's Health*. Garden City, N.Y.: Doubleday, 1980.

Kowalski, Karen, ed. *Nursing Dimensions* [Women's Health Care] 7.1 (Spring 1979).

Lanson, Lucienne. *From Woman to Woman: A Gynecologist Answers Questions About You and Your Body*. New York: Knopf, 1975.

Laversen, N., and S. Whitney. *It's Your Body: A Woman's Guide to Gynecology*. New York: Grosset and Dunlap, 1978.

Lewis, Charles E., and Mary Ann Lewis. "The potential impact of sexual equality on health." *New England Journal of Medicine* (Oct. 1977): 863-869.

Llewellyn-Jones, Derek. *Every Woman and Her Body*. New York: Taplinger, 1971.

Lynch, Peggy and Esther Rome. "Venereal disease (VD) also known as sexually transmitted diseases (STD) and how to avoid them." Boston Women's Health Book Collective, 1980. Available from: Boston Women's Health Book Collective, Department EP, Box 192, Somerville, MA 02144.

MacKeith, Nancy, ed. *The New Women's Health Handbook*. London: Virago, 1978.

Milio, Nancy. *The Care of Health in Communities: Access for Outcasts*. New York: Macmillan, 1975.

Montagu, M. F. *Reproductive Development of the Female*. Littleton, Mass.: PSG Publishing Co., 1979.

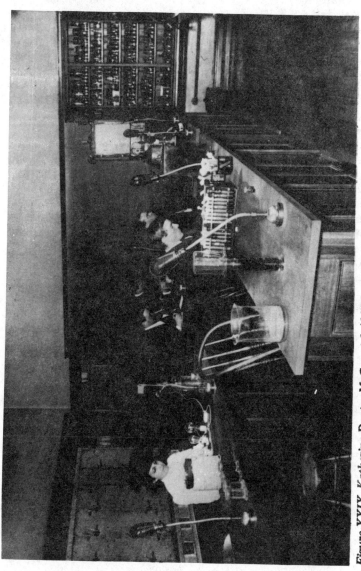

Figure XXIX. Katherine Dexter McCormick (1875–1967), early woman graduate of M.I.T. (B.S. in biology, 1904). (M.I.T. Historical Collections)

Moore, H. and S. Thompson. *The Surgical Beauty Racket*. Port Washington, N.Y.: Ashley Bks, 1978.

Motta, P.M., and E.S.E. Hafez, eds. *Biology of the Ovary*. The Hague: Nijhoff, 1980.

*National Women's Health Network. *Resource Guides 1–9*. Washington, D.C.: National Women's Health Network, 1980.

Notman, Malkah, and Carol C. Nadelson. *The Woman Patient, Medical and Psychological Interfaces. Vol. 1: Sexual and Reproductive Aspects of Women's Health Care*. New York: Plenum Pub, 1978.

Parvati, J. *Hygea: A Woman's Herbal*. Berkeley, Calif.: Distributed by Bookpeople, 1978.

Rennie, Susan, and Anna Rubin. "Women's survival catalog: Holistic healing." *Chrysalis* No. 1: 67–69, 1977.

Rosenkrantz, B. G. *Health of Women and Children* [Historical analysis]. New York: Arno, 1977.

Rowland, Beryl. *Medieval Woman's Guide to Health: The First English Gynecological Handbook*. Kent, Ohio: Kent State University Press, 1981.

Rush, Anne Kent. *Getting Clear: Body Work for Women*. New York: Random House, 1973.

Sandelowski, Margarete. *Women, Health, and Choice*. Englewood Cliffs, N.J.: Prentice-Hall, 1981.

*Sloane, Ethel. *Biology of Women*. New York: Wiley, 1980.

Social Policy. Special Issue on Women and Health. Sept./Oct. 1975.

Thompson, D. S., ed. *Everywoman's Health: A Complete Guide to Body and Mind*. Garden City, N.Y., 1980.

III. Women in Science and Health Care

SCIENCE

Acker, Fabian. "Two kids with every Ph.D." *New Scientist* 72 (December 16, 1976): 653.

"Adventures of women in science and labor." *Federation Proceedings* 35 (Sept. 1976): 11.

Arditti, Rita. "Women in science: Women drink water while men drink wine." *Science For the People* 8 (March 1976): 24.

Auvinen, Rita. "Women and work (II): Social attitudes and women's careers." *Impact of Science on Society* 20 (1970): 85–92.

AWIS Newsletter (Association for Women in Science). Available from: Suite 1122, 1346 Connecticut Ave., N. W., Washington, D. C. 20036.

AWM Newsletter (Association of Women in Mathematics). Available from: c/o Department of Mathematics, Wellesley College, Wellesley, MA 02181.

AWP Newsletter (Association for Women in Psychology). Available from: c/o Jill Bellinson, 114 W. 86th St., New York, NY 10024.

Bachtold, L. M., and E. E. Werner. "Personality characteristics of women scientists." *Psychological Reports* 31 (1972): 391–396.

Baizer, E. *Resources for Women in Science*. Washington, D.C.: Association for Women in Science, 1980.

Barranger, Elizabeth et al. "Goals for women in science." *Technology Review*, June 1973, pp. 48–57.

Bayer, A. E. and H. S. Astin. "Sex differences in academic rank and salary among science doctorates in teaching." *Journal of Human Resources* 3 (1968): 191.

Bayer, A. E., and J. Austic. "Sex differentials in the academic reward system." *Science* 188 (1975): 796.

Bernard, Jessie. *Academic Women*. New York: New American Library, 1974.

Bissell, Mina. "Equality for women scientists." *Grants Magazine* 1 (Dec. 1978): 331–334.

Boalt, G., H. Lantz, and H. Herlin. *The Academic Pattern: A Comparison Between Researchers and Non-Researchers. Men and Women*. Stockholm: Almquist and Wiksell, 1973.

Booth, Egbert Perry. *Women at War: Engineering*. London: J. Crowther, 1943.

Bregman, Elsie Oschrin. "A study of vocational traits of women secretaries, lawyers, chemists, statisticians, and women in positions of responsibility in department stores." *Psychological Bulletin* 24 (March 1926).

*Briscoe, Anne M., and Sheila M. Pfafflin, eds. *Expanding the Role of Women in the Sciences*. New York: The New York Academy of Sciences, 1979.

Brown, Janet Welsh. "Are women scientists getting a fair deal?" *Nature* 272 (20 April 1978): 658–659.

Chepelinsky, Ana Berta et al. "Women in chemistry—Part of the 51% minority." *Science for the People* 4 (1972).

Chinn, Phyllis Zweig, compiler. *Women in Science and Math*. Available from: Karen Ehrlich, AAAS, 1776 Massachusetts Ave., N.W., Washington, D. C. 20036. [Bibliography].

Cole, Jonathan R., and Stephen Cole. *Social Stratification in Science*. York: Free Press, 1979.

_____. *Woman's Place in the Scientific Community*. New York: Wiley, 1977.

_____. "Women in science." *American Scientist* 69 (July/Aug. 1981): 385–391.

Cole, Jonathon R., and Stephen Cole. *Social Stratification in Science*. Chicago: University of Chicago Press, 1973.

Committee on the Education and Employment of Women in Science and Engineering. *Climbing the Academic Ladder: Doctoral Women Scientists in Academe*, Washington, D.C.: National Academy of Sciences, 1979.

Conference on the Role of Women in Professional Engineering. A conference held under the auspices of the Executive Office of the President of the U.S. and sponsored by the University of Pittsburgh and the Society of Women Engineers. New York: Society of Women Engineers, 1962.

Couture-Cherki, Monique. "Women in Physics." In *Ideology of/in the Natural Sciences*. Hilary Rose and S. Rose, eds. Boston, Mass.: G. K. Hall; Cambridge, Mass.: Schenkman, 1980.

Crawford, H. Jean. "Report on the association to aid scientific research by women." *Science* 76 (1932): 492–493.

Creative Women Quarterly [Women in Science, special issue] 2.4 (Spring 1979).

Davis, Audrey B. *Bibliography on Women: With Special Emphasis on Their Roles in Science and Society*. New York: Science History Publications, 1974.

Dodge, N. T. *Women in the Soviet Economy: Their Role in Economic, Scientific and Technical Development*. Baltimore: Johns Hopkins Press, 1966.

Dumbar, M. "Women in science: how much progress have we really made." *Science Forum* 6 (18 April 1973).

Easlea, Brian. *Science and Sexual Oppression*. London: Weidenfeld and Nicolson, 1981.

Emberlin, Diane P. *Contributions of Women: Science*. Minneapolis, Minn.: Dillon, 1977.

Ernest, J. "Mathematics and sex." *The American Mathematical Monthly* 83 (1976): 595–614.

Fee, Elizabeth. "Is There a Feminist Science." *Science and Nature* 4 (1981): 46–57.

Ferriman, Annabel. "Women academics publish less than men." *Impact of Science on Society*. UNESCO 25 (1975): 153–154.

Fields, Cheryl. "Women in science: Breaking the barriers." *Chronicle of Higher Education* 15 (October 31, 1977): 7–8.

Fischer, Ann. "The position of women in anthropology." *American Anthropologist* 70 (1968): 338.

Fox, Lynn. "Women and the career relevance of mathematics and science." *School Science and Mathematics* 76 (1976): 357–365.

*Fox, Lynn, Linda Brody, and Dianne Tobin, eds. *Women and the Mathematical Mystique*. Baltimore, Md.: The Johns Hopkins University Press, 1980.

Frithiof, P. *Women in Science*. Lund, Sweden: University of Lund, 1967.

Goldman, R. D., R. M. Kaplan, and B. B. Platt. "Sex differences in the relationship of attitudes toward technology to choice of field of study." *Journal of Counseling Psychology* 20 (1973): 412.

Goldsmit, N. F. "Women in science: Symposium and job mart." *Science* 168 (1970): 1124.

Gray, M. "The mathematical education of women." *The American Mathematical Monthly* 84 (1977): 374–377.

Gruchow, Nancy. "Discrimination: Women charge universities, colleges with bias." *Science* 169 (Sept. 25, 1970): 1284–1290.

Haber, Julia Moesel. *Women in the Biological Sciences*. State College, Pennsylvania: Pennsylvania State University Press, 1939.

Halsey, S. D., ed. *Women in Geology*. Canton, N.Y.: Ash Lad Press, 1976.

Handler, Philip. "Women scientists: steps in the right direction." *The Sciences* 18 (March 1978): 6-9.

Hansen, R., and J. Neujahr. "Career development of males and females gifted in science." *The Journal of Educational Research* 68 (1974): 43, 45.

Haraway, D. J. "In the beginning was the word: The genesis of biological theory." *Signs* 6 (1981): 469-481.

Helson, Ravenna. "Women mathematicians and the creative personality." *Journal of Consulting and Clinical Psychology* 36 (1971): 210-220.

Hollingworth, Leta, and Robert H. Lowe. "Science and feminism." *Popular Science Monthly* (1916).

Holyrup, Else. *Women and Mathematics, Science, and Engineering. A Partially Annotated Bibliography with Emphasis on Mathematics and with References on Related Topics*. Vaerlose, Denmark: Roskilde University Library, 1978. Available from: author, Bogevej 8, DK 3500 Vaerlose, Denmark.

Hopkins, Nancy. "The high price of success in science." *Radcliffe Quarterly* 62 (June 1976): 16-18. See also Replies: *Radcliffe Quarterly* 62 (Sept. 1976): 31-32.

Howell, Mary. "Just like a housewife: Delivering 'human services.'" *Radcliffe Quarterly* 63 (Dec. 1977): 1-4.

Hubbard, Ruth. "Rosalind Franklin and DNA" [book review]. *Signs* 2 (1976): 229-236.

_____. "Sexism in science." *Radcliffe Quarterly* 62 (March 1976): 8-11.

_____. "When women fill men's roles. . . ." *Trends in Biochemical Sciences* 1 (1976): N 52-53.

Hudson, L. "Fertility in the arts and sciences." *Science Studies* 3 (1973): 305.

Inke, Lillian V., and Mildred S. Barker. *Employment Opportunities for Women in Professional Engineering*. Washington, D.C.: Government Printing Office, 1954.

Jacobs, Judith E., *Perspectives on Women and Mathematics*. Columbus, Ohio: ERIC Clearinghouse for Science, Mathematics, and Environmental Education, 1978.

*Jacobs, Karen Folger, and Carol S. Wolman. "Twenty-seven strategies of women academics." In *Women in the Organization*. Edited by Hal H. Frank. Philadelphia: University of Pennsylvania Press, 1977, pp. 256-262.

Kashket, Eva R. et al. "Status of women microbiologists: A study of microbiologists based on objective and subjective criteria is presented." *Science* 183 (1974): 488-494.

Keller, Evelyn Fox. "Feminism and Science." *Signs* 7 (1982): 589-602.

_____. "Feminist critique of science: A forward or backward move." *Fundamenta Scientiae* 1: 241-249.

*_____. "Gender and science." *Psychoanalytic and Contemporary Thought* 1 (1978): 409-433.

Magazine (October 1974).

Kelly, Alison. *Girls and Science: An International Study of Sex Differences in School Science Achievement.* Stockholm: Almquist and Wiksell, 1978.

———. *Girls and Science Education.* Manchester, U.K.: Manchester University Press, 1979.

———. "Women in physics and physics education." In *New Trends in Physics Teaching* 3: 241–266. Edited by J. L. Lewis. UNESCO: 1976.

Kistiakowsky, V. *Women in Engineering, Medicine and Science.* Conference on Women in Science and Engineering. Washington, D.C.: National Research Council, June 1973 (Revised: Sept. 1973).

Kreinberg, Nancy. *I'm Madly in Love with Electricity and Other Comments about Their Work by Women in Science and Engineering.* Berkeley, Calif.: Lawrence Hall of Science, 1977.

*Kundsin, Ruth B., ed. "Successful women in the sciences: An analysis of determinants." *Annals of the New York Academy of Science* 208 (1973).

*———, ed. *Women and Success: The Anatomy of Achievement.* New York: Morrow, 1974.

Law, Margaret E., ed. *Goals for Women in Science.* Boston: Women in Science and Engineering, 1972.

Lonsdale, K. "Women in science: Reminiscences and reflections." *Impact of Science on Society* 20 (1970): 45.

Lowry, Nancy. *Science Education at Hampshire College with Special Emphasis on Women and Science.* Amherst, Mass.: Hampshire College, 1981.

Lubkin, Gloria. "Women in physics." *Physics Today* 24 (1971): 24.

Maccoby, E. E. "Feminine intellect and the demands of science." *Impact of Science on Society* 20 (1970): 13.

Malcom, Shirley, Paul Q. Hall, and J. W. Brown. *The Double Bind: The Price of Being a Minority Woman in Science.* AAAS Report No. 76-R-3. Washington, D.C. (April 1976).

Math/Science [women's] *Network Broadcast.* Available from: Jan MacDonald, Math/Science Resource Center, Mills College, Oakland, CA 94613.

Mattfeld, J. A., and C. G. van Aken. *Women and the Scientific Professions: The MIT Symposium on American Women in Science.* Cambridge, Mass.: MIT Press, 1965.

Menninger, Sally Ann and Clare Rose. "Women scientists and engineers in American academia." *International Journal of Women's Studies* 3 (May-June 1980): 292–299.

Miller, Helen H. "Science careers for women." *Atlantic Monthly* 1957, pp. 123–128.

Motz, A. B. "The roles of the married woman in science." *Marriage and Family Living* 23 (1961): 374.

Muir, Helen. "Emphasize quality not sex." *Trends in Biochemical Sciences* 2 (June 1977): N 121–123.

Bibliography 355

Murphy, Mary Claire, and E. S. Spiro. *Careers for Women in the Biological Sciences.* U.S. Dept. of Labor, Women's Bureau. Washington, D.C.: Government Printing Office, 1961.

National Research Council, Committee on the Education and Employment of Women in Science and Engineering. *Climbing the Academic Ladder: Doctoral Women Scientists in Academe.* Washington, D.C.: National Academy of Sciences, 1979.

National Women's Anthropology Newsletter. Available from: San Francisco State University, 1600 Holloway Ave., San Francisco, CA 94132.

Nelson, R. "Women mathematicians and the creative personality." *Journal of Consulting and Clinical Psychology* 35 (1970): 210.

Nichols, Robert. "Women in science and engineering: Are jobs really sexless?" *Technology Review* (June 1973).

Noble, Iris. *Contemporary Women Scientists of America.* New York: Julian Messner, 1979.

Parish, J. B., and J. S. Block. "The future for women in science and engineering." *Bulletin of the Atomic Scientist* 24 (May 1968): 46.

Perrucci, C. C. "Minority status and the pursuit of professional careers: Women in science and engineering." *Social Forces* 49 (1970): 245.

Ramaley, J. *Covert Discrimination and Women in the Sciences.* Washington, D.C.: American Association for the Advancement of Science Symposium, 1977.

Reskin, Barbara F. "Scientific productivity, sex, and location in the institution of science." *American Journal of Sociology* 83 (March 1978): 1235-1243.

_____. "Sex differences in status attainment in science: The case of the postdoctoral fellowship." *American Sociological Review* 41 (1976): 597-612.

Roark, Anne. "Women in science: Unequal pay, unsold ideas, and sometimes, unhappy marriages." *Chronicle of Higher Education* 20 (April 21, 1980): 3-4.

Robinson, Gail. "A woman's place is in the [environmental] movement." *Environmental Action* 9 (25 March 1978): 12-23.

Roe, A. "Women in science." *Personnel and Guidance Journal* 44 (1966): 748.

Rossi, A. S. "Women in science: Why so few?" *Science* 148 (1965): 1196.

*Ruddick, Sara, and Pamela Daniels, eds. *Working It Out: 23 Women Writers, Artists, Scientists, and Scholars Talk About Their Lives and Work.* New York: Pantheon, 1977.

Schilling, Gerhard F., and M. Kathleen Hunt. *Women in Science and Technology: US/USSR Comparisons.* Santa Monica, Calif.: The Rand Paper Series, 1974.

Schwartz, B. *Women in Scientific Careers.* Washington, D.C.: National Science Foundation, 1961.

Scientific Manpower Commission, AAAS. *Professional Women and Minorities: A Manpower Data Resource Service.* Washington, D.C.: AAAS, 1975.

_____. "Women in science: An analysis of a social problem." *Harvard*

Shapley, Deborah. "Obstacles to women in science." *Impact of Science on Society* 25 (1975): 105-114.

Smith, Judy. *Something Old, Something New, Something Borrowed, Something Due: Women and Appropriate Technology.* Missoula, Mont.: Women and Technology Project (315 S. 4th St.) 1981.

Stehelin, Liliane. "Sciences, women and ideology." In *Ideology of/in the Natural Sciences.* Edited by Hilary Rose and S. Rose. Boston, Mass.: G. K. Hall; Cambridge, Mass.: Schenkman, 1980.

Tereshkan-Nikolayeva, V. "Women in space." *Impact of Science on Society* 20 (1970): 5-12.

Tobias, Sheila. *Overcoming Math Anxiety.* Boston: Houghton Mifflin, 1978.

Tossi, Lucia. "Women's scientific creativity." *Impact of Science on Society,* 25 (1975): 105-114.

Undercurrents [Women and Energy, special issue]. Available from: 27 Clerkenwell Close, London EC1R OAT, England.

Vetter, B. M. "The outlook for women in science." *The Science Teacher* 40 (1973): 22.

*_____. "Sex discrimination in the halls of science." [Review of J. Cole's *Fair Science*] *Chemical and Engineering News* (March 1980): 37-38.

_____. "Women in the natural sciences." *Signs* 1 (Spring 1976): 713-720.

_____. "Working women scientists and engineers." *Science* 207 (4 Jan. 1980): 26-34.

Walberg. "Physics, femininity and creativity." *Development Psychology* 1 (1969): 47.

Weisstein, Naomi. "Adventures of a Woman in Science." In this volume.

White, Martha S. "Psychological and social barriers to women in science." In *Toward a Sociology of Women.* Edited by C. Safilios-Rothschild. Xerox College Publications, 1972.

_____. "Psychological and social barriers to women in science." *Science* 170 (1970): 413.

Women and Mathematics Education. Available from: Judith Jacobs, WME, George Mason University, 4400 University Drive, Fairfax, VA 22030.

Women and Technology Project. *Women and Technology: Deciding What's Appropriate.* Missoula, Mont.: Women and Technology Project (315 S. 4th St.), 1981.

Women Authors in Science and Engineering, Boston, Mass. "Goals for women in science." *Technology Review* (June 1973).

Women in Research. "Women in science: Working together to vitalize research." *Women's Studies Program: Papers in Women's Studies.* The University of Michigan 1 (1976): 1.

Zinberg, Dorothy. "The past decade for women scientists—win, lose, or draw?" *Trends in Biochemical Sciences* 2 (June 1977): N 123-126.

Zuckerman, Harriet, and Jonathan Cole. "Women in American science." *Minerva* 13 (Spring 1975): 82-102.

PATIENTS AND PROVIDERS

Advances in Nursing Science [Special Issue *Women's Health*]. (January 1981).
American Psychiatric Association. *Women in Psychiatry*. Washington, D.C.:
APA, 1973.
Ashley, JoAnn. *Hospitals, Paternalism and the Role of the Nurse*. New York:
Teachers College Press, 1976.
Bellaby, Paul and Patrick Oribabor. "Determinants of the occupational
strategies adopted by British hospital nurses." *International Journal of
Health Services* 10.2 (1980): 291-310.
Benedict, Thomas G. "The changing relationship between midwives and
physicians during the Renaissance." *Bulletin of the History of Medicine*
51 (Winter 1977): 550-564.
Beshiri, Patricia H. *The Woman Doctor: Her Career in Modern Medicine*.
New York: Cowles Book Co., 1969.
Boston Nurses Group. *The False Promise: Professionalism in Nursing*.
Somerville, Mass.: New England Free Press. Reprinted from *Science for
the People*, 1978.
Brown, Carol A. "Women workers in the health service industry." *International Journal of Health Services* 5 (1975): 173-184.
Bullough, Bonnie. "Barriers to the nurse practitioner movement." *International Journal of Health Services* 5 (1975): 225-234.
Bullough, Vern L. and Bonnie Bullough. *The Care of the Sick*. New York:
Produst, 1979.
Campbell, J. E. "Women dentists: An untapped resource." *Journal of the
American College of Dentistry* 36 (1970): 265-269.
*Campbell, Margaret [Mary C. Howell]. *"Why Would a Girl Go Into
Medicine?" Medical Education in the United States: A Guide for Women*.
Old Westbury, N.Y.: Feminist Pr, 1973.
Cannings, Kathleen, and William Lazonick. "The Development of the nursing
labor force in the United States: A basic analysis." *International Journal
of Health Services* 5 (1975): 185-216.
Chaff, S. L. et al. *Women in Medicine: A Bibliography of the Literature on
Women Physicians*. Metuchen, N.J.: Scarecrow, 1977.
Conference on Meeting Medical Manpower Needs. *The Fuller Utilization of
the Woman Physician: Report*. Washington, D.C.: American Medical
Women's Association, 1968.
*Corea, Gena. *The Hidden Malpractice: How American Medicine Treats
Women as Patients and Professionals*. New York: Morrow, 1977.
Coste, Chris. "Women in medicine: Progress and prejudice." *The New
Physician* 24 (1975): 25-31.
Crawford, Robert. "Healthism and the medicalization of everyday life."
International Journal of Health Services 10 (1980): 365-388.
Fee, Elizabeth. "Women and health care: A comparison of theories." *International Journal of Health Services* 5 (1975): 397-415.

Feldman, Jacqueline. "The savant and the midwife." *Impact of Science on Society* 24 (1975): 105–115.

Fields, Cheryl. "Problems, ironies plague nursing." *Chronicle of Higher Education* 19 (February 4, 1980): 9.

Fitzpatrick, M. Louise. "Nursing." *Signs* 2 (Summer, 1977): 818–834.

Fogel, Catherine and Nancy Woods., eds. *Health Care of Women: A Nursing Perspective.* St. Louis: Mosby, 1981.

Gershon, Michael and Henry B. Biller. *The Other Helpers: Paraprofessionals and Nonprofessionals in Mental Health.* Lexington, Mass.: Lexington Bks, 1977.

Grissum, M., and C. Spengler. *Womanpower and Health Care.* Boston: Little, Brown, 1976.

Halas, Mary A. "Sexism in women's medical care." *Frontiers: a Journal of Women's Studies* 4 (July 1979): 11–15.

Hare, Daphne. "The victim is guilty." *Federation Proceedings* 35 (1975): 2223–2225.

Haseltine, Florence, and Yvonne Yaw. *Woman Doctor: The Internship of a Modern Woman.* Boston, Mass.: Houghton Mifflin, 1976.

Health Group Collective. "Women: The right to choose what's right for your health: conflicts with the NHS." *Science for People* [Health, special issue]. 30 (Winter 1977–1978): 13–16.

Howell, Mary. "What medical schools teach about women." *New England Journal of Medicine* 291 (8 Aug. 1974): 304–307.

Hurst, Marsha and Ruth E. Sambrana. "The health careers of urban women: A study of East Harlem." *Signs* 5.3 (supplement) (Spring 1980).

ISIS Collective. "Who controls women's health. Drug industry, doctors and governments." *ISIS International Bulletin.* 7 (Spring 1978): 5–12.

Jacobson, Arthur C. "A medical view of women's lib." *Medical Times* 18 (July 1972).

Jefferys, Margot. *Women in Medicine: The Results of an Inquiry Conducted by the Medical Practitioners' Union in 1962–63.* London: Office of Health Economics, 1966.

Journal of the American Medical Women's Association. Available from: American Medical Women's Association, Inc., 1740 Broadway, New York NY 10019.

Kerrick, Margaret B. *Night Supervisor: An Irreverent Memoir.* West Roxbury, Mass.: Mrs. Margaret B. Kerrick, 1977.

Kjervik, Diane K. *Women in Stress: The Nursing Perspective.* New York: Appleton-Century-Crofts, 1979.

Leeson, Joyce and Judith Gray. *Women and Medicine.* London: Tavistock Publications, 1978.

Linn, Edwin. "Women dentists: Career and family." *Social Problems* 18 (1971): 394–395.

_____. "Women dentists: Some circumstances about their choice of a career." *Journal of the Canadian Dental Association* 10 (1972): 364–409.

Litoff, Judy B. *American Midwife: 1860 to the Present.* Contributions in Medical History No. 1. Westport, Conn.: Greenwood, 1978.

Lopate, C. *Women in Medicine.* Baltimore: Johns Hopkins Press, 1964.

*Lorber, Judith. "Women and medical sociology: Invisible professionals and ubiquitous patients. In *Another Voice.* Edited by M. Millmann and R. M. Kanter. Garden City, N.Y.: Anchor Pr, 1975, pp. 75–105.

Lutzker, Edythe. *Women Gain a Place in Medicine.* New York: McGraw-Hill, 1969.

Marieskind, Helen. "The women's health movement." *International Journal of Health Services* 5 (1975): 217–224.

*_____. *Women in the Health System: Patients, Providers, and Programs.* St. Louis: Mosby, 1980.

Melnick, Vijaya, and Franklin D. Hamilton. *Minorities in Science: The Challenge for Change in Biomedicine.* New York: Plenum Pr, 1977.

Morgan, Elizabeth. *The Making of a Woman Surgeon.* New York: Putnam, 1980.

Mundinger, M. O. *Autonomy in Nursing.* Germantown, Md.: Aspen Systems, 1980.

Nathanson, C. A. "Illness and the feminine role: A theoretical review." *Social Science Medicine* 9 (1975): 57–62.

Navarro, Vicente. "Women in health care." *New England Journal of Medicine* 292 (1975): 398–402.

Notman, Malka T., and Carol C. Nadelson, eds. *The Woman Patient: Medical and Psychological Interfaces.* New York: Plenum Pr, 1978.

Oakley, Ann. "Wisewoman and medicine man: Changes in the management of childbirth." In *The Rights and Wrongs of Women.* Edited by Juliet Mitchell and Ann Oakley. Harmondsworth, England: Penguin Books, 1977.

Practicing Midwife. Available from: 156 Drakes Lane, Summertown, TN 38483.

Psychology of Women Quarterly [Woman as Patient, special issue] 4.3 (Spring 1980).

Rodriguez-Trias, Helen. *Women and the Health Care System and Sterilization Abuse: Two Lectures.* New York: The Women's Center, Barnard College and CARASA, 1978.

Romney, Seymour L. et al. *Gynecology and Obstetrics: The Health Care of Women.* New York: McGraw-Hill, 1975.

*Scully, Diane. *Men Who Control Women's Health:.The Miseducation of Obstetrician-Gynecologists.* Boston: Houghton Mifflin, 1980.

Scully, Diane, and Pauline Bart. "A funny thing happened on the way to the orifice: Women in gynecology textbooks." *American Journal of Sociology* 78 (1973): 1045.

Slade, Margot. "The women in white." *The New Physician* 24 (1975): 34–35.

Spieler, Carolyn, ed. *Women in Medicine–1976.* New York: Independent Publishers Group, 1977.

Stark, Evan et al. "Medicine and patriarchal violence, the social construction of a 'private' event." *International Journal of Health Services* 9.3 (1977): 461–494.

Steppacher, R. C., and J. S. Mausner. "Suicide in male and female physicans." *Journal of the American Medical Association* 228 (1974): 323–328.

Tillman, Randi S. "Women in dentistry—a review of the literature." *Journal of the American Dental Association* 91 (1975): 1214–1215.

U.S. Dept. of Health, Education, and Welfare. *Minorities and Women in the Health Fields: Applicants, Students and Workers* (DHEW Pub. #MRA 76-776-22). Washington, D.C.: Government Printing Office, 1975.

_____. Women's Action Program Office of Special Concerns. *An Exploratory Study of Women in the Health Professions Schools, Vol. IV. Dentistry.* California Urban and Rural Systems Association, 1976.

Wagner, David. "The proletarianization of nursing in the United States, 1932–1946." *International Journal of Health Services* 18.2 (180): 271–290.

*Walsh, Mary Roth. *Doctors Wanted: No Women Need Apply. Sexual Barriers in the Medical Profession. 1835–1975.* New Haven: Yale University Press, 1977.

_____. "The Quirls of a Woman's Brain." In this volume.

_____. "The rediscovery of the need for a feminist medical education." *Harvard Educational Review* 49.4 (1979).

Williams, J. J. "The woman physician's dilemma." *Journal of Sociological Issues* 6 (1950): 38.

Women: A Journal of Liberation. [Women as Hospital Workers, special issue] 2.3. Available from: 3028 Greenmount Ave., Baltimore, MD 21218.

Women Workers at Two Abortion Clinics. *Getting Stronger: Women Workers Organize the Abortion Clinics.* Cambridge, Mass.: Red Sun Pr, 1978.

NONTRADITIONAL APPROACHES TO HEALTH CARE

Chamberlin, Jean. *On Our Own: Patient-Controlled Alternatives to the Mental Health System.* New York: McGraw-Hill, 1978.

Elhai, Laurisa. "The quality of medical care delivered by lay practitioners in a feminist clinic." *American Journal of Public Health* 71 (Aug. 1981): 853–856.

The Federation of Feminist Women's Health Centers. "How to stay out of the gynecologist's office." *Women's Health in Women's Hands Series.* Available from: Box 551, 6520 Selma Avenue, Los Angeles, CA 90028.

Ferguson, Tom. *Medical Self-Care.* New York: Summit Bks, 1980.

Gottlieb, Naomi. *Alternative Social Services for Women.* New York: Columbia University Press, 1980.

*Howell, Mary. *Healing at Home.* Boston: Beacon Pr, 1979.

Kleiber, Nancy, and Linda Light. *Caring for Ourselves: An Alternative Structure for Nursing Care.* Vancouver: 1978. Available from: School of

Nursing, University of British Columbia, 2075 Wesbrook Crescent,
Vancouver, B.C., Canada V6T 1W5.

Lipnack, Jessica. "The women's health movement," *New Age* (March 1980).
*National Women's Health Network. *Self Help: Resource Guide 7*.
Washington, D.C.: National Women's Health Network, 1980.
Rennie, Susan and Anna Rubin. "Women's survival catalog: Holistic healing."
[a bibliography] *Chrysalis* 1.
Ruzek, Sheryl. *The Women's Health Movement: Feminist Alternatives to
Medical Control*. New York: Praeger, 1979.

BIOGRAPHIES

American Anthropologist [In Memoriam to Margaret Mead (1901-1978),
special issue] 82.2 (June 1980).
American Men and Women of Science. New York: Bowker, 1970.
Baker, Rachel. *The First Woman Doctor: The Story of Elizabeth Blackwell,
M.D.* New York: Julian Messner, 1944.
Baldwin, Richard. *The Fungus Fighters: Two Women Scientists and Their
Discovery*. Ithaca, N.Y.: Cornell University Press, 1981.
Balfour, Margaret Ida. *The Work of Medical Women in India*. London and
New York: Oxford University Press, 1929.
Barringer, Emily A. *Bowery to Bellevue: The Story of New York's First
Woman Ambulance Surgeon*. New York: Norton, 1950.
Basalla, George. "Mary Somerville: A neglected popularizer of science." *New
Scientist* (March 1973).
Bass, Elizabeth. "Dispensaries founded by women physicians in the South-
land." *Journal of the American Medical Women's Association* 2 (1947).
_____. "Pioneer women doctors in the South." *Journal of the American
Medical Women's Association* 2 (1947).
Bell, Enid Mobely. *Storming the Citadel: The Rise of the Woman Doctor*.
London: Constable, 1953.
Bennett, A. H. *English Medical Women: Glimpses of Their Work in Peace and
War*. London: Pitman, 1915.
Bluemel, Elinor. *Florence Sabin: Colorado Woman of the Century*. Boulder:
University of Colorado Press, 1957.
Bowen, Gertrude Maude. *I Have Lived* [autobiography of a doctor].
Grantham: Stanborough Press, 1973.
Breckenridge, Mary. *Wide Neighbors: A Story of the Frontier Nursing
Service*. New York: Harper and Bros., 1952.
Brooks, Paul. *The House of Life: Rachel Carson at Work*. Boston: Houghton
Mifflin, 1972.
Campbell, J. Menzies. "Bygone women dentists." *Journal of the Canadian
Dental Association* (April 1948).
Clark, Eugenie. *Lady With a Spear* [autobiography of an ichthyologist].
New York: Harper, 1953.

Clarke, Robert. *Ellen Swallow: The Woman Who Founded Ecology*. Chicago Follett, 1973.

Curie, Eve. *Madame Curie. A Biography*. Translated by Vincent Sheean. Garden City, N.Y.: Doubleday, Doran and Co., 1937.

Dakin, Susanna Bryant. *The Perennial Adventure: A Tribute to Alice Eastwood, 1859-1953*. San Francisco: California Academy of Science, 1954.

Dally, Ann G. *Cicely, The Story of a Doctor*. London: Victor Gollancz, 196.

Destreich-Levie, Nancy. "Women in early American anthropology." In *Pioneers of American Anthropology*. Edited by June Helm. AES Monograph 43. Seattle: University of Washington Press, 1966.

Douglass, Emily Taft. *Margaret Sanger: Pioneer of the Future*. New York: Holt, Rinehart, and Winston, 1970.

Dubreil-Jacobin, Marie Louise. "Women mathematicians." In *Great Currents of Mathematical Thought*. Edited by F. Le Lionnais. New York: Dover, 1971.

Dworkin, Susan. "The renewable energy of Lola Redford." *Ms.* 8.

Dykeman, Wilma. *Too Many People, Too Little Love: Edna Rankin McKinnon, Pioneer for Birth Control*. New York: Holt, 1974.

Edwards, R. W. "The first woman dentist: Lucy Hobbs Taylor, D.D.S. 1833-1910." *Bulletin of the History of Medicine* 25 (1915): 277-283.

Eagles, J. A. et al. *Mary Swartz Rose 1874-1944, Pioneer in Nutrition*. New York: Teachers College Press, 1979.

Edwards, Samuel. *The Divine Mistress*. [Emilie du Chatelet: mathematician, geometer, and physicist.] New York: McKay, 1970.

Elia, Joseph J. "Alice Hamilton, 1869-1970." *New England Journal of Medicine* 283 (1970).

Emberlin, Diane P. *Contributions of Women: Science*. Woodland, Calif.: Dillon-Tyler Pubs, 1977.

Emerson, Gladys A. "Agnes Fay Morgan and early nutrition discoveries in California." *Federation Proceedings* 36 (May 1977): 1911-1914.

Faucett, Mrs. Henry. *Some Eminent Women of our Times*. London: Macmillan, 1899.

Fleming, Alice. *Doctors in Petticoats*. Lippincott, 1964.

Fletcher, Maurine S. "Portrait for a western album." [Biography of Alice Eastwood, a botanist] *American West* 17 (January-February 1980).

Golde, Peggy, ed. *Women in the Field: Anthropological Experiences*. Chicago: Aldine, 1970.

Goodfield, June. *An Imagined World: A Story of Scientific Discovery* [The biography of Anna Brito, cancer biologist]. New York: Harper and Row, 1981.

Grant, Madeleine P. *Alice Hamilton, Pioneer Doctor in Industrial Medicine*. New York and London: Abelard-Schuman, 1967.

Green-Armytage, A. J. *Maids of Honor*. London: W. Blackwood and Sons, 1906.

Haber, Louis. *Women Pioneers of Science*. New York: Harcourt, Brace, Jovanovich, 1979.

Hamerstrom, Frances. *An Eagle to the Sky* [The autobiography of a wildlife biologist]. Trumansburg, N.Y.: Crossing Pr, 1970.

Hamilton, Alice. *Exploring the Dangerous Trades: Autobiography of an Industrial Toxicologist with Jane Addams at Hull House*. Boston, 1943.

_____. "Pioneering in industrial medicine." *Journal of the American Medical Women's Association* 2 (1947).

Hays, Elinor Rice. *Those Extraordinary Blackwells: The Story of a Journey to a Better World*. New York: Harcourt, Brace and World, 1967.

Hughes, Muriel Joy. *Women Healers in Medieval Life and Literature*. Freeport, N.Y.: Books for Libraries Press, 1968 (originally 1943).

Hume, Ruth Fox. *Great Women of Medicine*. New York: Random House, 1964.

Hunt, Caroline. *Ellen Richards* [Professor at MIT, founder of home economics]. Boston: Whitcomb and Borrows, 1912.

Iltis, Carolyn. "Madame du Chatelet: Metaphysics and mechanics." *Studies in History and Philosophy of Science* 8 (1977).

Ireland, N. O. *Index to Women from Ancient to Modern Times*. Westwood, Mass.: F. W. Faxon Co., 1970.

Jacobi, Mary Putnam. *Life and Letters of Mary Putnam Jacobi*. Edited by Ruth Putnam. New York: Putnam, 1925.

_____. *Mary Putnam Jacobi, M.D.: A Pathfinder in Medicine, With Selections from Her Writings and a Complete Bibliography*. Edited by the Women's Medical Association of New York City. New York: Putnam, 1928.

_____. "Women in medicine." In *Women's Work in America*. Edited by Annie Nathan Meyer. New York: Holt, 1891.

Johnston, Malcolm Sanders. *Elizabeth Blackwell and Her Alma Mater: The Story of the Documents*. Geneva, N.Y.: W. F. Humphrey Press, 1947.

Kendall, Phebe Mitchell. *Maria Mitchell, Life Letters and Journals*. Boston, 1896.

King-Salmon, Frances W. *House of a Thousand Babies: Experiences of an American Woman Physician in China 1922-1940*. New York: Exposition Press, 1968.

Knapp, Sally. *Women Doctors Today*. New York: Crowell, 1947.

Lonsdale, Kathleen. "Women in science: Reminiscenses and reflections." *Impact of Science on Society* 20 (1970): 45–59.

Lovejoy, Esther. *Women Doctors of the World*. New York: Macmillan, 1957.

Lurie, Alison. "Beatrix Potter: More than just Peter Rabbit." *Ms.*, Sept. 1977, pp. 42–45.

Lurie, N. O. "Women in early American anthropology." In *Pioneers of American Anthropology*. Edited by June Helm. Seattle: University of Washington Press, 1966.

Lutzker, Edythe. *Edith Peckey-Phipson, M.D.: The Story of England's Primary Woman Doctor*. New York: Exposition, 1973.

MacDermot, H. E. *Maude Abbott: A Memoir.* Toronto: Macmillan, 1941.

McFerran, Ann. *Elizabeth Blackwell, First Woman Doctor.* New York: Grosset and Dunlap, 1966.

McMaster, Gilbert Totten. "The first woman practitioner of midwifery and the care of infants in Athens. 300 B.C. [Agnodice]." *American Medicine* 18 (1912): 202–205.

Manson, Cecil, and Celia Manson. *Doctor Agnes Bennett.* London: Michael Joseph, 1960.

Manton, J. *Elizabeth Garrett Anderson* [first female English physician]. New York: Dutton, 1965.

Marks, Geoffrey, and William K. Beatty. *Women in White.* New York: Scribner's, 1972.

Mead, Margaret. *Blackberry Winter, My Earlier Years.* New York: Morrow, 1972.

_____, ed. *Writings of Ruth Benedict: An Anthropologist at Work.* New York: Avon, 1959.

Meyer, Gerald Dennis. *Science for Englishwomen 1650–1760: The Telescope, the Microscope, and the Feminine Mind.* Berkeley: University of California Press, 1955.

Michael, Helen Abott. *Studies in Plant and Organic Chemistry and Literary Papers. With Biographical Sketch.* Cambridge, Mass.: Riverside Press, 1907.

Moores, Richard G. "Isabel Bevier, lady with a mission." In *Fields of Rich Toil: The Development of the University of Illinois College of Agriculture.* Urbana: University of Illinois Press, 1970.

Morton, Rosalie Slaughter. *A Woman Surgeon: The Life and Work of Rosalie Slaughter Morton.* New York: Stokes, 1937.

Mozans, H. J. [John Augustine Zahn]. *Woman in Science.* Cambridge, Mass.: MIT Press, 1974 (originally 1913).

Nearing, Scott, and M. S. Nellie. *Women and Social Progress: A Discussion of the Biologic, Domestic, Industrial and Social Possibilities of American Women.* New York: Macmillan, 1912.

Nightingale, Florence. *Cassandra.* Westbury, N.Y.: Feminist Pr, 1979.

Noble, Ins. *First Woman Ambulance Surgeon: Emily Barringer.* New York: Julian Messner, 1962.

O'Hern, E. "Alice Evans, pioneer microbiologist." *ASM News* 39 (1975): 573.

_____. "Cora Mitchell Downs, pioneer microbiologist." *ASM News* 40 (1974): 862.

_____. "Rebecca Craighill Lancefield, pioneer in microbiology." *ASM News* 41 (1975): 805.

Osen, Lynn. *Women in Mathematics.* Cambridge, Mass.: MIT Press, 1974.

Overholser, Winfred. "Dorothea Lynde Dix: A note." *Bulletin of the History of Medicine* 9 (1941).

Owens, Helen B. *Early Scientific Work of Women and Women in Mathematics.* State College, Pa.: Pennsylvania State University Press, 1940.

Pearsall, Ronald. "A pioneer bone setter: Mrs. Sarah Mapp." *Practitioner* 195 (1965).

Perl, Teri. *Math Equals: Biographies of Women Mathematicians and Related Activities*. Menlo Park, Calif.: Addison Wesley, 1978.

Phalen, Mary Kay. *Probing the Unknown: The Story of Dr. Florence Sabin*. New York: Crowell, 1969.

Phillips, D. H. "Women in nineteenth-century Wisconsin medicine." *Wisconsin Medical Journal* 71 (1972).

Pirami, E. "An eighteenth-century woman physician." *World Medical Journal* 12 (1965): 154.

Power, Eileen. "Some women practitioners of medicine in the Middle Ages." *Proceedings of the Royal Society of Medicine* 14 (1921).

Rappaport, Karen D. "Women mathematicians: a bibliography." *Women's Studies Newsletter* 6.4 (Fall, 1978): 15–17.

Regnault, P., and K. Stephenson. "Dr. Suzanne Noen, the first woman to do esthetic surgery." *Plastic and Reconstructive Surgery* 48 (1971): 133–139.

Reid, Robert. *Marie Curie*. New York: Saturday Review Pr, Dutton, 1974.

Rizzo, P. V. "Early daughters of Urania: Review of women in astronomy." *Sky and Telescope* (Nov. 1954): 7–9.

Robb, Hunter, "Mm. Lachepelle, Midwife." *Bulletin of the Johns Hopkins Hospital* 2 (1891).

Robinson, Marion O. *Give My Heart: The Dr. Marian Hilliard Story*. Garden City, N.Y.: Doubleday, 1964.

Ross, Ishbel. *Child of Destiny: The Life Story of the First Woman Doctor*. New York: Harper Bros., 1949.

Rossiter, M. W. "Women scientists in America before 1920." *American Science* 62 (1974): 312–323.

Rubin, Vera. "Margaret Mead: An appreciation." *Human Organization* 38 (Summer 1979): 193–195.

Sanger, Margaret. *Margaret Sanger: An Autobiography*. New York: Dover, 1971 (originally 1938).

*Sayre, Anne. *Rosalind Franklin and DNA: A Vivid View of What it is Like to Be a Gifted Woman in an Especially Male Profession*. New York: Norton, 1975.

*Schacher, Susan, coordinator. *Hypatia's Sisters: Biographies of Women Scientists Past and Present*. Seattle, Wash.: Feminist Northwest, 1976.

Shryock, Richard H. "Women in American medicine." In *Medicine in America: Historical Essays*. Baltimore: The Johns Hopkins Press, 1966.

Singer, Charles. *From Magic to Science*. Chapter 6: "The Visions of Hildegard of Bingen" of Bingen," Chapter 7: "The Ladies of Salerno." 1928.

_____. "The Scientific views and visions of Saint Hildegard, 1098–1180." In *Studies on the History and Method of Science*. Oxford: Clarendon Press, 1917.

Smithsonian Institution. *Women in Science in Nineteenth-Century America* [pamphlet and exhibition catalog]. Washington, D.C.: Smithsonian Institution Press, 1978.

Snively, William D. "Discoverer of the cause of milk sickness [Anna Pierce Hobbs Bixby]." *Journal of the American Medical Association* 196 (1966).

Somerville, Martha. *Personal Recollections from Early Life to Old Age of Mary Somerville*. Boston: Roberts Bros., 1874.

Stage, Sarah. *Female Complaints: Lydia Pinkham and the Business of Women's Medicine*. New York: Norton, 1979.

Sterling, Philip. *Sea and Earth: The Life of Rachel Carson*. New York: Crowell, 1970.

Stern, Madeleine B. *So Much in a Lifetime: The Story of Dr. Isabel Burrows*. New York: Messner, 1964.

_____. *We the Women: Career Firsts of Nineteenth-Century America*. New York: Shulte Publishing, 1963.

Strasser, Judy. "Jungle law: Stealing the double helix." *Science for the People* 8 (Sept./Oct. 1976): 29–31.

Strohl, E. Lee. "The fascinating Lady Mary Wortley Montagu. 1689–1762." *Archives of Surgery* 89 (1964).

Sturgis, Katharine R. "First woman fellow of the College of Physicians of Philadelphia: Memoir of Catharine MacFarlane, 1877–1969." *Transactions and Studies of the College of Physicians of Philadelphia* 38 (1971).

Todd, Margaret. *The Life of Sophia Jex-Blake*. London: Macmillan, 1918.

Traux, Rhoda. *The Doctors Jacobi*. Boston: Little, Brown, 1952.

Trescott, Martha B. "Julia B. Hall and aluminum." *Journal of Chemical Education* 54 (1977): 24.

Truman, Margaret. *Women of Courage*. New York: Morrow, 1976.

Victor, Agnecci. *A Woman's Quest: The Life of Marie Zakrzewska, M.D.* New York: Appleton, 1924.

Visher, Stephen S. "Women starred in 'American Men of Science. 1903–1943.'" In *Scientists Starred*. Edited by Stephen S. Visher. Baltimore: Johns Hopkins Press, 1947.

Wade, Ira O. *Voltaire and Madame du Chatelet: An Essay on the Intellectual Activity at Circy*. Princeton, N.J.: Princeton University Press, 1941.

Wade, Nicholas. "Discovery of pulsars: a graduate student's story." *Science* 189 (1975): 358–364.

Warner, Deborah Jean. *Graceanna Lewis: Scientist and Humanitarian*. Washington, D.C.: Smithsonian Institution Press, 1979.

Wauchope, Gladys Mary. *The Story of A Woman Physician*. Bristol: John Wright, 1963.

Willard, Mary Louisa. *Pioneer Women in Chemistry*. State College, Pa.: Pennsylvania State University Press, 1940.

William, Henry Smith. *The Great Astronomers*. New York: Simon and Schuster, 1930.

485

Williams, Margot, and Paul Elliot. "Maria Martin: The brush behind Audubon's birds." *Ms.* 5 (1977): 14–18.

Wilson, Carol Green. *Alice Eastwood's Wonderland: The Adventures of a Botanist.* San Francisco, Calif.: California Academy of Science, 1955.

Wilson, Dorothy Clarke. *Lone Woman: The Story of Elizabeth Blackwell, The First Woman Doctor.* Boston: Little, Brown, 1970.

———. *Palace of Healing: The Story of Dr. Clara Swain, First Woman Missionary Doctor and the Hospital She Founded.* New York: McGraw-Hill, 1968.

Winkler, Karen. "Anthropologists take a new look at the work of the late Margaret Mead." *Chronicle of Higher Education* 19 (December 10, 1979): 3.

Wright, Helen. *Sweeper of the Sky: The Life of Maria Mitchell, First Astronomer in America.* New York: Macmillan, 1949.

Wupperman, Alice. "Women in 'American Men of Science.'" *Journal of Chemical Education* (1941): 120–121.

Yost, Edna. *American Women of Nursing.* Philadelphia: Lippincott, 1947.

———. *American Women of Science.* New York: Fred A. Stokes, 1943.

———. *Women of Modern Science.* New York: Dodd, Mead, 1959.

HISTORY

Alcott, William. "The Young Woman's Book of Health." In *Root of Bitterness.* Edited by N. F. Cott. New York: Dutton, 1972 (originally 1855).

Alsop, Gulielma Fell. *History of the Women's Medical College of Philadelphia, Pennsylvania, 1850–1950.* Philadelphia: Lippincott, 1950.

*Barker-Benfield, G. J. *The Horrors of the Half-Known Life: Male Attitudes Toward Women and Sexuality in Nineteenth-century America.* New York: Harper and Row, 1976.

Beecher, Catherine. "Letters to the people on health and happiness." In *Root of Bitterness.* Edited by N. F. Cott. New York: Dutton, 1972 (originally 1855).

Bennett, A. Hughes. "Hygiene in the higher education of women." *Popular Science Monthly* 26 (Feb. 1880): 519–529.

Blackwell, Elizabeth. *Address on the Medical Education of Women.* New York: Baptist and Taylor, 1864.

———. "The human element in sex: Being a medical inquiry into the relation of sexual physiology to Christian morality." In *Root of Bitterness.* Edited by N. F. Cott. New York: Dutton, 1972 (originally 1894).

———. "The influence of women in the profession of medicine." In *Essays in Medical Sociology.* New York: Arno Press and the New York Times, 1972 (originally 1902).

———. *Medicine as a Profession for Women.* New York: Trustees of the New York Infirmary for Women, 1860.

_____. *Opening the Medical Profession to Women*. Edited by Mary Roth Walsh. New York: Schocken, 1977 (originally 1895).

Blackwell, Emily. "The industrial position of women." *Popular Science Monthly* 23 (July 1883): 388-398.

Blake, John B. "Women and medicine in antebellum America." *Bulletin of the History of Medicine* 39 (1965).

Bolton, H. Carrington. "The early practice of medicine by women." *Popular Science Monthly* 18 (Dec. 1880): 191-201.

Bowditch, Henry I. "The medical education of women." *Boston Medical and Surgical Journal* 105 (August 1881).

Bridenthal, R. and Koonz, C., eds. *Becoming Visible: Women in European History*. Boston: Houghton Mifflin, 1977.

Burstyn, Joan N. "Education and sex: The medical case against higher education for women in England, 1870-1900." *Proceedings of the American Philosophical Society* 117 (April 1973): 2.

Chadwick, James Read. "The study and practice of medicine by women." *International Review* (October 1879).

Colden, Jane. *Jane Colden Botanic Manuscript*. Limited Edition. New York: Garden Club of Orange and Dutchess Counties, 1963.

Crawford, H. Jean. "Association to aid scientific research by women." *Science* 75 (1932).

Cumings, Elizabeth. "Education as an aid to the health of women." *Popular Science Monthly* 27 (Oct. 1880): 823-827.

Daughters of Aesculapius: Stories Written by Alumnae and Students of the Woman's Medical College of Pennsylvania. Philadelphia: George W. Jacobs, 1897.

Davis, Paulina W. "Female physician." *Boston Medical and Surgical Journal* 41 (1853).

Donegan, Jane B. *Women and Men Midwives: Medicine, Morality, and Misogyny in Early America*. Westport, Conn.: Greenwood, 1978.

Donnison, Jean. *Midwives and Medical Men: A History of Inter-Professional Rivalries and Women's Rights*. New York: Schocken, 1977.

*Ehrenreich, Barbara, and Deirdre English. *Complaints and Disorders: The Sexual Politics of Sickness*. Old Westbuty, N.Y.: Feminist Pr, 1973.

* _____. *For Her Own Good: 150 Years of the Experts' Advice to Women*. Garden City, N.Y.: Anchor Pr/Doubleday, 1978.

* _____. *Witches, Midwives, and Nurses: A History of Women Healers*. Old Westbury, N.Y.: Feminist Pr, 1973.

Figlio, Karl. "Chlorosis and chronic disease in 19th century Britain: The social constitution of somatic illness in a capitalist society." *International Journal of Health Services* 8.4 (1978): 589-618.

Forster, Emily L. *Analytical Chemistry as a Profession for Women*. London: C. Griffin and Co., 1920.

Gage, Matilda J. *Woman as Inventor*. Fayetteville, N.Y.: New York State Woman Suffrage Association, 1870.

Gamble, E. B. *The Sexes in Science and History.* New York: Putnam, 1916.

Gardener, Helen Hamilton. *Facts and Fictions of Life.* [Including article on Sex in Brain]. Fenno, 1893.

Graham, Davis. "The demand for medically educated women." *Journal of the American Medical Association* 6 (1886).

Gregory, George. *Medical Morals . . . and the Importance of Establishing Female Medical Colleges, and Educating and Employing Female Physicians for Their Own Sex.* New York: G. Gregory, 1853.

Haller, John S., Jr. "From maidenhood to menopause: Sex education for women in Victorian America." *Journal of Popular Culture* 6 (Spring 1972).

Hunt, Harriot K. [1805-1875]. "On medical education for women." In *Voices From Women's Liberation.* Edited by Leslie R. Tanner. New York: New American Library, 1970.

Jacobi, Dr. Mary Putnam. "Female invalidism: From a letter to Dr. Edis." In *Root of Bitterness.* Edited by N. F. Cott. New York: Dutton, 1972 (originally 1895).

Jex-Blake, Sophia. *Medical Women: A Thesis and a History.* New York: Source Books Press, 1970 (originally 1886).

Kalisch, Philip A. and Beatrice J. Kalisch. *The Advance of American Nursing.* Boston: Little, Brown, 1978.

Kinsler, Miriam S. "The American woman dentist: A brief historical review from 1855 through 1968." *Bulletin of the History of Dentistry* 17 (December 1967).

Kohlstedt, Sally Gregory. *The Formation of the American Scientific Community: The American Association for the Advancement of Science 1848-1860.* Urbana, Ill.: University of Illinois Press, 1976.

_____. "In from the periphery: American women in science, 1830-1880." *Signs* 4 (Autumn 1978): 81-96.

Lander, K. E. "Study of anatomy by women before the nineteenth century." In *Proceedings of the Third International Congress of the History of Medicine.* London, 1972.

Litoff, J. *American Midwives: 1860 to the present.* Westport, Conn.: Greenwood, 1978.

_____. "Forgotten women: American midwives at the turn of the twentieth century." *Historian* 40.

Lowie, Robert H., and Leta Stetter Hollingworth. "Science and Feminism." *Scientific Monthly* (Sept. 1916).

McGrew, Elizabeth A. "The history of women in medicine: a symposium: The present. *Bulletin of the Medical Library Association* 44 (1956).

Marshall, Clara. *The Women's Medical College of Pennsylvania: A Historical Outline.* Philadelphia: Blakiston, 1897.

Mead, Kate Hurd. *A History of Women in Medicine from Earliest Times to the Early Nineteenth Century.* Haddam, Conn.: Haddam Pr, 1938.

_____. "The seven important periods in the evolution of women in medicine." *Bulletin of the Women's Medical College of Pennsylvania* 81 (1931): 6-15.

370 Biological Woman—The Convenient Myth

*Merchant, Carolyn. *The Death of Nature: Women, Ecology, and the Scientific Revolution*. San Francisco: Harper and Row, 1980.

*Mosedale, Susan. "Science corrupted: Victorian biologists consider 'the woman question.'" *Journal of the History of Biology* 11 (1978): 1-55.

Munster, L. "Women doctors in medieval Italy." *CIBA Symposium* 10 (1962): 136-140.

Murray, Flora. *Women as Army Surgeons, Being the History of the Women's Hospital Corps in Paris, Winereux, and Endell Street, Sept. 1914-Oct. 1919*. London: Hodder and Stoughton, 1920.

Nichols, Mary Sargeant Gove. *Lectures to Women on Anatomy and Physiology With an Appendix on Water Cure*. New York: Harper Bros., 1846.

Rossiter, Margaret W. "'Women's Work'" in Science, 1880-1910." *ISIS: Official Journal of the History of Science Society* 71.258 (1980): 381-398.

Smith, Hilda. "Gynecology and ideology in seventeenth-century England." In *Liberating Women's History*. Edited by Berneice A. Carroll. Urbana, Ill.: University of Illinois Press, 1976, pp. 97-114.

Smith-Rosenberg, Carroll, and Charles Rosenberg. "The female animal: Medical and biological views of woman and her role in nineteenth-century America." *The Journal of American History* (Sept. 1973).

Stage, S. *Female Complaints: Lydia Pinkham and the Business of Women's Medicine*. New York: Norton, 1979.

Thorne, May. "Women in medicine: The early years." *Postgraduate Medical Journal* 27 (1951).

Verbrugge, M. H. "Women and medicine in nineteenth-century America." *Signs* 1 (1976): 957-972.

Vicinus, Martha. *Suffer and Be Still: Women and the Victorian Age*. Bloomington, Ind.: University of Indiana Press, 1972.

Waite, Frederick C. *History of the New England Medical College 1848-1874*. Boston, 1950.

Walsh, James J. Medical education for women." In *Medieval Medicine*. London: A. and C. Black, 1920.

_____. "Women in medicine." In *History of Medicine in New York: Three Centuries of Medical Progress*. New York: National Americana Society, 1919.

Warner, Deborah Jean. "Science education for women in antebellum America." *ISIS* 69:246 (March 1978): 58-66.

_____. "Women in Science in Nineteenth Century America." Washington, D.C.: Smithsonian Institution Press, 1978.

Weaver, Jerry L., and Sharon D. Garrett. "Sexism and racism in the American health care industry: a comparative analysis." *International Journal of Health Services* 8.4 (1978): 677-704.

Wood, Ann Douglas. "The fashionable diseases: Women's complaints and their treatment in nineteenth-century America." *Journal of Interdisciplinary History* (Summer 1973).

Wright, Katherine W. "History of women in medicine: a symposium: Nineteenth century or transitional period." *Bulletin of the Medical Library Association* 44 (1956).

GENERAL

*The Ann Arbor Science for the People Collective. *Biology as a Social Weapon.* Minneapolis, Minn.: Burgess, 1977.

Arditti, Rita. "Women's biology in a man's world: some issues and questions." *Science for the People* 4 (July 1973).

*Arditti, Rita, Pat Brennan, and Steve Cavrak. *Science and Liberation.* Boston: South End Press, 1979, 1980.

Beatty, Jerome. *The Girls We Leave Behind: A Terribly Scientific Study of American Women at Home.* Garden City, N.Y.: Doubleday, 1963.

Berger, Peter L., and Thomas Luckmann. *The Social Construction of Reality: A Treatise in the Sociology of Knowledge.* Garden City, N.Y.: Anchor Books, Doubleday, 1966.

Bird, Caroline. *Born Female.* New York: McKay, 1968.

Bleier, Ruth. "Difficulties of detecting sexist biases in the biological sciences." *Signs* 4 (1978): 159–162.

Braxton, Bernard. *Women, Sex and Race.* Washington, D.C.: Verta Pr, 1973.

Costa, M. Dalla, and S. James. *The Power of Women and the Subversion of the Community.* Bristol: Falling Wall Press, 1972.

*Cott, Nancy F., ed. *Root of Bitterness: Documents of the Social History of American Women.* New York: Dutton, 1972.

*Daly, M. *Beyond God the Father: Toward a Philosophy of Women's Liberation.* Boston: Beacon Pr, 1973.

_____. *Gyn/Ecology: The Metaphysics of Radical Feminism.* Boston: Beacon Pr, 1978.

*de Beauvoir, Simone. *The Second Sex.* New York: Knopf, 1953; Bantam, 1961.

Figes, Eva. *Patriarchal Attitudes.* New York: Fawcett, Premier Bks, 1970.

*Firestone, Shulamith. *The Dialectic of Sex.* New York: Bantam, 1970.

Freeman, J., ed. *Women: A Feminist Perspective.* Palo Alto, Calif.: Mayfield Pub, 1975.

*Gornick, Vivian, and B. K. Moran. *Woman in Sexist Society.* New York: Basic, 1971.

*Griffin, Susan. *Woman and Nature: The Roaring Inside Her.* New York: Harper and Row, 1978.

Hall, Diana Long. "The social implications of the scientific study of sex." *The Scholar and the Feminist IV: Connecting Theory, Practice and Values.* A Conference sponsored by the Barnard College Women's Center, Papers from the Morning Session, April 23, 1977. The Women's Center, Barnard College.

Hubbard, Ruth, Mary Sue Henifin, and Barbara Fried, editors. *Women Look at Biology Looking at Women: A Collection of Feminist Critiques.* Boston, Mass.: G. K. Hall; Cambridge, Mass.: Schenkman, 1979.

Lipman-Blumen, J. "How ideology shapes women's lives." *Scientific American* 226 (1972): 34–42.

Mill, John Stuart. *On the Subjection of Women.* New York: Collectors Editions, Source Book Press, 1970 (originally 1869).

Mill, John Stuart, and Harriet Taylor Mill. *Essays on Sex Equality.* Compiled by Alice S. Rossi. Chicago: University of Chicago Press, 1970.

Millman, Marcia and Rosabeth Moss Kanter. *Another Voice.* Garden City, N.Y.: Anchor Pr, 1975.

Morgan, Robin, ed. *Sisterhood Is Powerful.* New York: Vintage, 1970.

Redstockings. *Feminist Revolution.* New York: Random House, 1978.

*Roberts, Joan L., ed. *Beyond Intellectual Sexism: A New Woman, A New Reality.* New York: McKay, 1976.

Roberts, Helen, ed. *Doing Feminist Research.* Boston: Routledge and Kegan Paul, 1981.

*Rose, Hilary, and S. Rose, eds. *Ideology of/in the Natural Sciences.* Boston: G. K. Hall; Cambridge, Mass.: Schenkman, 1980.

*Rossi, Alice, ed. *The Feminist Papers.* New York: Bantam, 1973.

Roszak, Betty, and Theodore Roszak, eds. *Masculine/Feminine.* New York: Harper, 1969.

Rousseau, J. J. *Emile: or a Treatise on Education.* Translated by W. H. Payne. New York: Appleton, 1899.

Ryan, Mary P. *Womanhood in America: From Colonial Times to the Present.* New York: Franklin Watts, 1975.

Seavey, Carol A., Phyllis Katz, A. Zalk, and Sue Rosenberg. "Baby X: The effect of gender labels on adult responses to infants." *Sex Roles* 1 (June 1975): 103–109.

Signs 4:1 ["Women, Science, and Society"] (1978): 1–216.

Stoll, C. S. *Sexism: Scientific Debates.* Menlo Park, Calif.: Addison Wesley, 1973.

*Wollstonecraft, Mary. *A Vindication of the Rights of Women.* Dublin: James Moore, 1793.

Women's Studies International Quarterly [Women, Technology and Innovation, special issue] 4:3 (1981).

*Woolf, Virginia. "Professions for Women." In *The Death of the Moth and Other Essays.* New York: Harcourt Brace, 1942; Harvest, 1974.

*____. *A Room of One's Own.* New York: Penguin, 1964 (originally 1928).

*____. *Three Guineas.* London: Hogarth Press, 1968 (originally 1938).

Wortig, Rochelle. "The acceptance of the concept of the maternal role by behavioral scientists: Its effect on women." In *The Women's Movement.* Edited by Helen Wortig and Clara Rabinowitz. New York: AMS Press, 1972.

IV. Bibliographies and Periodicals

Ballou, Patricia K. *Women: A Bibliography of Bibliographies.* Boston: G. K. Hall, 1980.

The Barnard College Women's Center. *Women's Work and Women's Studies: A Bibliography.* Published Yearly. Old Westbury, N.Y.: Feminist Pr.

Boston Women's Health Book Collective. *Women and Health: Issues and Alternatives, Recommended Readings.* W. Somerville, Mass.: unpublished, 1980.

Bullough, Vern L., and Barrett Wayne Elcaro. *A Bibliography of Prostitution.* New York: Garland Pub, 1977.

Canadian Newsletter of Research on Women [includes all feminist publications, bibliography of current scholarship, etc.]. Dept. of Sociology, Ontario Institute for Studies in Education, 252 Bloor St., W., Toronto, Ontario, Canada.

Canadian Women's Studies: Les Cahiers de la Femme. P. O. Box 631, Station "A," Scarborough, Ontario M1K 5E9 Canada.

Chrysalis: A Magazine of Women's Culture. 635 South Westlake Ave., Los Angeles, CA 90057.

*Davis, Audrey B. *Bibliography on Women: With Special Emphasis on Their Sex Roles in Science and Society.* New York: Science History Publications, 1974.

Feminist Issues. The Feminist Forum, Berkeley and Transaction Periodicals Consortium, Rutgers University, New Brunswick, N.J. 08903.

Feminist Studies. Available from: Managing Editor, *Feminist Studies*, c/o Women's Studies Program, University of Maryland, College Park, MD 20742.

Glenn, Sara et al. *Women and Society: Bibliography.* Boston University School of Social Work, Boston, Mass.

Haber, Barbara. *Women in America: A Guide to Books, 1963-1975.* Boston: G. K. Hall, 1974.

The Hastings Center Report. 360 Broadway, Hastings-on-Hudson, NY 10706.

Health/PAC Bulletin. Human Science Press, 72 5th Ave., New York NY 10011.

Healthright. 41 Union Square, Room 206-209, New York, NY 10003.

*Hughes, Marija Matich. *The Sexual Barrier: Legal, Medical, Economic and Social Aspects of Sex Discrimination* [an annotated bibliography with over 8,000 items]. Washington, D.C. 500 23rd St. N.W. Box B203, Washington, D.C. 20037.

International Journal of Health Services. Baywood Publishing Company, Inc., 120 Marine St., P.O. Box D, Farmingdale, NY 11735.

ISIS International Bulletin. Case Postale 301, 1327 Geneva, Switzerland.

Journal of the History of Biology. Kluwer Academic Publishers Group, Lincoln Building, 160 Old Derby St., Hingham, MA 02043.

Figure XXX. Investigators on collecting trip, Marine Biological Laboratory, Woods Hole, Massachusetts (1895).
(Society for the Preservation of New England Antiquities; Baldwin Coolidge, Photograph)

Key, Mary Ritchie. *Male/Female Language, with a Comprehensive Bibliography.* Metuchen, N.J.: Scarecrow, 1975.

Lesbian/Feminist Research Newsletter. c/o Julia P. Stanley, Department of English, University of Nebraska, Lincoln, NE 68588.

Meyer, Sharon I. "Bibliography of Bibliographies on Women and Health Care, Revised Listing." *Women and Health . . . the Journal of Women's Health Care* 5.2 (Summer 1980): 95–99.

Mother Jones. 1255 Portland Place, Boulder CO 80302.

Ms. Ms. Magazine Corporation, 370 Lexington Ave., New York, NY 10017.

National Women's Health Network News. 224 Seventh St. SE, Washington DC 20003.

New Scientist. IPC Magazines LTD, King's Reach Tower, Stamford St., London SE1 9LS.

Off Our Backs [Feminist Newspaper]. 1724 20th St. NW, Washington DC 20009.

Rosenberg, Marie B., and Len V. Bergstrom. *Women and Society: A Critical Review of the Literature with a Selected Annotated Bibliography.* Beverly Hills, Calif.: Sage, 1975.

Ruzek, Sheryl K. *Women and Health Care: An Annotated Bibliography.* Program on Women. Northwestern University.

Science for People. British Society for Social Responsibility in Science, 9 Poland St., London W1, England.

Science for the People. 897 Main St., Cambridge, MA 02139.

Science, Technology and Human Values. MIT Press, 28 Carleton St., Cambridge, MA 02142.

Signs: A Journal of Women in Culture and Society. Chicago: University of Chicago Press, 5801 Ellis Avenue, Chicago, Ill. 60637.

WIN, Women's International Network News. 187 Grant St., Lexington, MA 02173.

Women: A Journal of Liberation. 3028 Greenmount Ave., Baltimore, MD 21218.

Women and Environments International Newsletter. Faculty of Environmental Studies, York University, 4700 Keele Street, Downsview, Ontario M3J 2R2.

Women and Health. The Haworth Press, 149 Fifth Ave., New York, NY 10010.

Women and the Health System: Selected Annotated References. Health Planning Bibliography, Series #4, HRA #72629, National Health Planning Information Center.

Women's Studies International Quarterly. A Multidisciplinary Journal for the Rapid Publication of Research Communications and Review Articles in Women's Studies. Pergamon Press Ltd., Headington Hill Hall, Oxford, England.

Women's Studies Newsletter. A Publication of the Feminist Press and the National Women's Studies Association. c/o The Feminist Press, Box 334, Old Westbury, NY 11568.

Womenwise [New Hampshire Feminist Health Center Quarterly]. Women-Wise, c/o NHFHC, 38 S. Main St., Concord, NH 03301.

World Health: The Magazine of the World Health Organziation. Research in Family Planning.

[Joan Cindy Amatniek *is coediting a source book on the "Woman's Nature" debate of the late 1800s. She has been involved in studies concerning science, technology, and public policy in England, Norway, and Washington D.C. As an undergraduate at Harvard University, she helped teach the course, Biology and Women's Issues.*]

[Mary Sue Henifin *is Past Coordinator of the Women's Occupational Health Resource Center at Columbia University School of Public health where she also received her Master's of Public Health Degree in Environmental Science. She speaks and writes frequently about issues affecting women's health and has coproduced a monthly radio program, "The Women's Occupational Health Radio Hour." She is completing a book with Jeanne Stellman on the health hazards of office work.*]

TOMAS TRANSTRÖMER
COLLECTED POEMS

Tomas Tranströmer
COLLECTED POEMS

TRANSLATED BY ROBIN FULTON

BLOODAXE BOOKS

ISBN: 1 85224 23 7

First published 1987 by
Bloodaxe Books Ltd,
P.O. Box 1SN,
Newcastle upon Tyne NE99 1SN.

Bloodaxe Books Ltd acknowledges
the financial assistance of Northern Arts.

Typesetting by Bryan Williamson, Manchester.

Cover printing by
Tyneside Free Press Workshop Ltd, Newcastle upon Tyne.

Printed in Great Britain
at the University Printing House, Oxford